Symbiotic Association of Microorganisms with Medicinal and Herbal Plants

Research establishes that symbiotic association of microbes with medicinal and herbal plants enhance the growth and accumulation of bioactive materials and that species of microbes, including bacterial and fungal species, play a key role. *Symbiotic Association of Microorganisms with Medicinal and Herbal Plants* identifies the important symbiotic associations between microbes and medicinal plants, including perspectives in improving bioactive ingredients for the synthesis and preparation of pharmaceutical drugs.

FEATURES

- Provides a comprehensive overview of symbiotic associations of microorganisms with medicinal and herbal plants.
- Discusses the impacts of symbiotic associations on the diversity, growth, and chemical components of the medicinal and herbal plants.
- Elaborates on opportunities and future challenges regarding the understanding of symbiotic associations between microbes and medicinal plants.

A volume in the *Exploring Medicinal Plants* series, this book provides exciting information in a structured manner for scientists, researchers, and students working in various fields, including medicinal plants, microbiology, economic botany, chemistry, biotechnology, pharmacognosy, and pharmaceuticals.

EXPLORING MEDICINAL PLANTS

Series Editor: Azamal Husen, *Wolaita Sodo University, Ethiopia*

Medicinal plants render a rich source of bioactive compounds used in drug formulation and development; they play a key role in traditional or indigenous health systems. As the demand for herbal medicines increases worldwide, supply is declining as most of the harvest is derived from naturally growing vegetation. Considering global interests and covering several important aspects associated with medicinal plants, the Exploring Medicinal Plants series comprises volumes valuable to academia, practitioners, and researchers interested in medicinal plants. Topics provide information on a range of subjects including diversity, conservation, propagation, cultivation, physiology, molecular biology, growth response under extreme environment, handling, storage, bioactive compounds, secondary metabolites, extraction, therapeutics, mode of action, and healthcare practices.

Led by Azamal Husen, PhD, this series is directed to a broad range of researchers and professionals consisting of topical books exploring information related to medicinal plants. It includes edited volumes, references, and textbooks available for individual print and electronic purchases.

Antimalarial Medicinal Plants
Edited by Azamal Husen

Medicinal Spice and Condiment Crops
Edited by Azamal Husen

Therapeutic Medicinal Plants in Traditional Persian Medicine
Edited by Roja Rahimi and Roodabeh Bahramsoltani

Medicinal Plants for the Management of Neurodegenerative Diseases
Edited by Jamal Akhtar, Maryam Sarwat and Fouzia Bashir

Nutraceuticals Inspiring the Contemporary Therapy for Lifestyle Diseases
Edited by Mala Trivedi, Sachidanand Singh, Parul Johri, and Pedro López-Sánchez

Plants as Medicine and Aromatics: Uses of Botanicals
Edited by Mohd Kafeel Ahmad Ansari, Mushtaq Ahmad, and Gary Owens

Exploring Traditional Wild Edible Plants
Edited by Vibhor Agarwal, Sachidanand Singh and Rhaul·Datta

Propagation to Pharmacopeia: Modern Approaches in Medicinal Plants
Edited by Abha Manohar K, Gopal Shukla, Saikat Gantait, A P Das, and Sumit Chakravarty

Exploring Medicinal Orchids
Edited by Azamal Husen and Manu Pant

Symbiotic Association of Microorganisms with Medicinal and Herbal Plants
Edited by Allah Ditta, Sajid Mehmood, and Azamal Husen

For more information about this series, please visit: www.routledge.com/Exploring-Medicinal-Plants/book-series/CRCEMP

Symbiotic Association of Microorganisms with Medicinal and Herbal Plants

Edited by
Allah Ditta, Sajid Mehmood, and Azamal Husen

CRC Press
Taylor & Francis Group
Boca Raton London New York

CRC Press is an imprint of the
Taylor & Francis Group, an **informa** business

Designed cover image: Shutterstock

First edition published 2025
by CRC Press
2385 NW Executive Center Drive, Suite 320, Boca Raton FL 33431

and by CRC Press
4 Park Square, Milton Park, Abingdon, Oxon, OX14 4RN

CRC Press is an imprint of Taylor & Francis Group, LLC

ISBN: 978-1-032-35877-2 (hbk)
ISBN: 978-1-032-35879-6 (pbk)
ISBN: 978-1-003-32915-2 (ebk)

DOI: 10.1201/b23155

Typeset in Times
by Apex CoVantage, LLC

Contents

Preface

Medicinal plants have been used for health care since ancient times, as they are rich in various bioactive compounds such as flavonoids, alkaloids, terpenoids, sterols, and saponins. Most of these plants grow inland, in the terrestrial ecosystem. The rhizosphere of these plants contains a variety of microbes such as bacteria and fungi, which play an important role in nutrient dynamics and improving growth and productivity in the form of bioactive materials in medicinal and herbal plants. The microbes have a symbiotic association with medicinal and herbal plants, which plays an important role in nutrient acquisition and their growth and productivity. Recently, various researchers around the world have established that the symbiotic association of microbes with medicinal and herbal plants has played an important role in enhancing the growth and accumulation of bioactive materials. Various species of microbes, including bacterial and fungal species, have been reported in this regard. Despite these reports, no book is available highlighting the symbiotic association of microorganisms with medicinal and herbal plants with special emphasis on diversity, growth, and bioactive ingredients.

Symbiotic Association of Microorganisms with Medicinal and Herbal Plants aims to provide exciting and remarkable information in a structured manner to scientists, researchers, and students working in the fields of medicinal plants, microbiology, plant production, plant science, economic botany, chemistry, biotechnology, pharmacognosy, pharmaceuticals, industrial chemistry, and many other interdisciplinary subjects. This book will also be helpful in identifying important symbiotic associations between microbes and medicinal plants and their future perspectives in improving bioactive ingredients for the synthesis or preparation of modern drugs.

We are grateful to all contributors for readily accepting our invitation, sharing their knowledge in specialized areas of research, and readily adjusting to suggestions for improving the shape of their contributions. With great pleasure, we also extend our sincere thanks to Randy Brehm, Tom Connelly, and all associates at Taylor & Francis Group/CRC Press, for their active cooperation.

Allah Ditta
Sajid Mehmood
Azamal Husen

About the Editors

Allah Ditta, PhD, earned his BSc (Hons) in 2007 and MSc (Hons) in 2009 from University of Agriculture, Faisalabad, Pakistan. He earned his PhD from the Institute of Soil and Environmental Sciences, University of Agriculture, Faisalabad, Pakistan, in 2014. He has carried out research projects in the areas of eco-restoration by using integrated organic, inorganic, and microbial strategies. Dr. Ditta has served as an associate professor and Chairman of the Department of Environmental Sciences, Shaheed Benazir Bhutto University, Sheringal, Upper Dir, Khyber Pakhtunkhwa, Pakistan, since 2021. He has also served as an adjunct research fellow at the School of Biological Sciences, The University of Western Australia, Perth, Australia, since 2018. He has also served as an assistant professor at the Department of Environmental Sciences, PMAS, Arid Agriculture University Rawalpindi, Pakistan. His major research interests encompass broad, interdisciplinary fields of plant production, with a focus on eco-restoration under abiotic stresses. The breadth of his research spans degraded land ecological amelioration using integrated and sustainable approaches. He has supervised seven MPhil and eight undergraduate-level research theses. Dr. Ditta has published more than 150 research and review articles (impact factor 455) in internationally well-recognized journals, along with 19 book chapters. He has also edited two books: *Bio-Organic Amendments for Heavy Metal Remediation: Water, Soil and Plant Approaches and Technologies* (Plant Biology, Sustainability and Climate Change) with Elsevier, USA, and *Nanomaterials in Industrial Chemistry* with CRC Press, USA.

Sajid Mehmood completed his BSc (Hons) in agriculture from Arid Agricultural University of Rawalpindi, Pakistan, in 2012; he continued his MSc (Hons) in soil science and graduated in 2014. In 2015, he moved to Wuhan, China; finished his PhD under Prof. Dr. Shuxin Tu in June 2018 at the College of Resources and Environment, Huazhong Agricultural University, Wuhan, China; and examined the different types of risk management strategies and efficiency at the crop level using biochar, slag, and ferrous manganese ore. He is currently a postdoctoral fellow at Hainan University in the Department of Ecology and Environment, working on the remediation of fish pond sediments for the utilization of sediments as a fertilizer. He has three years of research experience on as-is and modified (magnetic and acidic) biochars. He has published his findings in science journals. Dr. Mehmood worked as a research associate in a project of the National Natural Science Foundation of China. Dr. Mehmood has been on the editorial boards and panels of reviewers of several reputed journals of Elsevier, Frontiers Media SA, Taylor & Francis, Springer Nature, RSC, Oxford University Press, Sciendo, the Royal Society, CSIRO, PLOS, and John Wiley & Sons.

Azamal Husen is a professor at Sankalchand Patel University, Visnagar, India, and adjunct professor at Graphic Era (Deemed to be University), Dehradun, Uttarakhand, India. He is also working as a visiting professor at University Putra Malaysia, Selangor, Malaysia, and acting as a foreign delegate at Wolaita Sodo University, Wolaita, Ethiopia. Previously, he served as Professor and Head of the Department of Biology, University of Gondar, Ethiopia. He also worked as a visiting faculty of the Forest Research Institute and the Doon College of Agriculture and Forest at Dehra Dun, India. His research and teaching experience of 25 years encompasses biogenic nanomaterial fabrication and application; plant responses to nanomaterials; plant adaptation to harsh environments at the physiological, biochemical, and molecular levels; herbal medicine; and clonal propagation for improvement of tree species. He has conducted research sponsored by the World Bank, the National Agricultural Technology Project, the Indian Council of Agriculture Research, the Indian Council of Forest Research Education, and the Japan Bank for International Cooperation. Dr. Husen has published extensively (~300 works) and served on the editorial board and as reviewer of reputed journals published by Elsevier, Frontiers Media, Taylor & Francis, Springer Nature, RSC, Oxford University Press, Sciendo, the Royal Society, CSIRO, PLOS, MDPI, John Wiley & Sons, and UPM Journals. He is on the advisory board of Cambridge Scholars Publishing, UK. He is a fellow of the Plantae group of the American Society of Plant Biologists and a member of the International Society of Root Research, Asian Council of Science Editors, and International Natural Product Sciences. He is Editor-in-Chief of the *American Journal of Plant Physiology* and a series editor of *Exploring Medicinal Plants* (Taylor & Francis Group, USA); *Plant Biology, Sustainability, and Climate Change* (Elsevier, USA); and *Smart Nanomaterials Technology* (Springer Nature, Singapore).

Contributors

Arusa Aftab
Department of Botany
Lahore College for Women University
Lahore, Pakistan

Zubair Ahmad
Department of Chemistry
University of Swabi, Pakistan

Asghar Ali
Department of Botany
Hazara University Mansehra
Khyber Pakhtunkhwa, Pakistan

Muhammad Shahbaz Ali
Department of Environmental Sciences
Government College University Faisalabad
Pakistan

Atefe Amirahmadi
Department of Plant Science
Damghan University
Damghan, Iran

Farrukh Azeem
Department of Bioinformatics & Biotechnology
Government College University Faisalabad
Pakistan

Muhammad Bilal
Department of Environmental Sciences
Government College University Faisalabad
Pakistan

Yasir Bilal
Department of Environmental Sciences
Government College University Faisalabad
Pakistan

Sri Wahyuni Budiarti
Research Centre for Food Crops
National Research and Innovation Agency (BRIN), Cibinong Science Centre
Indonesia

Indra Jeet Chaudhary
Department of Environmental Science
Savitribai Phule Pune University
Pune (Maharashtra) India

Allah Ditta
Department of Environmental Sciences
Shaheed Benazir Bhutto University
Sheringal, Dir (U)
Khyber Pakhtunkhwa, Pakistan

Hubert Olivier Dossou-Yovo
Laboratory of Applied Ecology
Faculty of Agronomic Sciences
University of Abomey-Calavi
Benin

Muhammad Esa
Department of Pharmacy
Shaheed Benazir Bhutto University
Sheringal, Dir (Upper)
Khyber Pakhtunkhwa, Pakistan

Shivani Gupta
Plant Virology Laboratory
Department of Plant Pathology
Dr. YS Parmar University of Horticulture and Forestry, Nauni
Solan, Himachal Pradesh, India

Samina Hanif
Department of Botany
Lahore College for Women University
Lahore, Pakistan

Wiwiek Harsonowati
Research Centre for Horticultural and Estate Crops
National Research and Innovation Agency (BRIN) Cibinong Science Centre
Indonesia

Sabir Hussain
Department of Environmental Sciences
Government College University Faisalabad
Pakistan

Sajjad Hyder
Department of Botany
Government College for Women University
Sialkot, Pakistan

Mizu Istianto
Research Centre for Horticultural and Estate Crops
National Research and Innovation Agency (BRIN), Cibinong Science Centre
Cibinong, Indonesia

Abdul Khaliq Jan
Department of Chemistry
Shaheed Benazir Bhutto University Sheringal
Pakistan

Zul Kamal
Department of Pharmacy
Shaheed Benazir Bhutto University, Sheringal, Dir (Upper)
Khyber Pakhtunkhwa, Pakistan
and
School of Pharmacy
Shanghai Jiao Tong University
China

Alevcan Kaplan
Department of Crop and Animal Production
Sason Vocational School
Batman University
Türkiye

Muhammad Nauman Khan
Department of Botany
Islamia College Peshawar
Pakistan
and
University Public School
University of Peshawar
Pakistan

Muhammad Talha Khokhar
Department of Environmental Sciences
Government College University Faisalabad
Pakistan

Amrender Kumar
Agricultural Knowledge Management Unit (AKMU)
ICAR-Indian Agricultural Research Institute
India

Sumit Kumar
Department of Mycology and Plant Pathology
Institute of Agricultural Sciences
Banaras Hindu University
India
and
Department of Plant Pathology
B.M. College of Agriculture, Khandwa
Rajmata Vijayaraje Scindia Krishi Vishwa Vidyalaya
India

Kumar M
ICAR-National Bureau of Agriculturally Important Microorganisms
Kushmaur, India

Faisal Mahmood
Department of Environmental Sciences
Government College University Faisalabad
Pakistan

Zainab Maqbool
Department of Botany
Lahore College for Women University Lahore
Pakistan

Dur Muhammad
Department of Chemistry University of Baluchistan Quetta
Pakistan

Shahneela Mushtaq
Department of Botany
Lahore College for Women University Lahore
Pakistan

Gulshan Negi
Plant Virology Laboratory
Department of Plant Pathology
Dr. YS Parmar University of Horticulture and Forestry
India

Bhavna Nigam
School of Environment and Sustainable Development
Central University of Gujarat
Gandhinagar, Gujarat, India

Vahid Poozesh
Department of Plant Science
Faculty of Biology, Damghan University
Iran

Ratna Prabha
Agricultural Knowledge Management Unit (AKMU)
ICAR-Indian Agricultural Research Institute
India

Dheeraj Rathore
School of Environment and Sustainable Development
Central University of Gujarat, Gandhinagar
Gujarat, India

Abdur Rauf
Department of Chemistry
University of Swabi
Khyber Pakhtunkhwa, Pakistan

Nadia Riaz
Department of Botany
Lahore College for Women University Lahore
Pakistan

Hasin Sahar
Department of Botany
University of Chitral, Chitral
Khyber Pakhtunkhwa, Pakistan

Wiwin Setiawati
Research Centre for Horticultural and Estate Crops
National Research and Innovation Agency (BRIN)
Cibinong Science Centre, West Java, Indonesia

Muhammad Shahid
Department of Bioinformatics & Biotechnology
Government College University Faisalabad
Pakistan

Tanvir Shahzad
Department of Environmental Sciences
Government College University Faisalabad
Pakistan

Zainab Shahzadi
Department of Botany
Lahore College for Women University Lahore
Pakistan

Uzma Sher
Department of Botany
University of Chitral, Chitral, Khyber Pakhtunkhwa
Pakistan

DP Singh
ICAR-Indian Institute of Vegetable Research
India

Brice Sinsin
Laboratory of Applied Ecology, Faculty of Agronomic Sciences, University of Abomey-Calavi, Benin

Deciyanto Soetopo
Research Centre for Horticultural and Estate Crops, National Research and Innovation Agency (BRIN), Cibinong Science Centre, West Java, Indonesia

Sakshi Suchita
Maize Pathology Laboratory
Department of Plant Pathology, Punjab Agricultural University, Ludhiana
Punjab, India

Suliasih
Research Centre for Applied Microbiology
National Research and Innovation Agency (BRIN), Cibinong Science Centre
West Java, Indonesia

Farman Ullah
Department of Chemistry
University of Swabi
Khyber Pakhtunkhwa, Pakistan

Wajahat Ullah
Department of Chemistry
Shaheed Benazir Bhutto University Sheringal
Pakistan

Dwinita Wikan Utami
Research Centre for Horticultural and Estate Crops
National Research and Innovation Agency (BRIN)
Cibinong Science Centre, Indonesia

Sri Widawati
Research Centre for Applied Microbiology
National Research and Innovation Agency (BRIN)
Cibinong Science Centre, Indonesia

Susi Wuryantini
Research Centre for Horticultural and Estate Crops
National Research and Innovation Agency (BRIN)
Cibinong Science Centre, Indonesia

Afifa Younas
Department of Botany
Lahore College for Women University, Lahore
Pakistan

Zubaida Yousaf
Department of Botany
Lahore College for Women University Lahore
Pakistan

1 Symbiotic Association of Microbes with Medicinal and Herbal Plants

Vahid Poozesh and Allah Ditta

1.1 INTRODUCTION

Nature always presents itself as a prime example of the exceptional phenomena of symbiosis. Natural products of plant, animal, and mineral origin have been the basis of treating human diseases. The use of herbal medicines and supplements has increased dramatically over the past three decades, with 80% of people worldwide relying on them as part of their primary health care (Ekor, 2014). The use of herbal medicines and phytonutrients or nutraceuticals is growing and expanding rapidly all over the world, and many people are now using these products for the treatment of various health and medical problems at different national health care centres (WHO, 2004).

In recent years, in developing and developed countries, we have seen a significant trend towards medicinal plants, so that today herbal medicines are available to the public in supermarkets and grocery stores in addition to pharmacies. Medicinal plants are an integral part of alternative health care and are of great value and importance in providing health and safety to society, both in terms of treatment and prevention of diseases. Traditional medicine and modern medicine are considered two main custodians of global public health (Sen *et al.*, 2011). Currently, 80% of the world's population, who live mainly in developing countries, use herbal medicine and its derivative compounds. During the last century, about 121 medicinal products have been obtained based on traditional knowledge and science. Since ancient times, traditional medicine has been a way of life that has its roots in the culture of tribes and nations and is a part of the local folklore in many different cultures of the world (Bodeker *et al.*, 2005; Bandaranayake, 2006; Verma and Singh, 2008; Heinrich and Jäger, 2015).

With the increase in the global consumption of herbal medicinal products and the introduction of many new products to the market, issues related to public health and concerns regarding the safety status of these drugs have increased. Although some herbal medicines have good potential and are widely used, many of them have not yet been subjected to clinical trials, and there is not enough supervision on how to use them.

This severely limits awareness of their potential adverse effects and makes it more difficult to identify the safest and most effective treatments and promote their rational use (WHO, 2002b). With the increasing demand for the use of herbal medicine in developed countries, the use of complementary and alternative medicine (CAM) has become conventional in North America, Europe, and Australia (Calapai, 2008; Braun *et al.*, 2010; Anquez-Traxler, 2011). In these countries, they believe that the use of herbal medicinal helps one lead a healthier and more useful life. Therefore, the use of medicinal plants is often considered a modest and balanced procedure for treatment and recovery, and huge amounts (more than billions of dollars) are spent on buying and consuming these plants. This partly explains the boom in herbal medicine sales and accounts for a significant share of the global medicine market (Roberts and Tyler, 1997; WHO, 2002a, 2005; Kong *et al.*, 2003; Bandaranayake, 2006). However, side effects caused by herbal medicine are less than synthetic drugs because allopathic medicines have a high potential to cause side effects. Perhaps a simple

DOI: 10.1201/b23155-1

explanation of the reason for this is that plants and humans have had mutual relationships and interactions over many centuries, which leads to more compatible metabolic processes in medicinal plants than synthetic drugs in the human body (Firenzuoli and Gori, 2007). This issue has led to more activation of the medicinal plant market and its high demand from consumers, which resulted in a sudden increase in the number of herbal medicine manufacturers (Agarwal, 2005).

Many articles have been published in fields related to the traditional medicinal knowledge of different ethnic and cultures to date. Nevertheless, despite spending a lot of time and money in this regard, so far no significant successes have been seen in the field of treating diseases with the help of traditional medicine or new drugs based on nature (Rezaei *et al.*, 2022). Various environmental factors such as temperature, humidity, light, soil type, and characteristics of soil microbes can significantly affect the quantity and quality of medicinal plants (Namdar *et al.*, 2019). It has been well demonstrated that specific bacterial endophytes can affect medicinal plants (Ek-Ramos *et al.*, 2019). All plant-associated microenvironments are colonized in excessive abundance by using microorganisms, in particular the nutrient-rich rhizosphere. This region is a microecological area influenced via plant roots and hosts a lot of microbes that are very important for plant nutrition, health, and quality (Berg, 2009; Mendes *et al.*, 2011, 2013).

These rhizosphere microorganisms are absorbed from the diverse reservoir of soil microbes by plant root exudates of the rhizosphere (Compant *et al.*, 2010) and other rhizosphere sediments and are guided through the diversity of compounds of these exudates (Doornbos *et al.*, 2012). Each plant species has a specific rhizosphere microbiome that depends on the existing soil community (Smalla *et al.*, 2001). In addition to plant species, the composition and diversity of microbial communities in the rhizosphere are formed through soil kind and pedoclimate, plant health and growth stages, climate and season, pesticide treatments, herbivores and animals, and countless different biotic and abiotic factors (Berg and Smalla, 2009; Barnard *et al.*, 2013). Some genera are present everywhere and are found scattered throughout plants, such as Bacillus and Pseudomonas, which are among the known genera related to plants (Berg *et al.*, 2011). Nevertheless, excessive specificity for every microenvironment was once discovered with the aid of a comparison of microbial colonization patterns of different microsites (Fürnkranz *et al.*, 2012; Köberl *et al.*, 2013). Certain plant-associated microorganisms that include beneficials and pathogens are additionally handed down from preceding generations through the seed (Hardoim *et al.*, 2012; Hirsch and Mauchline, 2012), and likewise, transmission between flora thru pollen grains was recently observed (Fürnkranz *et al.*, 2012). The interesting point is that the phylogenetically oldest land plants, mosses, move a highly diverse core microbiome of primarily potential beneficial bacteria from the sporophyte to the gametophyte and vice versa (Bragina *et al.*, 2012). Medicinal vegetation has a distinct microbiome due to its special and structurally different bioactive secondary metabolites, which possibly account for the excessive specificity of the related microorganisms (Qi *et al.*, 2012).

We discuss and evaluate current reports on how the plant-associated microbial community could be fashioned by using hosts and how the plant microbiome influences plant growth, productivity, and host survival in a variety of symbiotic associations. This understanding will inform efforts to enhance agricultural practices and predict how environmental elements will affect the microbial community and plant diversity.

1.2 ENVIRONMENTAL AND HOST PLANTS' EFFECTS ON THE MICROBIAL COMMUNITY

The various types of microorganisms related to a greater organism (human, animal, vegetation, etc.) are together described as its microbiome. All greater organisms studied to date, including plants, insects, fish, mice, monkeys, and humans, harbour microbiomes (Berg *et al.*, 2015; McFall-Ngai *et al.*, 2013). Research on the human microbiome has advanced very rapidly. Recently, researchers have additionally paid significant attention to elucidating the composition and function of plant and soil microbiomes. It is now believed that plants are not separate entities; however, as a substitute they stay in affiliation with a massive range of microbes. These microbes stay either inside (endosphere) or outside (episphere) of plant tissues.

Among these microorganisms, bacteria and fungi dominate. Approximately several thousand species of bacteria and fungi have been reported in plant tissues (Bulgarelli *et al.*, 2012; Berendsen *et al.*, 2012). They play essential roles such as expanded nutrient availability, plant uptake, and extended plant stress tolerance. Therefore, plant health (growth and survival) is the result of the physical and physiological features of the plant as much as the related microbiome, which collectively is recognized as a plant holobiont (Zilber-Rosenberg *et al.*, 2008). It is important to study the relationship between plants and microorganisms, especially the role of microbes in absorbing nitrogen and phosphorus. Some of these microbes cannot live in the absence of a host plant that supplies space, oxygen, protein, and carbohydrates to the microorganisms (Koide and Mosse, 2020). Traditionally, culturable microbes have not been explored for studies of plant–microbe interactions with large non-culturable microbes, and therefore our knowledge of the role of these non-culturable microbes is limited. The discovery of the kinds and consequences of plant–microbe interactions has attracted the attention of ecologists, evolutionary biologists, plant biologists, and agronomists (Bulgarelli *et al.*, 2013; Turner *et al.*, 2013).

1.2.1 Interactions between Environment and Microbes

Soil microorganisms represent the greatest share of global biodiversity, play a key role in biogeochemical cycles, and are essential drivers of soil fertility and primary productiveness (Figure 1.1) (van der Heijden *et al.*, 2008). Environmental elements (such as the seasonality of rainfall, temperature, and vegetation type) and soil characteristics (such as pH, texture, organic matter, and nutrient content) tend to generate seasonal and spatial changes in soil microbial abundance and composition (Barnard *et al.*, 2013; Pajares *et al.*, 2016), and these changes can affect ecosystem procedures (Strickland *et al.*, 2009). Pollution of soil has seriously hampered the healthful and sustainable development of agricultural production (Figure 1.1). Hence, it is necessary to take effective and useful measures to solve this problem. Therefore, understanding the elements that cause changes

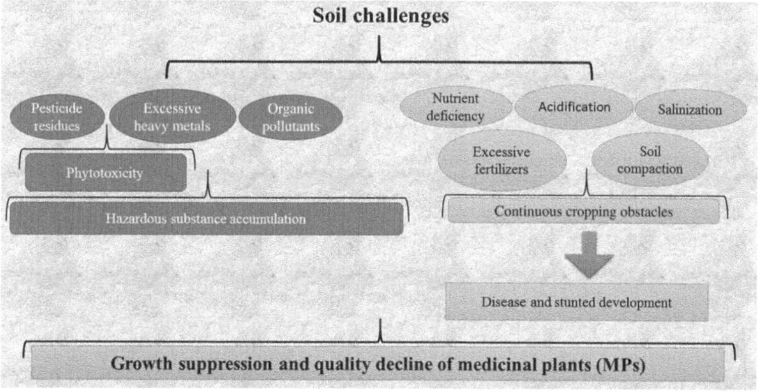

FIGURE 1.1 Negative and destructive effects of soil on the cultivation of medicinal plants. Nutrient deficiency and imbalance, soil compaction, salinization, and acidification retard the growth and development of MPs, and as a result, serious obstacles are created in the way of continuous cultivation of crops. Harmful substances, such as chemical pesticides and herbicides, heavy metals, and organic soil pollutants, can be toxic to plants (phytotoxicity), and they are absorbed and accumulated by medicinal plants. Finally, they can pose many health risks to consumers through food chains. (Adapted from Wang *et al.*, 2022.)

in the diversity of soil microbial communities is significant in predicting ecosystem response to a changing environment (Singh *et al.*, 2010).

The microbial community refers to all the microbial populations in a habitat. The action of complex microbial communities influences biogeochemical changes in natural, managed, and human-made ecosystems. Microbial communities are crucial for serious advances in agriculture that increase crop production rates. The microbial community can be terrestrial or aquatic (Gupta *et al.*, 2016). A community of microbes and their environment that occurs on the continents' landmasses and islands constitutes an earthy microenvironment. Terrestrial microenvironments are distinguished from aquatic microenvironments by less access to water and therefore the significance of water as a limiting agent.

Rainforests are the most diverse and productive terrestrial microenvironment; however, their soil is poor in nutrients due to heavy leaching through rainwater. Topsoil is still rich in native populations of bacteria (including actinomycetes), fungi, algae, and protozoa. Moreover, human and animal activities also introduce certain microbes into the soil in different ways. Human activities directly add bacteria to agricultural fields as agents of biodegradation or through the use of sewage sludge. Animals introduce microbes through bird droppings or excretion (Gupta *et al.*, 2016). The atmosphere is an unsuitable climate for microbes due to stress from dehydration. This results in a restricted amount of time for the microbes to become active. Also, some microbes resist these stresses through specific mechanisms that reduce their biological activity.

Hence, survival is highly affiliated with the environment, the time spent in the environment, and the type of microbes. Nevertheless, many other factors such as humidity, temperature, oxygen content, specific ions, UV radiation, diverse pollutants, and other air-related factors (AOFs) affect microbial viability. Relative humidity of the air is very important for the survival of airborne microbes. Microbes' ability to survive in aerosols is linked to the biochemistry of the organism. Temperature is an important factor that affects the microbial activity. Generally, high temperatures cause deactivation due to drying and denaturation of the protein, while lower temperatures lead to longer survival. At very low temperatures, some microbes lose their liveability due to the formation of ice crystals on their surface because of freezing. Most short-wavelength radiation, such as ultraviolet radiation and ionizing radiation (X-rays), is detrimental to microbes due to damage and destruction of DNA.

This radiation targets DNA by creating single- or double-strand breaks and by modifying the structure of the bases of nucleic acids. UV radiation causes damage by forming intrastrand thymidine dimers, which inhibit biological activity like genome replication, transcription, and translation (Gupta *et al.*, 2016). In addition, certain reactive forms of oxygen, including superoxide radicals, hydrogen peroxide, and hydroxide radicals, are produced by light, UV rays, or pollution and damage DNA by causing mutations. Likewise, OAFs (a mixture of ozone and hydrocarbons) also inactivate microbes by damaging nucleic acids and enzymes.

1.2.2 Interactions between Host Plants and Microbes

Microbial interactions play an important role in the ecology and function of plant communities. How do plants have special microbial communities? How do plants form a special microbial community in the rhizosphere? These are the questions to be answered. Modern genomic technologies (e.g., excessive throughput sequencing) can provide clues to the response. Lundberg *et al.* tried to describe the root microbiome of *Arabidopsis thaliana* (2012). They gathered over 600 *A. thaliana* plants and compared bacterial communities using 16S rRNA gene sequencing. They located that vegetation's root microbial communities were sufficiently dependent on host genotype to rang between populations of *A. thaliana* will congenerate. Nevertheless, the mechanisms were unclear but involved differences in host physiology and immune responses. The terrestrial plant community induces dynamic changes in soil microbial ecology that can result in modifications in ecosystem functioning. Bever *et al.* observed evidence that microbial-mediated positive and negative feedback can play an important role throughout the plant ecosystem and contribute to these plant–plant interaction mechanisms (2010).

During evolution, plants have developed complex strategies to protect themselves against biotic and abiotic stresses and nutrient availability. One example is the establishment of mutually beneficial symbiotic relationships with soil microbes, whether bacteria (such as rhizobia) or fungi (such as arbuscular mycorrhizal fungi). Beneficial microbes that are symbiotic with the roots of plants are considered 'biofertilizers' that provide nutrients to plants. Therefore, due to the beneficial effects of these fertilizers, which can be fruitful in reducing the use of chemical fertilizers and thus protecting the environment, it has been noticed by humankind. Therefore, as a general rule and not an exception, plant roots can be colonized by different microorganisms (Partida-Martínez and Heil, 2011).

Sometimes plants can benefit from a symbiotic relationship with microbes; for example, by providing fixed nitrogen and other nutrients for microorganisms, they are protected against diseases. There is documented evidence that microbes can affect, directly and indirectly, the health of plants by affecting their function (Figure 1.2). This influence is through the effect on things such as the state of nutrients, the performance of the photosynthetic apparatus, the growth and development of the plant, and the tolerance of salt stress (Friesen *et al.*, 2011; van der Heijden *et al.*, 2008). Plants have lately been identified as a meta-organism that has a definite microbiome and symbiotic relationships with related microorganisms (Mendes *et al.*, 2013). Plant ecology is influenced by complicated interactions with plant-associated microbes. Because plants are sessile, they have co-evolved with microbes and obtained a wide variety of mechanisms that modulate the consequence of their interactions (Oldroyd, 2013).

Roots can consistently synthesize, accumulate, and secrete a vast variety of compounds into the soil (Jones and Dangl, 2006). Recent research has confirmed that plant root secretions shape the soil bacterial community (Lareen *et al.*, 2016). According to Ciccazzo *et al.*'s experiments (2014), a plant species can select a specific bacterial community in the rhizosphere zone. The change in microbial composition creates feedback in relative plant performance that determines the long-term outcomes of soil microbes on their symbiosis with that plant species (Bever, 2013). How can microbiome diversity and function affect host plant function? The presence of microbial foci in plant microbiome networks performs an important function between the plant and its microorganism community (van der Heijden and Hartmann, 2016).

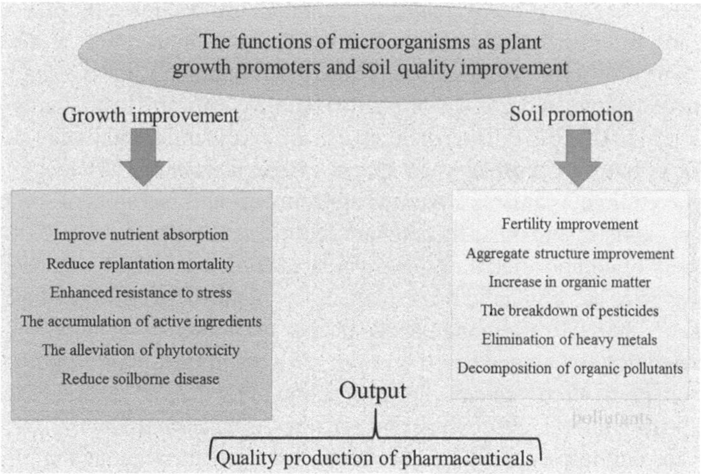

FIGURE 1.2 The performance of microorganisms in soil amendment and plant growth enhancement. (Adapted from Wang *et al.*, 2022.)

1.2.2.1 Plant Genes Responsible for Defence Affecting the Microbial Community

Various studies have confirmed that the plant genotype has a small but sizable impact on the composition of endophyte, rhizosphere, or phyllosphere microbial communities (Horton *et al.*, 2014; Bodenhausen *et al.*, 2014). Quantitative trait locus (QTL) evaluation and genome-wide association study (GWAS) have been used to discover species associated with host genes in humans, mice, plants, and flies (Horton *et al.*, 2014; Chaston *et al.*, 2016). It was proposed by a genome-wide association study of the *A. thaliana* leaf microbial community that the loci of *A. thaliana* are accountable for defence, and the composition of the community is influenced by the integrity of the cell wall (Horton *et al.*, 2014).

In addition, host genetic changes could influence the species richness of microbial communities. Horten *et al.* stated that the genes responsible for the defence system against pathogens fashioned the ecosystems of microorganisms. Matthew reported that 196 accessions of *A. thaliana* were planted in a greenhouse and then transferred to land. Data from a field experiment showed that the structure of plant tissues (i.e., cell wall integrity) could influence the microbial community of leaves. To find the influence of host plant genetic factors on the population composition of related microorganisms, Bodenhausen *et al.* utilized a candidate gene approach to examine host effects on phyllosphere community composition and abundance of *A. thaliana* (2014).

A panel including 55 mutants of the *A. thaliana* plant was made; in this panel, changes were constructed in secondary metabolism, defence signalling, cell wall, surface structure, and identification of pathogens. This was done to show the effect of these changes on the composition of microbiota and/or the frequency of a small number of individual host mutants. The outcomes confirmed that lacs and pecl mutations affect cuticle formation, leading to elevated bacterial abundance and community composition. Additionally, the ethylene signalling gene ein2 was discovered as a host agent that modulates community composition. Peiffer *et al.* additionally stated that about 19% of the interline variation in species richness may be attributed to host genotypes in various maize lines grown in an equal geographic area (2013).

1.3 MICROORGANISMS AND MEDICINAL PLANTS

The focus of research on herbal plants was mostly in the field of bioactive plant chemicals, but currently, research is directed toward the issue of whether phytotherapeutic compounds are produced with the help of related microbes or through interaction with their host. Medicinal plants are a massive biological resource for potential utilization in modern medicine and agriculture, but their microbiome is mostly not known. All plant microhabitats are occupied by different microorganisms, especially the nutrient-rich rhizosphere zone, which is the environment around plant roots and can host a large number of types of microbes. These microorganisms have a significant impact on the health, nutritional status, and growth of plants (Hiltner, 1904 in Hartmann *et al.*, 2008; Berg, 2009; Mendes *et al.*, 2011, 2013). Rhizospheres can host 1011 microbial cells per gram of plant roots, with over 30,000 different prokaryotic species (Berendsen *et al.*, 2012).

These rhizosphere microorganisms are taken up from various reservoirs of soil microbes by rhizosphere plant root exudates and other rhizosphere sediments (Compant *et al.*, 2010) and are guided through the diversity of compositional of these exudates (Bais *et al.*, 2006; Doornbos *et al.*, 2012). Each of the plant species has a specific rhizosphere microbiome dependent on the current soil community (Smalla *et al.*, 2001). In addition to plant species, the composition and diversity of microbial rhizosphere communities are formed based on soil type and climate, plant health and growth stage, weather and season, pesticide treatments, herbivores and animals, and several other biotic and abiotic factors (Singh and Mukerji, 2006; Berg and Smalla, 2009; Barnard *et al.*, 2013).

Some genera are omnipresent and can be found scattered throughout the plant, as the well-known genera associated with Bacillus and Pseudomonas plants (Berg *et al.*, 2011). Nonetheless, a high grade of specificity for every microhabitat was also observed when comparing the microbial colonization patterns of various microhabitats (Berg *et al.*, 2005b; Fürnkranz *et al.*, 2012; Köberl *et al.*, 2013). Some plant-associated microorganisms, including beneficial and pathogens, are also transmitted by the seeds of previous generations (Hardoim *et al.*, 2012; Hirsch and Mauchline, 2012), and, more recently, interplant transmission via pollen grains has been observed (Fürnkranz *et al.*, 2012). It is noteworthy that mosses, as the phylogenetically oldest land plants, move a widely varied core microbiome of potentially advantageous bacteria from the sporophyte to the gametophyte and vice versa (Bragina *et al.*, 2012). Medicinal plants have an exclusive microbiome because of their unique and structurally different bioactive secondary metabolites, which are likely responsible for the high specificity of associated microorganisms (Qi *et al.*, 2012).

The analysis of several microbiomes of Chinese medicinal plants showed interesting results (such as *Ainsliaea henryi* Diels, *Dioscorea opposite*, *Potentilla discolor* Bge, *Stellera chamaejasme* L., and others; Li *et al.*, 2008; Zhao *et al.*, 2011) because each of them harboured a community-specific actinobacterial. They also showed a high and diverse endophytic and rhizosphere colonization by actinobacteria that have antimicrobial and antitumour attributes (Zhao *et al.*, 2012). Gram-positive and mostly spore-forming bacteria that are promising biological control agents (BCAs), such as the genus Streptomyces, are a well-known and special origin of new antibiotics (Goodfellow and Fiedler, 2010; Niraula *et al.*, 2010; Nachtigall *et al.*, 2011; Raaijmakers and Mazzola, 2012).

1.3.1 Medicinal Plants: Compounds Derived from Plants and Microbes

Plants include many biologically active compounds, and plant-derived medicines have been a section of traditional health care in almost all parts of the world for thousands of years. Now, the global pharmaceutical market is worth approximately US $1.1 trillion per year. About 35% of these drug treatments come directly or indirectly from natural products, including plants (25%), microorganisms (13%), and animals (about 3%). Natural derivatives are a very important source for pharmaceutical companies that work on the improvement and development of new drugs (Calixto, 2019).

Therefore, herbal productions have a very important position in the drug discovery and development process. Earlier, research on medicinal plants focused mostly on their constituents, but currently, the centre of attention has shifted to the structure and function of various microbiomes of medicinal plants (Köberl *et al.*, 2013). Surprisingly, not only are plants themselves able to produce substances with phytotherapeutic attributes, but the microbes associated with them, specifically the endophytes, can also produce such substances (Table 1.1; adapted from Köberl *et al.*, 2013).

Current research indicates that a considerable number of natural products are produced by way of microbes and/or microbial interactions with the host from which they had been removed, and for numerous medicinal vegetation. It is hypothesized that the microbiome associated with plants, especially the complicated endomicrobial community, is directly or indirectly concerned with the manufacturing of bioactive phytochemicals (Gunatilaka, 2006; Wang *et al.*, 2022). However, at present, only a small subset of potential microbial strains can be definitively attributed to herbal medicine (Castronovo *et al.*, 2021; Miller *et al.*, 2012a, 2012b), and their relative contribution to the known biological activity of medicinal plants is still unclear.

Due to the alarming emergence of antibiotic resistance in medicine-related bacteria, medicinal plants with antibacterial properties are of fundamental importance as biological sources of new active metabolites (Keita *et al.*, 2022). In addition, there is a growing need for more and better antifungal drugs to help people with attenuated immune systems who are susceptible to fungal infections, like in the AIDS epidemic, cancer treatment, or organ transplants (Strobel *et al.*, 2004). For centuries, various herbal remedies have also been known for their antiphlogistic properties, but despite advances in medical research, chronic inflammatory diseases like asthma, arthritis, and rheumatism remain some of the biggest problems in health care in the world (Li *et al.*, 2003).

Hence, nature must still harbour many currently unknown active agents that may serve as leads and scaffolds for the development of desperately needed drugs for a multitude of diseases (Newman and Cragg, 2012). Today, the phenomenon of globalization has caused more countries to connect, which may affect the use of herbal medicines. Therefore, it has been proven that this phenomenon can be the availability of more useful medicinal plants.

TABLE 1.1
Examples of Bioactive Phytometabolites That Microorganisms Are Involved in Their Production

Host Plant	Producing Microbe	Compounds of Bioactive	Reference
Kennedia nigricans	*Streptomyces* sp.	Munumbicins	Castillo *et al.*, 2002
grevillea pteridifolia	*Streptomyces* sp.	Kakadumycins	Castillo *et al.*, 2003
Monstera sp.	*Streptomyces* sp.	Coronamycins	Ezra *et al.*, 2004
Salvia officinalis	*Chaetomium* sp.	Cochliodinol	Debbab *et al.*, 2009
Bidens pilosa	*Botryosphaeria rhodina*	Botryorhodines	Abdou *et al.*, 2010
Taxus brevifolia; Ginkgo biloba; Aloe vera	*Taxomyces andreanae*; *Alternaria* sp.; *Phoma* sp.	Paclitaxel (Taxol, Onxal)	Wani *et al.*, 1971; Stierle *et al.*, 1993; Kim *et al.*, 1999; Immaculate *et al.*, 2011
Dysoxylum binectariferum	*Fusarium proliferatum*	Rohitukine	Mohana Kumara *et al.*, 2012
Putterlickia verrucosa	*Actinosynnema pretiosum*	Maytansine	Wings *et al.*, 2013
Podophyllum hexandrum; Juniperus communis	*Alternaria* sp.; *Aspergillus fumigatus*	Podophyllotoxin	Yang *et al.*, 2003; Kusari *et al.*, 2009a
Artemisia annua	*Colletotrichum* sp.	Artemisinin	Wang *et al.*, 2001
Nothapodytes foetida; Camptotheca acuminate	*Entrophospora infrequens; Fusarium solani*	Camptothecin	Puri *et al.*, 2005; Amna *et al.*, 2006; Kusari *et al.*, 2009b
Erythrina crista-galli	*Phomopsis* sp.	Phomol	Weber *et al.*, 2004
Terminalia morobensis	*Pestalotiopsis microspora*	Pestacin and Isopestacin	Strobel *et al.*, 2002; Harper *et al.*, 2003
Artemisia mongolica	*Colletotrichum gloeosporioides*	Colletotric acid	Zou *et al.*, 2000

Source: Adapted from Köberl *et al.*, 2013

1.4 PROMOTION OF PLANT GROWTH AND MICROBE CONTROL

Various rhizosphere microbes interact beneficially with their host plants through different mechanisms. They can have a plant growth–stimulating effect based on enhancing nutrient uptake or hormonal incitement or indirectly affect plant fitness by suppressing plant pathogens (Berg, 2009; Lugtenberg and Kamilova, 2009). Plant growth–promoting rhizobacteria (PGPR) can produce a complicated combination of unstable substances, which are divided between bacterial species and other closely relevant species (Garbeva *et al.*, 2014). Studies have shown that certain bacterial volatiles can stimulate plant growth (Peñuelas *et al.*, 2014); suppress disease-causing ISR (Yi *et al.*, 2013); or fight plant pathogens (Pineda *et al.*, 2013), nematodes, or insects (Bray *et al.*, 2000). Biofertilizers are microorganisms that provide nutrients to plants; for instance, the symbiotic rhizobia in root nodules is of the utmost importance among nitrogen-fixing microbes. Other microbial biofertilizers, like mycorrhizal fungi and various rhizobacteria, can solubilize plant-available phosphate from organic or inorganic phosphate (Lugtenberg *et al.*, 2002).

Plant stimulants or phytostimulators are microbes that hormonally improve plant growth; for example, auxin, as a plant hormone, can be mentioned. This hormone is produced by fluorescent pseudomonads and is one of the best-known examples of these stimuli (Kamilova *et al.*, 2006; Khare and Arora, 2010). Several rhizobacteria, such as *Burkholderia cepacia*, *Staphylococcus epidermidis*, and strains of the *Bacillus subtilis* group, stimulate vegetation growth by releasing volatile organic compounds (VOCs; Ryu *et al.*, 2003; Kai *et al.*, 2007; Effmert *et al.*, 2012; Bitas *et al.*, 2013). Plant growth can be improved indirectly through the biological control of plant pathogens. In the meantime, the spread of the pathogen can be prevented using antibiotics or volatile organic compounds, toxins, biosurfactants, or enzymes that destroy the extracellular cell wall. Microbial antagonism can occur by destruction of pathogen factors such as toxins or occur solely through competition for nutrients, minerals, or colonization sites (Berg, 2009).

Another feasible way to minimize the activity of pathogenic microorganisms is to activate plant defence mechanisms like induced systemic resistance (ISR), which is caused by non-pathogenic rhizobacteria. Flagella, lipopolysaccharides, siderophores, VOCs, and numerous different bacterial factors are involved in the activation of the ISR signalling pathway by non-pathogenic rhizobacteria (van Loon *et al.*, 1998; Lugtenberg and Kamilova, 2009). The biological control of plant pathogens and the improvement of plant growth with microorganisms have been studied for some time and are becoming real alternatives to pesticides and chemical fertilizers in sustainable agriculture (Weller, 2007). Although various microbial inoculants have been successfully exploited commercially (Berg, 2009, 2013), a clear strategy to combat various soil-borne pathogens of medicinal plants has not yet been provided. Although special biological control agents are needed for better management of medicinal plants, the existence of microbiomes related to them, which have significant metabolic activities, can be proposed as suitable and promising sources for use as new biological control agents.

1.4.1 Medicinal Plants and Potential Human Pathogens: Incidence and Feasible Biological Control

In most cases, medicinal plants are used in raw form, such as edible fruits, or dried, such as tea or other herbal brews. Therefore, when it comes to the direct consumption of herbal and medicinal plants, it is very important to pay attention to the harmful effects of plant-related microorganisms or any type of biological control agent used in human health and well-being. For example, lately, bacterial strains related to *Stenotrophomonas maltophilia* and *Rhodococcus* sp. were isolated from the oregano plant roots (*Origanum vulgare* L.) grown in the sub-Himalayan area (Bafana, 2013). Similarly, Ochrobactrum and Rhodococcus microbes were also identified by studying and investigating Egyptian medicinal plants (*Calendula officinalis* L., *Matricaria chamomilla* L., and *Solanum distichum* Schumach. & Thonn.; Köberl *et al.*, 2011). Among the studied genera, these bacterial genera have unknown interactions with eukaryotic hosts, so the processes that cause plant growth are similar to the processes that cause infectious diseases in humans and other animals (Berg *et al.*, 2005a). Moreover, suppressing plant pathogens and antagonistic activity against potentially dangerous human pathogens need to be viewed in a biological control strategy.

In contrast, ethanolic extracts of Chinese herbal medicines *Mallotus yunnanensis* Pax et. Hoffm., *Schima sinensis* (Hemsl. et. Wils) Airy-Shaw., *Garcinia morella* Desr., *Evodia daneillii* (Benn) Hemsl., *Meliosma squamulata* Hance., *Skimmia arborescens* Anders., and *Brandisia hancei* Hook. f. can be highly active in opposition to the clinical pathogens *Staphylococcus aureus*, *Escherichia coli*, *Pseudomonas aeruginosa*, and *Candida albicans*, consistent with their traditional uses in skin and other infections (Zuo *et al.*, 2012). Promising antimicrobial activities against multidrug-resistant human pathogens have also been discovered for Mexican medicinal plants (Jacobo-Salcedo Mdel *et al.*, 2011). As mentioned for herbal medicinal properties, suppression of human pathogens can often be attributed to microbes associated with medicinal plants and their secondary metabolites (Miller *et al.*, 2012b; Mousa and Raizada, 2013).

Therefore, medicinal plants should be considered meta-organisms that constitute both the plant itself and its microbiome. As meta-organisms, they constitute a largely unused and immense biological resource for bioactive compounds and potential microorganisms for use in modern medicine, agriculture, and the pharmaceutical industry. In this way, more research is needed to tap this huge reservoir for humanity.

1.5 MICROBES AND REDUCING BIOTIC AND ABIOTIC STRESSES

Global production of crops is affected by biotic and abiotic stressors, resulting in millions of dollars in losses (Suzuki *et al.*, 2014). The beginning of the Industrial Revolution in 1750 and the increase of human activities such as the burning of fossil fuels and deforestation became the beginning of global warming and climate change. An increase in carbon dioxide and temperature accelerates the life cycle of grains and oilseeds (Backlund *et al.*, 2008). According to Lesk *et al.*, excessive heat waves and droughts have caused a 10% decrease in the global harvest of cereals like corn, wheat, and rice over a 50-year period, which has grown to be a grave problem for a variety of governments (2016). The effect of global warming on the spring phenology of plants is obvious (Allstadt *et al.*, 2015). A longer growing season can increase carbon sequestration and potentially reduce climate change (Dragoni *et al.*, 2011), leaf greening (Jeong *et al.*, 2011), fruiting (Xia and Wan, 2013), and germination (Orsenigo *et al.*, 2015).

Abiotic stressors encompass temperature extremes, drought, water ingress, light, and salinity as essential parameters that affect plant growth (Majeed *et al.*, 2015). Plant-associated microbes or PGPR have been discovered to provide advantages to plants by generating indole acetic acid (Otieno *et al.*, 2015) and 1-aminocyclopropane-1-carboxylate (ACC) deaminases (Glick, 1995), solubilizing phosphate (Glick *et al.*, 1999), and increasing the uptake of nutrients from the environment (Egamberdieva *et al.*, 2015). They also help increase the plant's resistance to harmful environmental stresses such as drought, heavy metals, salts, and lack of nutrients (Table 1.2) (Zhang *et al.*, 2008). Biological stressors include interaction with other organisms and infection by pathogens or damage from insect pests, and some plant growth-promoting bacteria have been used as biological control agents against pathogens of plants (Ho *et al.*, 2015).

1.6 RHIZOSPHERE AND MICROBIAL COMMUNITY DIVERSITY

The dynamic density and combination of rhizosphere microbes can influence the coexistence of plant species in plant populations through indirect feedback (i.e., competition or inhibition of symbionts) (Pietikäinen and Kytöviita, 2007). Studies conducted on plants have found root symbionts that can increase the efficiency of nutrient uptake and allow the host to survive in a nutrient-poor environment, thus directly contributing to the competitive exclusion of other plants (O'Connor *et al.*, 2002). Rhizosphere microbes can alter the availability of different types of nitrogen or phosphorus in the

TABLE 1.2
Different Strategies of Microorganism–Plant Interaction to Withstand Abiotic Stresses in Plants)

Abiotic Stress	Plants	Microbial Inoculation	References
Salt	*Oryza sativa, Triticum aestivum, Zea mays, Gossypium hirsutum*	Cyanobacteria and cyanobacterial extracts	El-Tarabily *et al.*, 2020
Salt	*Vitis vinifera, Capsicum annuum*	Burkholderia, *Arthrobacter Bacillus*	Mathew *et al.*, 2020
Osmotic stress	*Zea mays*	*Bacillus megaterium*	Neubauer *et al.*, 2002
Drought	*Zea mays*	*Burkholderia phytofirmans Enterobacter* sp. FD17	Camerini *et al.*, 2008
Drought	*Arabidopsis thaliana*	*Pseudomonas chlororaphis* O6	Arshad *et al.*, 2007
Heat	*Triticum aestivum*	*Bacillus amyloliquefaciens, Azospirillum brasilence*	Kapoor *et al.*, 2013
Salinity	*Arabidopsis*	*Bacillus subtilis*	Guo *et al.*, 2015
Salinity	*Phaseolus vulgaris*	*Azospirillum brasilense* strain	Glick, 2012
Cd, As, Cu, Pb, and Zn toxicity	*Miscanthus sinensis*	*Pseudomonas koreensis* AGB-1	Laxa *et al.*, 2019
Arsenic toxicity	*Brassica juncea*	*Staphylococcus arlettae*	Balestrini and Lumini, 2018
Zn toxicity	*Triticum aestivum*	*Pseudomonas aeruginosa*	Posta and Duc, 2020
Hg toxicity	*Phragmites australis*	*Photobacterium* spp.	Ouledali *et al.*, 2019

Source: Adapted from Munir *et al.*, 2022

soil and influence plant–plant interactions through useful resource sharing (Reynolds *et al.*, 2003). Also, soil resources can be transported by using common symbiotic fungi called common mycorrhizal networks (CMNs) (Simard and Durall, 2004). In nature, various plant species usually share mycorrhizal fungi. Simard and Durall indicated the direct transition of resources from one plant to another by CMN with labelled carbon, nitrogen, and phosphorus (2004). The dynamics of plant communities are managed through the microbial mediation of dividing and sharing soil resources.

Lorenz Hiltner described the term 'rhizosphere' as the zone of chemical, biological, and physical in the soil under the influence of plant roots (Hartmann *et al.*, 2008). Soil microorganisms are chemotactically absorbed by plant root secretions and volatile and deposited organic carbon and then multiply in these carbon-rich surroundings (Lugtenberg and Kamilova, 2009). The type of plant root secretions differs between different plant species, so these differences should be observed in the rhizosphere microbiomes of different plant species (Dhungana *et al.*, 2023). Recent studies have clearly shown that plant species have their own microbiomes (Howard *et al.*, 2020). In addition, plants can form a microbial community through root secretions. Root secretions can be classified into sugars, amino acids, organic acids, nucleotides, flavonoids, antimicrobial compounds, and enzymes (Dhungana *et al.*, 2023).

A prior report confirmed that organic acids modulate colonization and increase biofilm formation of the root microbiome. It was proven that organic acids obtained from banana root exudates facilitate root colonization through *Bacillus amylolymphaciens* (Yuan *et al.*, 2015). Fumaric acid remarkably induced biofilm formation, while malic acid elicited a considerable chemotactic reaction. The outcomes demonstrated that organic acids obtained from banana root secretions performed an important role in the uptake and initiation of plant growth–promoting rhizobacteria colonization on the roots of plants. Secretions of rice, which mainly contain amino acid residues histidine, proline, valine, alanine, and glycine and carbohydrates glucose, arabinose, mannose, galactose, and glucuronic acid, can induce more chemotactic reactions via endophytic bacteria *Corynebacterium flavescens* and *Bacillus pumilus* (Bacilio-Jiménez *et al.*, 2003). Secretions of phenolic root not only have antimicrobial activity but also beneficially absorb soil microorganisms that affect the native soil microbial community (Clocchiatti *et al.*, 2021). Hence, plants can form a particular microbial community of the rhizosphere through root exudates.

1.7 INCREASED BIOACTIVE COMPOUND ACCUMULATION AND SECONDARY METABOLITE PRODUCTION IN MEDICINAL PLANTS

Naturally, plants generate various products with several chemical characteristics that are used for plant growth and development. Primary metabolites supply the resources needed for processes such as photosynthesis, movement, and respiration. Products obtained from primary metabolites that are not directly involved in growth and development are called secondary metabolites. Generally, primary metabolites produce secondary metabolites and are obtained from changes in biosynthesis, including methylation, glycosylation, and hydroxylation. Secondary metabolites are more complex than primary plant metabolites in terms of structural composition and side chains (Figueredo *et al.*, 2008).

Microorganisms of soil have an important function in the growth and development of plants and plant secretions. Microorganisms can affect secretion by affecting root cell permeability and root metabolism. On the other hand, microorganisms can also take up some compounds in root secretions and excrete other compounds. Microorganisms in soil can produce secondary metabolites that influence plant signalling and metabolism and can be considered a 'plant secondary genome' that prepares microbe-derived compounds to host plants (Rout and Southworth, 2013; Naamala and Smith, 2021). Certain microbes, as well as certain antibiotics (such as penicillin and polymyxin), increase the secretion of organic substances, modify cell permeability, and increase leakage (Harmsen and Jager, 1963; Norman, 1960). Microbes of soil can also induce the secretion of phenolic compounds to enhance iron uptake by plants in iron-deficient soils (Pii *et al.*, 2015; Chandwani *et al.*, 2022).

Some of these metabolites are also obtained from bacteria of endophytes. Some medicinal plants harbour endophytes capable of generating a various group of metabolites with high commercial value (Gouda *et al.*, 2016). These microbes' ability to synthesize secondary metabolites analogous to their host plants, and in some cases even to increment the production of plant secondary metabolites (Tiwari *et al.*, 2013, offers a tremendous opportunity to explore these great resources as new manufacturers of alkaloids and terpenoids. Additionally, compared to host plants, bacterial endophytes can convert some alkaloids and terpenoids to strong newer derivatives (Chu and Bae, 2022).

1.8 CONCLUSION AND PERSPECTIVES

Against the pressure caused by climate change, polluted environments, and pathogens, agricultural products, food materials, and sources are currently under threat. On the other hand, symbiotic and plant-associated microorganisms can have significant effects on the health and performance of the host. However, improving production and increasing product performance permanently through using microorganisms requires more study and investigation. Understanding of the interactions between plants and microbial symbiotic communities, the ecological outcomes of plant-associated microorganisms, and the microbial metabolic dynamics of plants is limited. The emergence of genomic approaches has greatly contributed to the comprehension of plant-bacteria interactions; however, genomic approaches are not yet sufficient to explain interactions among plants and pathogens (Bender, 2005). Procedures regarding the use of metagenomics and amplicon sequencing, along with other omics technologies (Quirino *et al.*, 2010) and database development (based on PHI) (Winnenburg *et al.*, 2008) and metabolomics, have improved our comprehension of interactions between plants and bacteria.

Microorganism–vegetation ecological communities are influenced by factors of plant genotype and environment. The difference between various genotypes causes several physiological and immune responses and causes the formation of host-specific microbial communities. Plant root secretions (e.g., sugars, amino acids, organic acids, nucleotides, flavonoids, antimicrobial compounds, and enzymes) form specific communities, promoting colonization of plant growth and attracting pest pathogenic infections. In the complicated microbial community, we discovered that particular species could significantly influence the structure of the microbial community.

Through a systemic substructure for the analysis of microbial networks, we can recognize 'core species' and 'tipping point species' that are candidate microbial communities for disease management. Models of plant-associated microbiome networks offer new opportunities to improve disease management. Therefore, understanding the various microbe–plant and microbe–microbe interactions provides a useful future perspective as a regulatory microbiome to minimize disease occurrence and increase the gross productivity of plants. Moreover, beneficial bacteria associated with plants can act as homologs against pathogens in the ecosystem of microorganisms. In addition, they provide the possibility of stabilizing the ecosystem, enhancing biodiversity, preventing the spread of pathogens, and, ultimately, increasing the productivity of plants. Well-studied collaboration with plant microorganisms in the future will also help increase crop productivity at little cost, and this shift could guide a 'Third Agricultural Revolution'.

REFERENCES

Abdou, R., Scherlach, K., Dahse, H.M., Sattler, I., Hertweck, C. 2010. Botryorhodines A-D, antifungal and cytotoxic depsidones from *Botryosphaeria rhodina*, an endophyte of the medicinal plant Bidens pilosa. *Phytochemistry* 71: 110–116. doi: 10.1016/j.phytochem.2009.09.024.

Agarwal, A. 2005. *Pharma Times* 37(6): 9–11.

Allstadt, A.J., Vavrus, S.J., Heglund, P.J., Pidgeon, A.M., Thogmartin, W.E., Radeloff, V.C. 2015. Spring plant phenology and false springs in the conterminous US during the 21st century. *Environ. Res. Lett.* 10(10): 104008.

Amna, T., Puri, S.C., Verma, V., Sharma, J.P., Khajuria, R.K., Musarrat, J., *et al.* 2006. Bioreactor studies on the endophytic fungus Entrophospora infrequens for the production of an anticancer alkaloid camptothecin. *Can. J. Microbiol.* 52: 189–196. doi: 10.1139/W05-122.

Anquez-Traxler, C. 2011. The legal and regulatory framework of herbal medicinal products in the European Union: A focus on the traditional herbal medicines category. *Drug Inf. J.* 45: 15–23. doi: 10.1177/009286151104500102.

Arshad, M., Saleem, M., Hussain, S. 2007. Perspectives of bacterial acc deaminase in phytoremediation. *Trends Biotechnol.* 25: 356–362.

Bacilio-Jiménez, M., Aguilar-Flores, S., Ventura-Zapata, E., Pérez-Campos, E., Bouquelet, S., Zenteno, E. 2003. Chemical characterization of root exudates from rice (*Oryza sativa*) and their effects on the chemotactic response of endophytic bacteria. *Plant Soil* 249(2): 271–277.

Backlund, P., Janetos, A., Schimel, D. 2008. *The Effects of Climate Change on Agriculture, Land Resources, Water Resources, and Biodiversity in the United States*. Washington, DC: United States Climate Change Science Program.

Balestrini, R., Lumini, E. 2018. Focus on mycorrhizal symbioses. *Appl. Soil Ecol.* 123: 299–304.

Bandaranayake, W.M. 2006. Quality control, screening, toxicity, and regulation of herbal drugs. In *Modern Phytomedicine. Turning Medicinal Plants into Drugs*, eds. Ahmad, I., Aqil, F., Owais, M. Weinheim: Wiley-VCH GmbH & Co. KGaA, 25–57. https://doi.org/10.1002/9783527609987.ch2

Barnard, R.L., Osborne, C.A., Firestone, M.K. 2013. Responses of soil bacterial and fungal communities to extreme desiccation and rewetting. *ISME J.* 7(11): 2229–2241. doi: 10.1038/ismej.

Bender, C.L. 2005. The post-genomic era: New approaches for studying bacterial diseases of plants. *Australas. Plant Pathol.* 34(4): 471–474.

Berendsen, R.L., Pieterse, C.M.J., Bakker, P.A.H.M. 2012. The rhizosphere microbiome and plant health. *Trends Plant Sci.* 17: 478–486.

Berg, G. 2009. Plant-microbe interactions promoting plant growth and health: Perspectives for controlled use of microorganisms in agriculture. *Appl. Microbiol. Biotechnol.* 84: 11–18. doi: 10.1007/s00253-009-2092-7.

Berg, G., Rybakova, D., Grube, M., Köberl, M. 2015. The plant microbiome explored: Implications for experimental botany. *J. Exp. Bot.* 67: 995–1002.

Bever, J.D. 2003. Soil community feedback and the coexistence of competitors: Conceptual frameworks and empirical tests. *New Phytol.* 157(3): 465–473.

Bever, J.D., Dickie, I.A., Facelli, E., Facelli, J.M., Klironomos, J., Moora, M., *et al.* 2010. Rooting theories of plant community ecology in microbial interactions. *Trends Ecol. Evol.* 25(8): 468–478.

Bodeker, C., Bodeker, G., Ong, C.K., Grundy, C.K., Burford, G., Shein, K. 2005. *WHO Global Atlas of Traditional, Complementary and Alternative Medicine*. Geneva, Switzerland: World Health Organization.

Bodenhausen, N., Bortfeld-Miller, M., Ackermann, M., Vorholt, J.A. 2014. A synthetic community approach reveals plant genotypes affecting the phyllosphere microbiota. *PLoS Genet.* 10(4): e1004283.

Braun, L.A., Tiralongo, E., Wilkinson, J.M., Spitzer, O., Bailey, M., Poole, S., *et al.* 2010. Perceptions of use and attitudes of pharmacy customers on complementary medicines and pharmacy practice. BMC Complement. *Altern. Med.* 10: 38. doi: 10.1186/1472-6882-10-38.

Bray, E.A., Bailey-Serres, J., Weretilnyk, E. 2000. Responses to abiotic stresses. In *Biochemistry and Molecular Biology of Plants*, eds. Gruissem, W., Jones, R. Rockville: American Society of Plant Physiologists, 1158–1203.

Bulgarelli, D., Rott, M., Schlaeppi, K., van Themaat, E.V.L., Ahmadinejad, N., Assenza, F., *et al.* 2012. Revealing structure and assembly cues for Arabidopsis root-inhabiting bacterial microbiota. *Nature* 488(7409): 91–95.

Bulgarelli, D., Schlaeppi, K., Spaepen, S., van Themaat, E.V.L., Schulze-Lefert, P. 2013. Structure and functions of the bacterial microbiota of plants. *Annu. Rev. Plant Biol.* 64: 807–838.

Calapai, G. 2008. European legislation on herbal medicines: A look into the future. *Drug Saf.* 31: 428–431. doi: 10.2165/00002018-200831050-00009.

Calixto, J.B. 2019. The role of natural products in modern drug discovery. *An. Acad. Bras. Cienc.* 91(Suppl 3): e20190105.

Camerini, S., Senatore, B., Lonardo, E., Imperlini, E., Bianco, C., Moschetti, G., Rotino, G.L., Campion, B., Defez, R. 2008. Introduction of a novel pathway for IAA biosynthesis to rhizobia alters vetch root nodule development. *Arch. Microbiol.* 190: 67–77.

Castillo, U., Harper, J.K., Strobel, G.A., Sears, J., Alesi, K., Ford, E., *et al.* 2003. Kakadumycins, novel antibiotics from Streptomyces sp. NRRL 30566, an endophyte of *Grevillea pteridifolia*. *FEMS Microbiol. Lett.* 224: 183–190. doi: 10.1016/S0378-1097(03)00426-9.

Castillo, U., Strobel, G.A., Ford, E.J., Hess, W.M., Porter, H., Jensen, J.B., *et al.* 2002. Munumbicins, wide-spectrum antibiotics produced by Streptomyces NRRL 30562, endophytic on *Kennedia nigriscans*. *Microbiology* 148: 2675–2685.

Castronovo, L.M., Vassallo, A., Mengoni, A., Miceli, E., Bogani, P., Firenzuoli, F., Fani, R., Maggini, V. 2021. Medicinal plants and their bacterial microbiota: A review on antimicrobial compounds production for plant and human health. *Pathogens* 10(2): 106.

Chandwani, S., Chavan, S.M., Paul, D., Amaresan, N. 2022. Bacterial inoculations mitigate different forms of iron (Fe) stress and enhance nutrient uptake in rice seedlings (*Oryza sativa* L.). *Environ. Technol. Innov.* 26: 102326.

Chaston, J.M., Dobson, A.J., Newell, P.D., Douglas, A.E. 2016. Host genetic control of the microbiota mediates the Drosophila nutritional phenotype. *Appl. Environ. Microbiol.* 82(2): 671–679.

Chu, L.L., Bae, H. 2022. Bacterial endophytes from ginseng and their biotechnological application. *J. Ginseng Res.* 46: 1–10. doi: 10.1016/j.jgr.2021.04.004.

Ciccazzo, S., Esposito, A., Rolli, E., Zerbe, S., Daffonchio, D., Brusetti, L. 2014. Different pioneer plant species select specific rhizosphere bacterial communities in a high mountain environment. *SpringerPlus* 3(1): 391.

Clocchiatti, A., Hannula, S.E., van den Berg, M., Hundscheid, M.P.J., de Boer, W. 2021. Evaluation of phenolic root exudates as stimulants of saprotrophic fungi in the rhizosphere. *Front. Microbiol.* 12: 644046.

Debbab, A., Aly, A.H., Edrada-Ebel, R.A., Müller, W.E., Mosaddak, M., Hakiki, A., *et al.* 2009. Bioactive secondary metabolites from the endophytic fungus *Chaetomium* sp. isolated from *Salvia officinalis* growing in Morocco. *Biotechnol. Agron. Soc. Environ.* 13: 229–234.

Dhungana, I., Kantar, M.B., Nguyen, N.H. 2023. Root exudate composition from different plant species influences the growth of rhizosphere bacteria. *Rhizosphere* 25: 100645. doi: 10.1016/j.rhisph.2022.100645.

Dragoni, D., Schmid, H.P., Wayson, C.A., Potter, H., Grimmond, C.S.B., Randolph, J.C. 2011. Evidence of increased net ecosystem productivity associated with a longer vegetated season in a deciduous forest in south-central Indiana, USA. *Glob. Change Biol.* 17(2): 886–897.

Egamberdieva, D., Jabborova, D., Hashem, A. 2015. Pseudomonas induces salinity tolerance in cotton (Gossypium hirsutum) and resistance to Fusarium root rot through the modulation of indole-3-acetic acid. *Saudi J. Biol. Sci.* 22(6): 773–779.

Ekor, M. 2014. The growing use of herbal medicines: Issues relating to adverse reactions and challenges in monitoring safety. *Front. Pharmacol.* 4: 177. doi: 10.3389/fphar.2013.00177.

Ek-Ramos, M.J., Gomez-Flores, R., Orozco-Flores, A.A., Rodriguez-Padilla, C., Gonzalez-Ochoa, G., Tamez-Guerra, P. 2019. Bioactive products from plant-endophytic gram-positive bacteria. *Front. Microbiol.* 10: 463. doi: 10.3389/fmicb.2019.00463.

El-Tarabily, K.A., ElBaghdady, K.Z., AlKhajeh, A.S., Ayyash, M.M., Aljneibi, R.S., El-Keblawy, A., AbuQamar, S.F. 2020. Polyamine-producing actinobacteria enhance biomass production and seed yield in *Salicornia bigelovii*. *Biol. Fertil. Soils* 56: 499–519.

Ezra, D., Castillo, U.F., Strobel, G.A., Hess, W.M., Porter, H., Jensen, J.B., *et al.* 2004. Coronamycins, peptide antibiotics produced by a verticillate *Streptomyces* sp. (MSU-2110) endophytic on *Monstera* sp. *Microbiology* 150: 785–793. doi: 10.1099/mic.0.26645-0.

Figueredo, A.C., Barroso, J.G., Pedro, L.G., Scheffer, J.J. 2008. Factors affecting secondary metabolite production in plants: Volatile and essential oils. *Flavour Fragr. J.* 23: 213–226.

Firenzuoli, F., Gori, L. 2007. Herbal medicine today: Clinical and research issues. *Evid. Based Complement. Altern. Med.* 4: 37–40.

Friesen, M.L., Porter, S.S., Stark, S.C., von Wettberg, E.J., Sachs, J.L., Martinez-Romero, E. 2011. Microbially mediated plant functional traits. *Annu. Rev. Ecol. Evol. Syst.* 42: 23–46.

Garbeva, P., Hordijk, C., Gerards, S., de Boer, W. 2014. Volatile-mediated interactions between phylogenetically different soil bacteria. *Front. Microbiol.* 5: 285–290.

Glick, B.R. 1995. The enhancement of plant growth by free-living bacteria. *Can. J. Microbiol.* 41(2): 109–117.

Glick, B.R. 2012. Plant growth-promoting bacteria: Mechanisms and applications. *Scientifica*. 963401. doi: 10.6064/2012/963401.

Glick, B.R., Patten, C.L., Holguin, G., Penrose, D. 1999. *Biochemical and genetic mechanisms used by plant growth promoting bacteria*. London, UK: Imperial College Press.

Gouda, S., Das, G., Sen, S.K., Shin, H.S., Patra, J.K. 2016. Endophytes: A treasure house of bioactive compounds of medicinal importance. *Front. Microbiol.* 7: 1538. doi: 10.3389/fmicb.2016.01538.

Guo, J.K., Ding, Y.Z., Feng, R.W., Wang, R.G., Xu, Y.M., Chen, C., Wei, X.L., Chen, W.M. 2015. *Burkholderia metalliresistens* sp. Nov. multiple metal-resistant and phosphate-solubilizing species isolated from heavy metal-polluted soil in southeast China. *Antonie Van Leeuwenhoek* 107: 1591–1598.

Gupta, A., Gupta, R., Singh, R.L. 2016. Microbes and environment. In *Principles and Applications of Environmental Biotechnology for a Sustainable Future. Applied Environmental Science and Engineering for a Sustainable Future*, ed. Singh, R. Singapore: Springer. doi: 10.1007/978-981-10-1866-4_3

Harmsen, G., Jager, G. 1963. Determination of the quantity of carbon and nitrogen in the rhizosphere of young plants. In Doeksen, J., van der Drift, J., eds. *Soil Organisms*. Amsterdam: North Holland, 245–251.

Harper, J.K., Arif, A.M., Ford, E.J., Strobel, G.A., Porco, J.A., Tomer, D.P., *et al*. 2003. Pestacin: A 1,3-dihydro isobenzofuran from *Pestalotiopsis microspora* possessing antioxidant and antimycotic activities. *Tetrahedron* 59: 2471–2476. doi: 10.1016/S0040-4020(03)00255-2.

Hartmann, A., Rothballer, M., Schmid, M., Hiltner, L. 2008. A pioneer in rhizosphere microbial ecology and soil bacteriology research. *Plant Soil* 312(1–2): 7–14.

Heinrich, M., Jäger, A.K. 2015. *Ethnopharmacology*. John Wiley & Sons Ltd. 464 pp.

Ho, Y-N., Chiang, H-M., Chao, C-P., Su, C-C., Hsu, H-F, Guo, C-t., *et al*. 2015. In planta biocontrol of soil-borne Fusarium wilt of banana through a plant endophytic bacterium, *Burkholderia cenocepacia* 869T2. *Plant Soil* 387(1–2): 295–306.

Horton, M.W., Bodenhausen, N., Beilsmith, K., Meng, D., Muegge, B.D., Subramanian, S., *et al*. 2014. Genome-wide association study of *Arabidopsis thaliana's* leaf microbial community. *Nat. Commun*. 5: 5320.

Howard, M.M., Muñoz, C.A., Kao-Kniffin, J., Kessler, A. 2020. Soil microbiomes from fallow fields have species-specific effects on crop growth and pest resistance. *Front. Plant Sci*. 11: 1171.

Immaculate, N.R., Kumar, M., Srimathi, S., Muthumary, J., Kalaichelvan, P.T. 2011. Isolation of Phoma species from Aloe vera: An endophyte and screening the fungus for Taxol production. *World J. Sci. Technol*. 1: 23–31.

Jeong, S.J., Ho, C.H., Gim, H.J., Brown, M.E. 2011. Phenology shifts at start vs. end of growing season in temperate vegetation over the Northern Hemisphere for the period 1982–2008. *Glob. Chang. Biol*. 17(7): 2385–2399.

Jones, J.D., Dangl, J.L. 2006. The plant immune system. *Nature* 444(7117): 323–329.

Kapoor, R., Evelin, H., Mathur, P., Giri, B. 2013. *Plant Acclimation to Environmental Stress*. Springer: Berlin/Heidelberg, Germany, 359–401.

Keita, K., Darkoh, C., Okafor, F. 2022. Secondary plant metabolites as potent drug candidates against antimicrobial-resistant pathogens. *SN Appl. Sci*. 4(8): 209.

Kim, S.U., Strobel, G.A., Ford, E. 1999. Screening of taxol-producing endophytic fungi from Ginkgo biloba and *Taxus cuspidata* in Korea. *Agric. Chem. Biotechnol*. 42: 97–99. doi: 10.1042/BA20080110.

Koide, R.T., Mosse, B. 2004. A history of research on arbuscular mycorrhiza. *Mycorrhiza* 14: 145–163.

Kong, J.M., Goh, N.K., Chia, L.S., Chia, T.F. 2003. Recent advances in traditional plant drugs and orchids. *Acta Pharm. Sin*. 24: 7–21.

Köberl, M., Müller, H., Ramadan, E.M., Berg, G. 2011. Desert farming benefits from microbial potential in arid soils and promotes diversity and plant health. *PLoS ONE* 6: e24452. doi: 10.1371/journal.pone.0024452.

Köberl, M., Schmidt, R., Ramadan, E.M., Bauer, R., Berg, G. 2013. The microbiome of medicinal plants: Diversity and importance for plant growth, quality and health. *Front. Microbiol*. 4: 400. doi: 10.3389/fmicb.2013.00400.

Kusari, S., Lamshöft, M., Spiteller, M. 2009a. Aspergillus fumigatus Fresenius, an endophytic fungus from *Juniperus communis* L. Horstmann is a novel source of the anticancer pro-drug deoxypodophyllotoxin. *J. Appl. Microbiol*. 107: 1019–1030. doi: 10.1111/j.1365-2672.2009.04285.x.

Kusari, S., Zühlke, S., Spiteller, M. 2009b. An endophytic fungus from *Camptotheca acuminata* that produces camptothecin and analogs. *J. Nat. Prod*. 72: 2–7. doi: 10.1021/np800455b.

Lareen, A., Burton, F., Schäfer, P. 2016. Plant root-microbe communication in shaping root microbiomes. *Plant Mol. Biol*. 90(6): 575–587.

Laxa, M., Liebthal, M., Telman, W., Chibani, K., Dietz, K.J. 2019. The role of the plant antioxidant system in drought tolerance. *Antioxidants* 8(4): 94. doi: 10.3390/antiox8040094.

Lesk, C., Rowhani, P., Ramankutty, N. 2016. Influence of extreme weather disasters on global crop production. *Nature* 529(7584): 84–87.

Li, R.W., Myers, S.P., Leach, D.N., Lin, G.D., Leach, G. 2003. Across-cultural study: Anti-inflammatory activity of Australian and Chinese plants. *J. Ethnopharmacol*. 85: 25–32.

Lugtenberg, B., Kamilova, F. 2009. Plant-growth-promoting rhizobacteria. *Ann. Rev. Microbiol*. 63: 541–556.

Lugtenberg, B.J., Chin-A-Woeng, T.F., Bloemberg, G.V. 2002. Microbe-plant interactions: Principles and mechanisms. *Antonie Van Leeuwenhoek* 81(1–4): 373–383. doi: 10.1023/a:1020596903142.

Lundberg, D.S., Lebeis, S.L., Paredes, S.H., Yourstone, S., Gehring, J., Malfatti, S., *et al*. 2012. Defining the core *Arabidopsis thaliana* root microbiome. *Nature* 488(7409): 86–90.

Majeed, A., Abbasi, M.K., Hameed, S., Imran, A., Rahim, N. 2015. Isolation, and characterization of plant growth-promoting rhizobacteria from wheat rhizosphere and their effect on plant growth promotion. *Front. Microbiol.* 6: 198.

Mathew, B.T., Torky, Y., Amin, A., Mourad, A.I., Ayyash, M.M., El-Keblawy, A., Hilal-Alnaqbi, A., AbuQamar, S.F., El-Tarabily, K.A. 2020. Halotolerant marine rhizosphere-competent actinobacteria promote *Salicornia bigelovii* growth and seed production using seawater irrigation. *Front. Microbiol.* 11: 552. doi: 10.3389/fmicb.2020.00552.

McFall-Ngai, M., Hadfield, M.G., Bosch, T.C., Carey, H.V., Domazet-Lošo, T., Douglas, A.E., *et al.* 2013. Animals in a bacterial world, are a new imperative for the life sciences. *Proc. Natl. Acad. Sci. U. S. A.* 110: 3229–3236.

Mendes, R., Garbeva, P., Raaijmakers, J.M. 2013. The rhizosphere microbiome: Significance of plant beneficial, plant pathogenic, and human pathogenic microorganisms. *FEMS Microbiol. Rev.* 37: 634–663. doi: 10.1111/1574-6976.12028.

Mendes, R., Kruijt, M., de Bruijn, I., Dekkers, E., van der Voort, M., Schneider, J.H., *et al.* 2011. Deciphering the rhizosphere microbiome for disease-suppressive bacteria. *Science* 332: 1097–1100. doi: 10.1126/science.1203980.

Miller, K.I., Qing, C., Sze, D.M., Neilan, B.A. 2012a. Investigation of the biosynthetic potential of endophytes in traditional Chinese anticancer herbs. *PLoS One* 7: e35953.

Miller, K.I., Qing, C., Sze, D.M., Roufogalis, B.D., Neilan, B.A. 2012b. Culturable endophytes of medicinal plants and the genetic basis for their bioactivity. *Microb. Ecol.* 64: 431–449.

Mohana Kumara, P., Zuehlke, S., Priti, V., Ramesha, B.T., Shweta, S., Ravikanth, G., Vasudeva, R., Santhoshkumar, T.R., Spiteller, M., Uma Shaanker, R. 2012. Fusarium proliferatum, an endophytic fungus from *Dysoxylum binectariferum* Hook. f, produces rohitukine, a chromane alkaloid possessing anticancer activity. *Antonie Van Leeuwenhoek* 101(2): 323–329. doi: 10.1007/s10482-011-9638-2.

Munir, N., Hanif, M., Abideen, Z., Sohail, M., El-Keblawy, A., Radicetti, E., Mancinelli, R., Haider, G. 2022. Mechanisms and strategies of plant microbiome interactions to mitigate abiotic stresses. *Agronomy* 12: 2069. doi: 10.3390/agronomy12092069.

Naamala, J., Smith, D.L. 2021. Microbial derived compounds, a step toward enhancing microbial inoculants technology for sustainable agriculture. *Front Microbiol.* 12: 634807.

Namdar, D., Charuvi, D., Ajjampura, V., Mazuz, M., Ion, A., Kamara, I., Hinanit, K. 2019. LED lighting affects the composition and biological activity of Cannabis sativa secondary metabolites. *Ind. Crop. Prod.* 132: 177–185. doi: 10.1016/j. indcrop.2019.02.016.

Neubauer, U., Furrer, G., Schulin, R. 2002. Heavy metal sorption on soil minerals affected by the siderophore desferrioxamine b: The role of Fe (iii)(hydr) oxides and dissolved Fe (iii). *Eur. J. Soil Sci.* 53: 45–55.

Newman, D.J., Cragg, G.M. 2012. Natural products as sources of new drugs over the 30 years from 1981 to 2010. *J. Nat. Prod.* 75: 311–335.

Norman, A.G. 1960. Microbial products affecting root development. *Transactions of the 7th Congress Wisconsin International Soil Science Society* 2: 531–536.

O'Connor, P.J., Smith, S.E., Smith, F.A. 2002. Arbuscular mycorrhizas influence plant diversity and community structure in a semiarid herbland. *New Phytologist.* 154(1): 209–218.

Oldroyd, G.E. 2013. Speak, friend, and enter: Signalling systems that promote beneficial symbiotic associations in plants. *Nat. Rev. Microbiol.* 11(4): 252–263.

Orsenigo, S., Abeli, T., Rossi, G., Bonasoni, P, Pasquaretta, C., Gandini, M., Mondoni, A. 2015. Effects of autumn and spring heat waves on seed germination of high mountain plants. *PLoS One* 10(7): e0133626. doi: 10.1371/journal.pone.0133626

Otieno, N., Lally, R.D., Kiwanuka, S., Lloyd, A., Ryan, D., Germaine, K.J., Dowling, D.N. 2015. Plant growth promotion induced by phosphate solubilizing endophytic Pseudomonas isolates. *Front. Microbiol.* 6: 745. doi: 10.3389/fmicb.2015.00745.

Ouledali, S., Ennajeh, M., Ferrandino, A., Khemira, H., Schubert, A., Secchi, F. 2019. Influence of *arbuscular mycorrhizal* fungi inoculation on the control of stomata functioning by abscisic acid (aba) in drought-stressed olive plants. *S. Afr. J. Bot.* 121: 152–158.

Pajares, S., Escalante, A.E., Noguez, A.M., García-Oliva, F., Martínez-Piedragil, C., Cram, S.S., Eguiarte, L.E., Souza, V. 2016. Spatial heterogeneity of physicochemical properties explains differences in microbial composition in arid soils from Cuatro Cienegas, Mexico. *PeerJ.* 4: e2459. doi: 10.7717/peerj.2459.

Partida-Martínez, L.P., Heil, M. 2011. The microbe-free plant: Fact or artifact? *Front Plant Sci.* 2: 100. doi: 10.3389/fpls.2011.00100.

Peiffer, J.A., Spor, A., Koren, O., Jin, Z., Tringe, S.G., Dangl, J.L., Buckler, E.S., Ley, R.E. 2013. Diversity and heritability of the maize rhizosphere microbiome under field conditions. *Proc. Natl. Acad. Sci.* 110(16): 6548–6553. doi: 10.1073/pnas.1302837110.

Peñuelas, J., Asensio, D., Tholl, D., Wenke, K., Rosenkranz, M., Piechulla, B., Schnitzler, J.P. 2014. Biogenic volatile emissions from the soil. *Plant Cell Environ.* 37(8): 1866–1891. doi: 10.1111/pce.12340.

Pietikäinen, A., Kytöviita, M.M. 2007. Defoliation changes mycorrhizal benefits and competitive interactions between seedlings and adult plants. *J. Ecol.* 95(4): 639–647.

Pineda, A., Dicke, M., Pieterse, C.M., Pozo, M.J. 2013. Beneficial microbes in a changing environment: Are they always helping plants to deal with insects? *Functional Ecology* 27(3): 574–586.

Pii, Y., Mimmo, T., Tomasi, N., Terzano, R., Cesco, S., Crecchio, C. 2015. Microbial interactions in the rhizosphere: Beneficial influences of plant growth-promoting rhizobacteria on nutrient acquisition process. A review. *Biol. Fertil. Soils* 51(4): 403–415.

Posta, K., Duc, N.H. 2020. Benefits of arbuscular mycorrhizal fungi application to crop production under water scarcity. In *Drought Detections and Solutions*. Rijeka, Croatia: InTechOpen.

Puri, S.C., Verma, V., Amna, T., Qazi, G.N., Spiteller, M. 2005. An endophytic fungus from *Nothapodytes foetida* that produces camptothecin. *J. Nat. Prod.* 68: 1717–1719. doi: 10.1021/np0502802.

Quirino, B., Candido, E., Campos, P., Franco, O., Krüger, R. 2010. Proteomic approaches to study plant-pathogen interactions. *Phytochemistry* 71(4): 351–362.

Raynor, D.K., Dickinson, R., Knapp, P., Long, A.F., Nicolson, D.J. 2011. Buyer beware? Does the information provided with herbal products available over the counter enable safe use? *BMC Med.* 9: 94. doi: 10.1186/1741-7015-9-94.

Reynolds, H.L., Packer, A., Bever, J.D., Clay, K. 2003. Grassroots ecology: Plant-microbe – soil interactions as drivers of plant community structure and dynamics. *Ecology*. 84(9): 2281–2291.

Rezaei, A., Amirahmadi, A., Poozesh, V. 2022. Herbal research: Gaps affecting the quality and validity of research findings. *J. Chem. Health Risks* 12(0): 0–0.

Roberts, J.E., Tyler, V.E. 1997. *Tyler's Herbs of Choice. The Therapeutic Use of Phytomedicine*. New York: The Haworth Press.

Rout, M.E., Southworth, D. 2013. The root microbiome influences scales from molecules to ecosystems: The unseen majority. *Am. J. Bot.* 100(9): 1689–1691.

Sen, S., Chakraborty, R., De, B. 2011. Challenges and opportunities in the advancement of herbal medicine: India's position and role in a global context. *J Herb. Med.* 1(3–4): 67–75.

Simard, S.W., Durall, D.M. 2004. Mycorrhizal networks: A review of their extent, function, and importance. *Canad. J. Bot.* 82(8): 1140–1165.

Singh, B.K., Bardgett, R.D., Smith, P., Reay, D.S. 2010. Microorganisms and climate change: Terrestrial feedbacks and mitigation options. *Nat. Rev. Microbiol.* 8(11): 779–790. doi: 10.1038/nrmicro2439.

Stierle, A., Strobel, G., Stierle, D. 1993. Taxol and taxane production by *Taxomyces andreanae*, an endophytic fungus of Pacific yew. *Science* 260: 214–216. doi: 10.1126/science.809706.1.

Strickland, M.S., Lauber, C., Fierer, N., Bradford, M.A. 2009. Testing the functional significance of microbial community composition. *Ecology* 90(2): 441–451. doi: 10.1890/08-0296.1.

Strobel, G., Daisy, B., Castillo, U., Harper, J. 2004. Natural products from endophytic microorganisms. *J. Nat. Prod.* 67: 257–268.

Strobel, G., Ford, E., Worapong, J., Harper, J.K., Arif, A.M., Grant, D.M., Fung, P.C., Ming Wah Chau, R. 2002. Isopestacin, an isobenzofuranone from *Pestalotiopsis microspora*, possesses antifungal and antioxidant activities. *Phytochemistry* 60: 179–183. doi: 10.1016/S0031-9422(02)00062-6.

Suzuki, N., Rivero, R.M., Shulaev, V., Blumwald, E., Mittler, R. 2014. Abiotic and biotic stress combinations. *New Phytol.* 203(1): 32–43. doi: 10.1111/nph.12797.

Tiwari, R., Awasthi, A., Mall, M., Shukla, A.K., Srinivas, K.V., Syamasundar, K.V., Kalra, A. 2013. Bacterial endophyte-mediated enhancement of in planta content of key terpenoid indole alkaloids and growth parameters of *Catharanthus Roseus*. *Ind. Crops Prod.* 43: 306–310. doi 10.1016/j.indcrop.2012.07.045.

Turner, T.R., James, E.K., Poole, P.S. 2013. The plant microbiome. *Genome Biol.* 14(6): 209. doi: 10.1186/gb-2013-14-6-209.

van der Heijden, M.G., Bardgett, R.D., van Straalen, N.M. 2008. The unseen majority: Soil microbes as drivers of plant diversity and productivity in terrestrial ecosystems. *Ecol Lett.* 11(3): 296–310. doi: 10.1111/j.1461-0248.2007.01139.x.

van der Heijden, M.G., Hartmann, M. 2016. Networking in the plant microbiome. *PLoS Biol.* 14(2): e1002378.

Verma, S., Singh, S.P. 2008. Current and future status of herbal medicines. *Vet. World* 1(11): 347.

Wang, G., Ren, Y., Bai, X., Su, Y., Han, J. 2022. Contributions of beneficial microorganisms in soil remediation and quality improvement of medicinal plants. *Plants (Basel)* 11(23): 3200.

Wang, J.W., Zhang, Z., Tan, R.X. 2001. Stimulation of artemisinin production in *Artemisia annua* hairy roots by the elicitor from the endophytic *Colletotrichum* sp. *Biotechnol. Lett.* 23: 857–860. doi: 10.1023/A:1010535001943.

Wani, M.C., Taylor, H.L., Wall, M.E., Coggon, P., McPhail, A.T. 1971. Plant antitumor agents. VI. The isolation and structure of Taxol, a novel antileukemic and antitumor agent from *Taxus brevifolia*. *J. Am. Chem. Soc.* 93: 2325–2327. doi: 10.1021/ja00738a045.

Weber, D., Sterner, O., Anke, T., Gorzalczancy, S., Martino, V., Acevedo, C. 2004. Phomol, a new anti-inflammatory metabolite from an endophyte of the medicinal plant Erythrina crista-galli. *J. Antibiot. (Tokyo)* 57: 559–563. doi: 10.7164/antibiotics.57.559.

WHO. 2002a. *WHO Monographs on Selected Medicinal Plants.* Vol. 2. Geneva, Switzerland: World Health Organization.

WHO. 2004. *WHO Guidelines on Safety Monitoring of Herbal Medicines in Pharmacovigilance Systems.* Geneva, Switzerland: World Health Organization.

WHO. 2005. WHO Global Atlas of traditional, complementary and alternative medicine. In *Map Volume*, eds. Ong, C.K., Bodeker, G., Grundy, C., Burford, G., Shein, K. Geneva, Switzerland: World Health Organization.

Wings, S., Müller, H., Berg, G., Lamshöft, M., Leistner, E. 2013. A study of the bacterial community in the root system of the maytansine-containing plant *Putterlickia verrucosa*. *Phytochemistry* 91: 158–164. doi: 10.1016/j.phytochem.2012.06.016.

Winnenburg, R., Urban, M., Beacham, A., Baldwin, T.K., Holland, S., Lindeberg, M., *et al.* 2008. PHI-base update: Additions to the pathogen-host interaction database. *Nucleic Acids Res.* 36(Suppl 1): D572–D576. doi: 10.1093/nar/gkm858.

Xia, J., Wan, S. 2013. Independent effects of warming and nitrogen addition on plant phenology in the Inner Mongolian steppe. *Ann. Bot.* 111(6): 1207–1217.

Yang, X., Guo, S., Zhang, L., Shao, H. 2003. Selection of producing podophyllotoxin endophytic fungi from podophyllin plant. *Nat. Prod. Res. Dev.* 15: 419–422.

Yi, H.S., Yang, J.W., Ryu, C.M. 2013. ISR meets SAR outside: Additive action of the endophyte Bacillus pumilus INR7 and the chemical inducer, benzothiadiazole, on induced resistance against a bacterial spot in field-grown pepper. *Front. Plant Sci.* 4: 122. doi: 10.3389/fpls.2013.00122.

Yuan, J., Zhang, N., Huang, Q., Raza, W., Li, R., Vivanco, J.M., Shen, Q. 2015. Organic acids from root exudates of bananas help root colonization of the PGPR strain Bacillus amyloliquefaciens NJN-6. *Sci. Rep.* 5: 13438. doi: 10.1038/srep13438.

Zhang, H., Kim, M.S., Sun, Y., Dowd, S.E., Shi, H., Paré, P.W. 2008. Soil bacteria confer plant salt tolerance by tissue-specific regulation of the sodium transporter HKT1. *Mol. Plant Microbe Interact.* 21(6): 737–744. doi: 10.1094/MPMI-21-6-0737.

Zilber-Rosenberg, I., Rosenberg, E. 2008. Role of microorganisms in the evolution of animals and plants: The hologenome theory of evolution. *FEMS Microbiol. Rev.* 32: 723–735.

Zou, W.X., Meng, J.C., Lu, H., Chen, G.X., Shi, G.X., Zhang, T.Y., Tan, R.X. 2000. Metabolites of Colletotrichum gloeosporioides, an endophytic fungus in *Artemisia mongolica*. *J. Nat. Prod.* 63: 1529–1530. doi: 10.1021/np000204t.

2 Mycorrhiza – A Symbiotic Association between Root-Inhabiting Fungi and Medicinal Plants with Potential Applications in Pharmacology and Biotechnology

Zainab Shahzadi, Zubaida Yousaf, Zainab Maqbool, Arusa Aftab, Shahneela Mushtaq, and Nadia Riaz

2.1 SYMBIOTIC ASSOCIATION BETWEEN ROOT-INHABITING FUNGI AND PLANTS

Symbiosis, the coexistence of dissimilar organisms, has played a crucial role in the development of complex life on Earth. Among the significant symbiotic relationships found on land, one stands out: the association between approximately 85% of land plants and fungi from the basal phylum Mucoromycota. This phylum comprises two subphyla, Glomeromycotina and Mucoromycotina, alongside the higher categories of the fungal phyla Basidiomycota and Ascomycota, making it an essential terrestrial symbiosis (Smith *et al.*, 2008). The fungi establish a mutually beneficial association with nonvascular plants, which lack true roots, benefiting both organisms. This symbiotic relationship is referred to as mycorrhiza. According to Read *et al.* (2000), mutualism is defined as a reciprocal exchange of resources that benefits the survival of both partners (Raven and Allen, 2003).

More than 25,000 plants have symbiotic relationships with various environmental microorganisms, and these interactions, particularly with fungi, take place in mycorrhiza, the portions of plant roots. To complete their life cycle, defend themselves against infections, and survive in hostile environments, plants interact with other living things in nature. The association of plant roots and other living organisms is collectively called the rhizosphere. It is the region of mycorrhizal association (Pérez-Brocal *et al.*, 2011).

2.2 MYCORRHIZA – A MUTUALISTIC BENEFICIAL SYMBIOTIC ASSOCIATION BETWEEN FUNGI AND THE ROOTS OF MEDICINAL PLANTS

Mycorrhiza represents a prime illustration of the mutually beneficial bond between plants and fungi. Its name originates from the Greek words 'mycosis', meaning fungus, and 'rhiza', meaning roots, combining to create the term 'mycorrhiza'. In the natural world, mycorrhizal associations are prevalent, with over 80% of angiosperms and nearly all gymnosperms engaging in such relationships. Two primary types of mycorrhizal associations exist: endomycorrhizae or arbuscular-mycorrhizae (AM), including species like *Endogone* and *Rhizophagus*, and ectomycorrhizae (EM), represented by species such as *Laccaria bicolor* and *Amanita muscaria* (Singh *et al.*, 2011). The association of

DOI: 10.1201/b23155-2

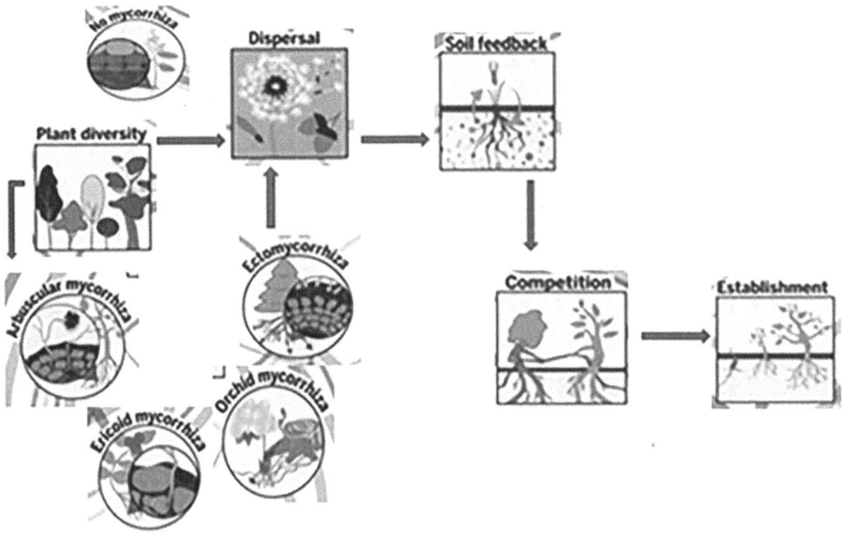

FIGURE 2.1 Mycorrhizal types and their effect on plant population.

mycorrhiza is beneficial to the host plants to survive in unsuitable ecological circumstances such as nutrient-deficient soil conditions and water stress such as drought situations. It helps the plants by extension of the roots' surface area and effective mineral uptake. The threats of climate change, such as an increase in temperature, drought conditions, and soil infertility by nutrient deficiency, are important factors in agriculture, and these must be eliminated to ensure the security of the world's food supply. In this regard, crop production by mycorrhiza-based methods is one of the essential elements of environmentally friendly farming methods (Turk *et al.*, 2006; Latef *et al.*, 2016). In the natural environment, over 6,000 fungal species are recognized as having the ability to create this beneficial association, with approximately 240,000 species of different families (Singh, 2007).

2.3 TYPES OF MYCORRHIZA ASSOCIATION

The two major reported types of mycorrhiza are ectomycorrhizae and endomycorrhizae. Both types can be distinguished based on their mode of penetration into the host roots (Figure 2.1). Ectomycorrhizae do not penetrate their hyphae within the roots of individual cells, while endomycorrhizae penetrate their hyphae within the cell wall and cell membrane of host cells (Imhof *et al.*, 2013). These two fungi can be further classified into various types, such as ericoid, orchid, and arbuscular mycorrhiza. Additionally, the arbutoid type can be categorized as ecto-endomycorrhiza. Within this diverse range, monotropoid mycorrhizae hold a special and distinct place (Kohout, 2017; Schneider-Maunoury *et al.*, 2020).

2.4 ECTOMYCORRHIZAE

2.4.1 Diversity and Distribution

Ectomycorrhizae form mutually beneficial partnerships among the roots of approximately 10% of plant families, predominantly woody plants like eucalyptus, pine, birch, rose families, dipterocarps, orchids, and oak. These associations involve fungi from the Zygomycota, Ascomycota, and Basidiomycota groups (Peterson *et al.*, 2004; Barman *et al.*, 2016). Certain ectomycorrhizal fungi, such as *Suillus* and *Leccinum*, have a symbiotic relationship with specific plant genera, while others like *Amanita* are more versatile, forming mycorrhizal connections with various plant species.

A single tree can host simultaneous partnerships with 15 or more different fungal ectomycorrhizae. The realm of ectomycorrhizal fungi encompasses over 200 genera, housing thousands of species. According to a recent study, the conservative estimate of the global richness of ectomycorrhizal fungal species is around 7,750, but considering the known and unknown diversity of macromycetes, the final estimation is likely to range between 20,000 and 25,000 species (Behie and Bidochka, 2014).

2.5 COMPOSITION AND INTERACTION WITH THE HOST

Ectomycorrhiza consists of a mantle, a protective layer made up of hyphae, enveloping the root tip, along with a hyphae network known as the Hartig net, which encircles the plant cells present in the cortex of the root. In some instances, the hyphae may penetrate the plant cells, forming an ecto-endo-mycorrhiza. Apart from the root, the extramatrical mycelium of ectomycorrhizal fungi forms a vast network throughout the leaf litter and soil (Turk *et al.*, 2006; Gardes, 2003).

The fungal network facilitates the transfer of nutrients among different plants (Figure 2.2). For instance, carbon can move from two trees like paper birch to Douglas fir, thus promoting ecosystem succession. Additionally, a specific type of fungus called *Laccaria bicolor* has been found to attract and eradicate springtails to acquire nitrogen, some of which can be transferred to the plant to form a symbiotic relationship (Liu *et al.*, 2022). When *L. bicolor* was introduced to an eastern white pine, it obtained up to 25% of its nitrogen from springtails. Comparatively, ectomycorrhizae often contain significantly higher amounts of trace elements, including potentially harmful substances and toxic metals like chlorine, silver, or cadmium, compared to fine roots (non-mycorrhizal) (Turk *et al.*, 2006).

2.6 THE FIRST GENOMIC SEQUENCE OF ECTOMYCORRHIZA

In 2008, the first genetic sequence of a symbiotic fungus known as the ectomycorrhizal basidiomycete *L. bicolor* was published. This fungus displayed an expansion in several families of genes, indicating that gene duplication played a role in its adaptation to symbiosis. Among the genes specific to this lineage, those responsible for coding symbiosis-regulated secreted proteins exhibited increased expression in the tips of ectomycorrhizal roots, suggesting their involvement in communication with the partner organism (Li *et al.*, 2020; Liu *et al.*, 2022). *L. bicolor* lacks the enzymes necessary for breaking down components of plant cell walls such as cellulose, hemicellulose, pectins, and pectates. This absence prevents the symbiotic fungus from breaking down host cells during root colonization. However, *L. bicolor* has expanded families of genes associated with the breakdown of polysaccharides and proteins found in bacteria and microfauna. This genome analysis uncovered the mycorrhiza's dual lifestyle as both a saprotroph and a biotroph (Li *et al.*, 2020).

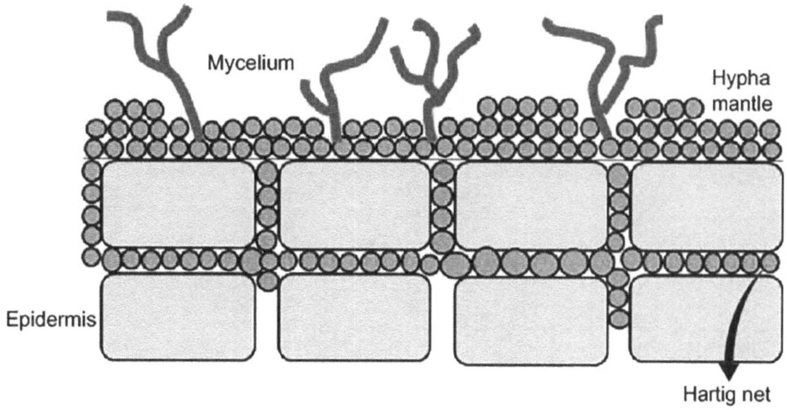

FIGURE 2.2 Cell penetration pattern of ectomycorrhiza.

2.7 ARBUTOID MYCORRHIZAE

2.7.1 Diversity and Distribution

Plants from the Ericaceae subfamily Arbutoideae are involved in this form of mycorrhiza. Nevertheless, it distinctively differs from ericoid mycorrhiza while sharing functional and fungal similarities with ectomycorrhiza. Truffles, which are classified under the genus *Tuber* (Ascomycota, Pezizales), are subterranean fungi that yield edible fruiting bodies. *Tuber* encompasses at least 180 species, several of which possess significant economic importance (Bonito *et al.*, 2010).

2.8 COMPOSITION AND INTERACTION WITH THE HOST PLANT

Arbutoid mycorrhiza can be distinguished from ectomycorrhiza by the penetration of some hyphae into the root cells. This unique characteristic classifies it as an ecto-endo-mycorrhizal fungus, allowing it to thrive within both soil and living plant roots. In arbutoid associations, an intercellular Hartig net is formed, primarily limited to the outer layer of root cells. This ‘paraepidermal’ net is also observed in ectomycorrhizae found in the majority of angiosperms. The Hartig net’s deeper penetration is hindered by a hypodermis, created by suberin deposits and a casparian strip in the outer layer of cortical cells (Singh *et al.*, 2019). Arbutoid mycorrhizae differ from ectomycorrhizae in a noteworthy way: in arbutoid associations, the hyphae penetrate the outer cortical cells and fill them with coils, a characteristic absent in ectomycorrhizae. The presence of mantle sheath, intracellular coils, and the Hartig net serves as distinctive traits indicating the presence of arbutoid mycorrhizae (Ori *et al.*, 2020).

The root systems of *Pyrola* species differ from ectomycorrhizae in comparison to Arbustus and Arctostaphylos. *Pyrola* species, except for *P. uniflora*, possess a comprehensive network of white rhizomes with slender adventitious roots (Figure 2.3). When fungi colonize *Pyrola* roots, their diameter increases and branching also intensifies. Initially, the tips of colonized roots are usually transparent but turn brown or black as they age. Colonization takes place a few millimetres behind the root tip, where hyphae create a surface layer. A complete sheath is not formed, and the hyphae penetrate between cells, forming a Hartig net of the paraepidermal type. Additionally, hyphae enter cells intracellularly and extensively grow within the plant cells (Kitamura *et al.* 2023).

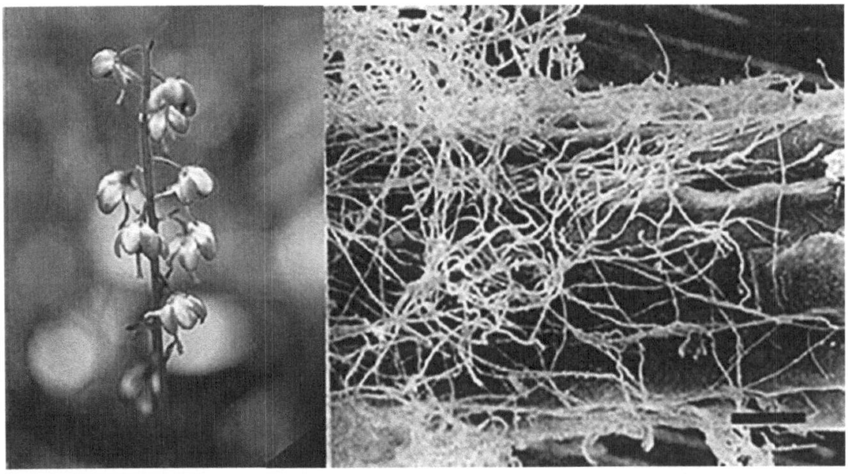

FIGURE 2.3 Pyrola roots and arbutoid formation.

2.9 ENDOMYCORRHIZAE

Endomycorrhizae exhibit variability and have been categorized into different types, such as ericoid, monotropoid, arbuscular, orchid, and arbutoid mycorrhizae (Roy *et al.*, 2020).

2.10 ARBUSCULAR MYCORRHIZAE

2.10.1 Diversity and Distribution

The Glomeromycota fungi are recognized for their method of asexual reproduction and are vital for promoting optimal plant growth. They play a crucial part in enabling arbuscular mycorrhizal symbiosis, which is the prevailing non-pathogenic symbiosis in soil and can be observed in the roots of 80% of vascular plant species. The presence of arbuscular mycorrhizae depends on suitable host plants and varies among different plant species. Fungal structures called arbuscules penetrate individual plant cells, and these arbuscular mycorrhizal fungi are present in nearly all phyla within the angiosperms (El Gabardi *et al.*, 2022).

2.11 COMPOSITION AND INTERACTION WITH THE HOST PLANT

Arbuscular mycorrhizae, previously known as vesicular-arbuscular mycorrhizae, establish a symbiotic relationship with plants by extending hyphae into plant cells. The hyphae create vesicles or arbuscules, which aid in nutrient exchange. It should be emphasized that the fungal hyphae do not physically penetrate the protoplast (the cell's interior) but instead fold the cell membrane inward (Figure 2.4). Arbuscules greatly enhance the contact area between the hyphae and the cell cytoplasm, leading to improved nutrient transfer. Arbuscular mycorrhizae are classified as obligate biotrophs, as they rely on the host plant for growth and reproduction. Approximately one-fifth of the plant's photosynthetic output is consumed by the fungi, while an equivalent amount of phosphate is reciprocally transferred from the fungi to the plant. This mutual exchange of carbon and phosphate occurs between the land-dwelling host plant and the fungi (Zou *et al.*, 2021).

2.12 FOSSIL EVIDENCE AND DNA SEQUENCE ANALYSIS

This symbiotic connection may have begun with the first colonization of the land by plants 400–460 million years ago, according to evidence from fossils and DNA sequence research. Eighty-five percent of plant families, including many agricultural species, have arbuscular mycorrhizae. These fungi

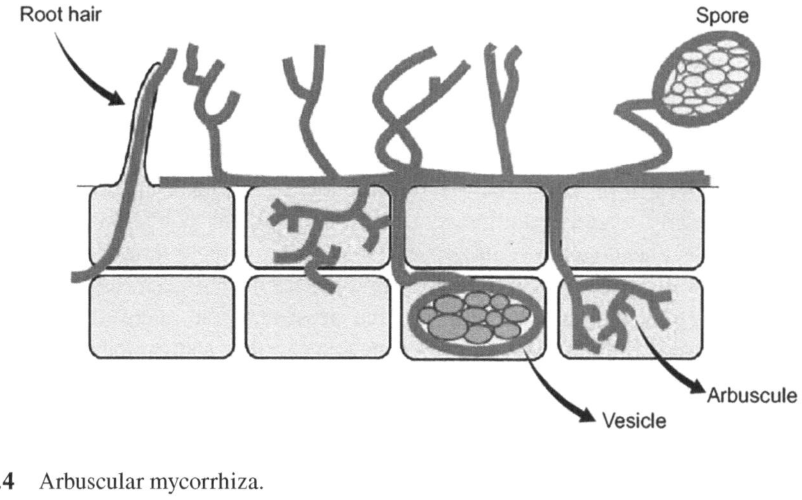

FIGURE 2.4 Arbuscular mycorrhiza.

create glomalin, a glycoprotein that may act as a sizable carbon storage facility in the soil, in their hyphae. Arbuscular mycorrhizal fungi have an unusual trait known as heterokaryosis, wherein individuals may harbour many genetically diverse nuclei and have probably reproduced asexually for millions of years (Nazir *et al.*, 2019).

2.13 ERICOID MYCORRHIZAE

Ericoid mycorrhizae are the third of the three ecologically relevant forms. In these mycorrhizae, the root cells go through a simple period of development that is distinguished by tightly coiled hyphae in the top layer. The soil around the root is devoid of any intermediary phases, and the hyphae are scarce and don't protrude very far (Pacheco *et al.*, 2022). Even though ericoid mycorrhizae have the potential to produce sporocarps that might resemble tiny cups, little is known about their reproductive biology. The noteworthy saprotrophic properties of these mycorrhizae have also been discovered, enabling plants to get nutrients from partially degraded materials through the decomposing actions of their ericoid partners (Ward *et al.*, 2022).

2.14 MONOTROPOID MYCORRHIZAE

This particular mycorrhizal type is found in the family Ericaceae and its subfamily Monotropoideae, as well as in various genera within the Orchidaceae family. Depending on the fungus partner, monotropoid mycorrhiza develops in non-chlorophyllous Monotropa species that grow beneath forest trees like Salix, Quercus, Fagus, and Pinus. Mycelium from the common mycorrhizal fungus Boletus connects Monotropa and its companion trees. These plants either have heterotrophic or mixotrophic growth patterns, and they get their carbon from the fungus they associate with. This symbiotic connection is categorized as a parasitic, non-mutualistic kind of mycorrhizal association (Bennett and Classen, 2020). One of the three ecologically relevant forms is ericoid mycorrhizae. They go through a straightforward intraradical phase in which the outermost layer of root cells contains thick coils of hyphae. No periradical phase exists, and the extraradical phase is made up of sparse hyphae that don't penetrate very deeply into the soil. Although little is known about their reproductive biology, it is plausible that they may produce sporocarps that would resemble little cups. It has also been shown that ericoid mycorrhizae exhibit strong saprotrophic properties. As a result of their ericoid fungus partners' decomposition processes, these plants may obtain nutrients from partially degraded materials (Tederosoo *et al.*, 2020).

2.15 ORCHID MYCORRHIZAE

During certain stages of their lifecycle, all orchids exhibit myco-heterotrophy, which means they can only survive by forming a mycorrhizal association with basidiomycete fungi. These fungi's hyphae penetrate the root cells and create coils called pelotons for exchanging nutrients. Myco-heterotrophic characteristics are present in various plant groups, primarily flowering plants. Full myco-heterotrophy is observed in all non-photosynthetic orchids, as well as in all monotropes and Cryptothallus, which are non-photosynthetic liverworts. Incomplete myco-heterotrophy is prevalent within the Gentian family, where certain genera like Voyria exhibit complete myco-heterotrophy (Yan *et al.*, 2022). This phenomenon is also observed in photosynthetic orchids and various other plant groups. Certain ferns and clubmosses possess myco-heterotrophic gametophyte stages. Myco-heterotrophs typically rely on parasitizing fungi that possess ample energy reserves, primarily mycorrhizal fungi. However, there is evidence indicating that they may also exploit parasitic fungi, such as *Armillaria*, which form extensive mycelial networks. The fungi targeted by myco-heterotrophic plants include arbuscular mycorrhizal, orchid mycorrhizal, and ectomycorrhizal fungi. The presence of myco-heterotrophic members in unrelated plant families, as well as the diversity of fungi they parasitize, suggests that myco-heterotrophs have evolved independently multiple times from mycorrhizal ancestors (Chaudhary *et al.*, 2022) (Figure 2.5).

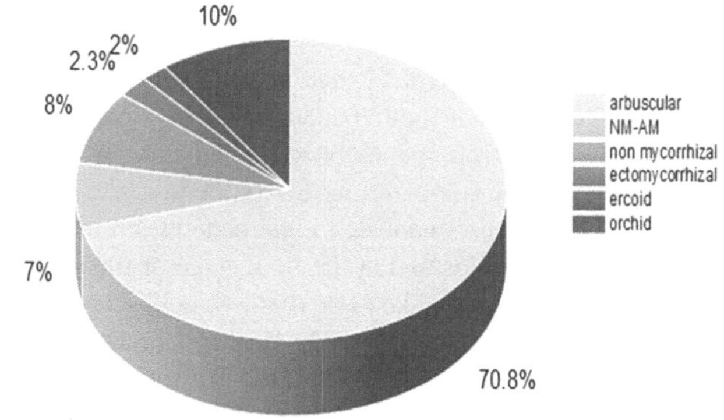

FIGURE 2.5 Worldwide relative diversity of mycorrhizal medicinal plant types.

2.16 HOW DID MYCORRHIZAE EVOLVE?

The fossil records and DNA-based phylogenetic analysis provide evidence that around 450–500 million years ago, mycorrhizal association emerged alongside the first land plants as an endomycorrhizal connection. Later, approximately 200–150 million years ago, with the increase in organic matter, ectomycorrhizal associations developed through co-evolution. The initial stages of mycorrhizal symbiosis likely involved reciprocal genetic exchanges between ancestral plants and free-living fungi. Both partners in both endophytic and ectophytic associations may have evolved simultaneously in response to environmental changes, eventually becoming reliant on sharing limited resources. As a result, the majority of present-day plants coexist in harmony with mycorrhizal fungi (Turk *et al.*, 2006; Barman *et al.*, 2016).

2.17 INTERACTION OF ROOT-INHABITING FUNGI WITH DIFFERENT COMMERCIALLY IMPORTANT MEDICINAL PLANT SPECIES

2.17.1 Mechanism

2.17.1.1 Initial Recognition of Host Plant Roots by Fungi

During the initial stages of establishing recognition between a fungal partner and the roots of a host plant, certain bioactive substances such as strigolactones are released by the roots (Figure 2.6). These strigolactones serve as beneficial agents that help the fungi identify their host plants. Additionally, strigolactones stimulate the growth and branching of arbuscular mycorrhizal fungi. In response to this signal, the fungi secrete a group of hypothetical substances called mycorrhizal factors. These factors also play a significant role in the interaction between arbuscular mycorrhizal fungi and nitrogen-fixing bacteria (Parniske, 2008).

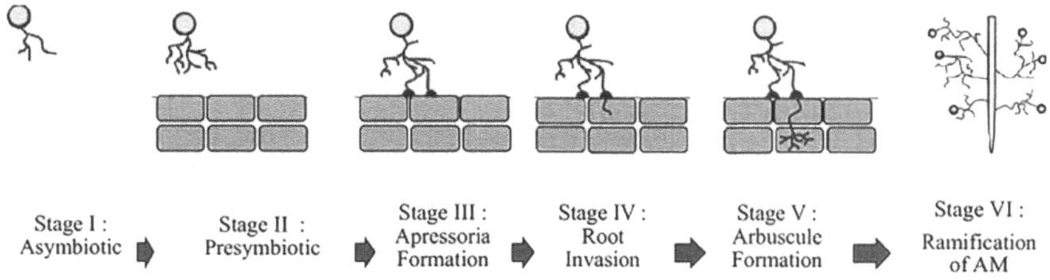

FIGURE 2.6 Mechanism of interaction of root-inhabiting fungi with medicinal plants.

2.18 PENETRATION AND ESTABLISHMENT OF MYCORRHIZAE

Following the chemical interaction, the fungal hyphae and the host root establish mutual interaction, leading to the gradual proliferation of the hyphae into the host root through the formation of the hyphopodium. Subsequently, numerous genes become activated as a result of hyphopodium formation, marking the initial stage of colonization. Subsequently, a crucial structure called the pre-penetration apparatus (PPA) develops, enabling fungal penetration into the plant without compromising cell integrity. Arbuscules are formed at the final stage of this symbiotic process. These arbuscules allow the fungi to enter the cytoplasm of the host cell. The arbuscular cells serve as mechanisms for nutrition uptake and transportation. The process of nutrient uptake involves a large number of genes and proteins, which ultimately aid in the achievement of symbiosis. The molecular mechanism used by ectomycorrhizal fungi is nearly identical to that of arbuscular mycorrhizal fungi, though not quite. To fully understand the mechanisms underlying ectomycorrhizal fungal interactions, more study is required (Paszkowski, 2006).

2.19 HOW THE PARTNERS BENEFIT

In the natural world, plants provide their microbial (bacteria) friends with the carbon supply. Hexose, a photosynthetic product, is transferred to the fungal cytoplasm's arbuscular region during that process, where it is transformed into glycogen and triacylglycerol (TAG). In the fungal network, these are the kinds of carbohydrates that can travel across great distances with ease. For plants, mycorrhiza increases the surface area of the roots to improve water and nutrient absorption. The plants take essential nutrients through diffusion. Mycorrhizal fungi can spread their exterior hyphae beyond the deficient zones in nutrient-poor tropical climates with high rainfall where vital nutrients are leached from soil surfaces. As a result, plant roots can access a larger amount of soil. As a result, plants with mycorrhizal connections are more effective at absorbing minerals like calcium, potassium, phosphorus, and nitrogen. Phosphorus is a very stationary element found in soil. Vascular-arbuscular fungi are primarily responsible for supplying phosphorus to plant roots through a phosphate transporter found in the hyphal membrane. The mycorrhizal association helps in the growth of plants in the soil in which the content of phosphate is low. Mycorrhizae also help plants absorb copper and zinc by similar mechanisms (Bonfante and Anca, 2009).

Plant growth relies heavily on the uptake of nitrogen, which exists in the soil as ammonium and nitrate. Fungi play a crucial role in this process by absorbing ammonium, nitrate, and amino acids through their mycelium. Ammonium, predominantly acquired through a protein transporter called AMT1 of fungal origin, is the primary form of nitrogen uptake. In terms of amino acids, arginine is commonly involved in nitrogen transportation. Additionally, arbuscular mycorrhizal fungal hyphae extend into the soil matrix, creating a framework that binds soil particles together and enhances soil structure and porosity. This structure becomes particularly important for retaining water, especially during arid periods. Furthermore, arbuscular mycorrhizal fungi can influence plant growth by altering hormonal signalling from the roots to the shoots and impacting stomatal responses under conditions of reduced soil water potential. It has been suggested that mycorrhizal connections promote hydraulic conductivity in roots, which enables plants to absorb more nutrients under water stress. Microorganisms engage in interactions with soil components and plant roots in the rhizosphere. Plant pathogen growth is competitively impeded by arbuscular mycorrhizal fungi's greater carbon demand. Moreover, by making up for the loss of root biomass brought on by pathogen attack and boosting the host plant's tolerance, the mycorrhizal fungus companion can also help the host plant's nutritional status. The detoxification processes include avoiding heavy metal uptake, binding of heavy metals to rhizodermal cell walls, and extracellular heavy metal chelation by root exudates (such as glycoprotein glomalin). Fungal hyphae's vast surface area serves as a significant sink site for heavy metals. Toxic substances can also be stored in fungal vesicles. Consequently, mycorrhizal fungi aid in phytoremediation, which enhances soil health (Barman *et al.*, 2016).

2.20 ROOT COLONIZATION AND MYCORRHIZAL PLANT INTERACTION MECHANISM

Mycorrhizae enable roots to access nutrients through two pathways: the plant uptake pathway (PP) and the mycorrhizal uptake pathway (MP) (Figure 2.7). In the PP pathway, nutrients are directly absorbed by the root hairs and epidermis through transporters. Conversely, in the MP pathway, nutrients are transferred indirectly through transporters found in the extraradical mycelium (ERM) of the fungus. These nutrients are then transported to either the Hartig net in ectomycorrhizal interactions or the intraradical mycelium (IRM) in arbuscular mycorrhizal interactions, as depicted in the mycorrhizal interface. The plant transporters induced by mycorrhiza facilitate nutrient uptake in the periarbuscular membrane located in the interfacial apoplast. The visual representation of fungal structures illustrates the colonization of a single host root by multiple fungal species, each exhibiting different levels of efficiency. Through these mechanisms, fungi can extract soil nutrients and transfer them to the host plants (Bücking *et al.*, 2012).

Roots rapidly absorb nutrients, such as P, resulting in the creation of a depletion zone. Fungal hyphae extend beyond these zones to penetrate and exploit the uptake of larger volumes of nutrients in the soil. Nutrients taken up by the hyphae are then transported within the fungal network. They move from the hyphae to the fungal sheath, intercellularly to the Hartig net, and, except for monotropoid mycorrhizae and ectomycorrhizae, intracellularly into plant cells. The fungal sheath can store nutrients, ensuring a continuous supply to the plant host even when soil nutrient levels decline. When additional nutrients are received, mycorrhizal plants sacrifice around 10 to 20% of their photosynthates, which are utilized by the mycorrhizal fungi and their associated structures for growth, maintenance, and functioning.

The fungal ERM utilizes transporters (depicted in blue) to move nutrients such as Pi, NO3$^-$, or NH^{4+}. In the case of arbuscular mycorrhizae, N is assimilated into Arg through the anabolic pathway of the urea cycle. Additionally, Pi is converted into polyP within the ERM. The polyP is then transported from the ERM to the IRM, where it is hydrolysed, resulting in the release of Arg and Pi. Subsequently, Arg undergoes breakdown into NH4$^+$ through the catabolic arm of the urea cycle,

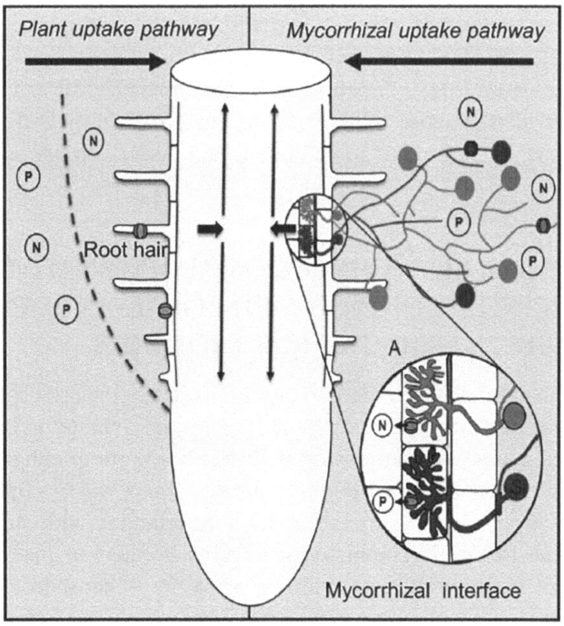

FIGURE 2.7 Nutrient uptake pathway of plants and mycorrhizae.

specifically observed in arbuscular mycorrhizae. Pi, $NH4^+$, and possibly amino acid (AA) efflux through the fungal plasma membrane (depicted in red) into the interfacial apoplast facilitates this process. The plant absorbs these nutrients from the mycorrhizal interface via mycorrhiza-inducible Pi or $NH4^+$ transporters. The enhanced nutrient supply stimulates photosynthesis, aided by sucrose efflux through the plant plasma membrane into the interfacial apoplast, sucrose hydrolysis via apoplastic plant invertase in the interfacial apoplast, and the uptake of hexoses by the mycorrhizal fungi through fungal monosaccharide transporters.

2.21 THREATS TO MYCORRHIZAL ASSOCIATIONS

Mycorrhizae play a vital role in soil ecosystems, making them essential for the survival of plant species. These symbiotic relationships also serve as indicators of plant well-being and soil toxicity. The agricultural practices employed by farmers have a significant impact on the functioning of mycorrhiza. Soil ploughing disrupts the networks of arbuscular mycorrhizal hyphae, resulting in a significant decline in root colonization and the uptake of phosphorus from the soil. Additionally, crop breeding can expedite the loss of AM fungal diversity if the selected genotype of a hybrid plant fails to form a symbiotic association with the previously partnered fungal species. Indiscriminate use of manures and insecticides can restrict the establishment and growth of both ectomycorrhiza and endomycorrhiza (Barman *et al.*, 2016).

2.22 POTENTIAL APPLICATIONS OF MEDICINAL PLANTS AND ROOT-INHABITING FUNGI IN PHARMACOLOGY AND BIOTECHNOLOGY

Medicinal plants or plant-based drugs have played a significant role in various traditional medical systems. The World Health Organization stated that about 80% of the world population prioritizes plant-based medicines as their primary treatment (Van Hunsel *et al.*, 2019). Plants are enriched with bioactive compounds; therefore, they are used as medicines. The global market value of medicinal plants has increased significantly (Nahin *et al.*, 2009). As a consequence, high-yielding plants with excess secondary metabolites are in demand. Until recently, more than 1% of global medicinal plant species were known because of their relation with root-inhabiting fungi, including arbuscular mycorrhizal fungi (AMF) (Figure 2.8). AMF have been observed to increase the performance of herbal species (Piszczek *et al.*, 2019). Inoculation of plants with AMF can improve growth (Chen *et al.*, 2017), which can also increase the content of bioactive compounds (Avio *et al.*, 2017). It can also enhance the resistance of plants to pathogens and insect herbivores (Dugassa *et al.*, 1996; Pineda *et al.*, 2010).

2.23 MYCORRHIZAE: A BIOTECHNOLOGICAL ALTERNATIVE FOR ENHANCING SECONDARY METABOLITES IN PLANTS AND VALUABLE PHARMACEUTICAL PRODUCTS

Various microorganisms exhibit harmonious relations with plant roots in the soil where they are found and play an active part in the decomposition of organic compounds, nutrient cycling, and plant health. Plant roots' symbiotic relationship with soil-borne fungi enhances plant development and productivity by increasing the absorption of nutrients and water. Higher plants grow faster and produce more active constituents through symbiotic relations with fungi (Ran *et al.*, 2022). Mycorrhiza increases the hydraulic conductivity of roots because of the greater surface area of mycelium. Hence, it can improve the water intake of plants (Khaosaad *et al.*, 2008; Jurkiewicz *et al.*, 2010). It also enhances the photosynthetic mechanism of the host plant by synchronizing the activity of the chloroplast enzyme, decreasing the decomposition rate of chlorophyll, and enhancing the synthesis of chlorophyll (Kapoor *et al.*, 2002; Abu-Zeyad *et al.*, 1999). Root-inhabiting fungi augment the antioxidant mechanism of plants and thus can overcome the negative impact of abiotic

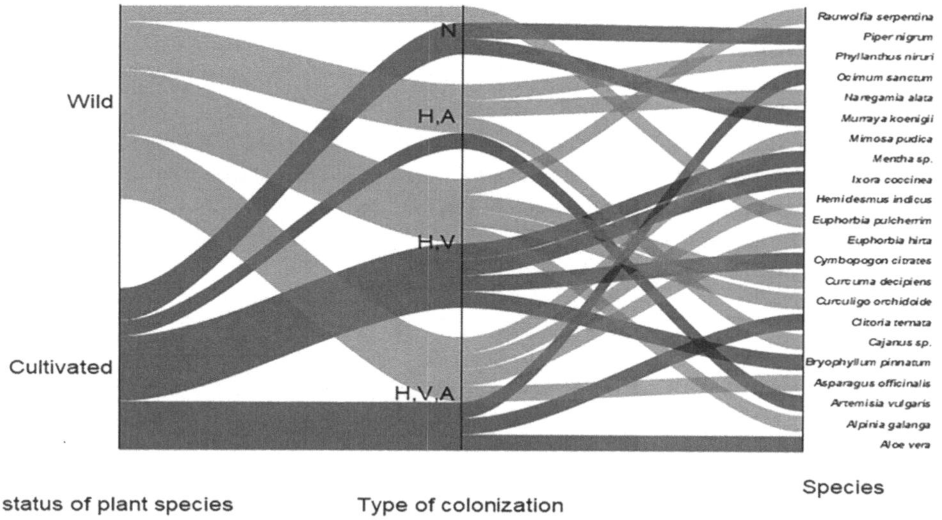

FIGURE 2.8 Mycorrhizal association with medicinal plant species.

Keynote: H: hyphal colonization, V: vesicular colonization, A: arbuscular colonization, N: no colonization

stressors on plants. Obligatory symbiotic arbuscular mycorrhizal fungus is among the beneficial soil microorganisms, as it can associate symbiotically with most of the terrestrial plant species. Medicinal plants are an important source of antioxidants, which may be enhanced by inoculating them with mycorrhizal fungi, which also increases the production of phytochemicals (Verma *et al.*, 2021). AMF helps the plant by enhancing the uptake of micronutrients and macronutrients from the soil through their extensive hyphal network (Maurich *et al.*, 2006; Pistelli *et al.*, 2017). Inoculation of soybean plants with AMF under drought stress increases the enzymatic antioxidant activity (ascorbate peroxidase: APX, superoxide dismutase: SOD, and guaiacol peroxidases: GPX) (Amani Machiani *et al.* 2021). An increase in antioxidant enzyme activities revealed that AMF plays an active role in protecting plants against oxidative damage imposed by abiotic and biotic stressors. AMF can also cause changes in the primary (amino acids, organic acids, sugars, etc.) and secondary metabolites (flavonoids, terpenoids, alkaloids, etc.) by reactivating the various pathways of metabolite synthesis. Regulation of the expression of defensive genes, including phenylalanine ammonia-lyase in the roots of *Medicago truncatula* and *Oryza sativa*, and chalcone synthase in the roots of *M. truncatula*, favoured the defensive role of mycorrhizal inoculation (Harrison and Dixon, 1993; Blilou *et al.*, 2000; Bonanomi *et al.*, 2001). Metabolites play an active role in protecting plants against stress. Mycorradicin (a yellow pigment), a derivate of glycosylated cyclohexanone, accumulated in the roots of mycorrhizal plants (Graminaceae, Liliopsida, and Rosopsida) (Klingner *et al.*, 1995). Roots and shoots of various mycorrhizal plants such as *Nicotiana attenuata* Torr. ex S.Watson, *Solanum tuberosum* L., *Lycopersicon esculentum* Mill., *Triticum aestivum* L., *Hordeum vulagare* L., *Brachypodium distachyon* (L.) P.Beauv., and *Medicago truncatula* Gaertn. were found to have blumenols (Wang *et al.*, 2018). Studies have shown that roots colonized with fungi went through various metabolic and cytological changes like regulation of biosynthetic pathways of plastids and Krebs cycle and increase in the production of fatty acids, amino acids, apocarotenoids (blumenols), and flavonoids (quercetin, polyphenols, rhamnetin, acacetin, polyketides, isoprenoids, etc.). (Fester *et al.*, 2001; Ponce, *et al.*, 2004; Lohse *et al.*, 2005; Copetta *et al.*, 2006; Schliemann *et al.*, 2008; Araim *et al.*, 2009; Caser *et al.*, 2019; Saleh *et al.*, 2020). Several studies revealed the regulation of genes encoding the enzymes involved in the production of bioactive compounds in mycorrhizal plants (Bonanomi *et al.*, 2001; Harrison, and Dixon, 1994; Liu *et al.*, 2007; Handa *et al.*, 2015; Shu *et al.*, 2016; Zhang *et al.*, 2020).

Approximately 50 medicinal and aromatic plant species have been evaluated to check the effect of inoculation of AMF on the concentration of bioactive compounds in various parts of plants (Shu *et al.*, 2016). *Ocimum basilicum* (sweet basil) inoculated with mycorrhiza was observed to have greater amount of rosmarinic acid (RA), caffeic acid Zhang *et al.*, 2020 and Zeng *et al.*, 2013), anthocyanins, and essential oils (Copetta *et al.*, 2006; Andrea *et al.*, 2007; Lee and Scagel, 2009; Rasouli-Sadaghiani *et al.*, 2010). Inoculation of *O. basilicum* also showed overexpression of the tyrosine aminotransferase (TAT) gene, which is involved in the production of RA (Table 2.1) [72]. Inoculation of AMF on strawberries had a substantial effect on anthocyanin and phenolic content (Cecatto *et al.*, 2016).

TABLE 2.1
Effect of Mycorrhizal Strains on Plant Secondary Metabolites

Plant Species	Mycorrhiza	Effect on Phyto-Active Constituents	References
Acorus calamus L.	*Funneliformis mosseae*	Enhance the production of phytochemicals	Verma *et al.* (2021)
Lactuca sativa L.	*Glomus fasciculatum*, *G. intraradices* + *Glomus mosseae* (Atens, Spain)	Regulate the production of phenolic compounds	Baslam *et al.* (2011).
	Rhizoglomus irregular *Funneliformis mosseae*		Avio *et al.* (2017)
Ocimum basilicum L.	*G. intraradices* NPI, USA *G. mosseae* BEG 12 *Rhizophagus intraradices* *Glomus caledonium* BEG 162 *G. mosseae* NBR 1–2	Regulate the concentration of phenolic compounds (total phenolic content, caffeic acid), hydroxycinnamic acids (rosmarinic acid, chicoric acid)	Lee and Scagel (2009) Toussaint *et al.* (2008) Battini *et al.* (2016) Toussaint *et al.* (2007)
Fragaria x ananassa Duch.	*G. intraradices* (Premier Tech Biotechnologies, Canada)	Flavanols, phenolic, quercetin, kaempferol, and catechin	Castellanos *et al.* (2010)
	*Glomus iranicum*var. *tenuihypharum* (Mycogrowth, Spain)		Cecatto *et al.* (2016)
	Mix[2] (Mybasol, Italy)	Anthocyanidins (pelargonidin 3-glucoside, pelargonidin 3-rutinoside)	Lingua *et al.* (2013)
	Septoglomus viscosum	Anthocyanidins (pelargonidin 3-glucoside)	Todeschini *et al.* (2018)
Cynara cardunculus L. var. *scolymus*	*G. mosseae* *G. intraradices*	Increase the concentration of phenolic compounds	Ceccarelli *et al.* (2010)
	F. mosseae 2W3 *F. mosseae* IMA1 *F. mosseae* IN101C		Avio *et al.* (2017)
	Claroideoglomus	Increase the concentration of phenolic acids (caffeic acid)	
Solanum lycopersicum L.	*F. mosseae* BEG12	Enhance the concentration of carotenes (ß-carotenes), lycopene,	Hart *et al.* (2015)
	R. irregularis BB-E (Agrauxine, F)	Enhance the concentration of carotenes (ß-carotenes), lycopene	
	R. irregularis (Premier Tech Inc., Canada)	Enhance the concentration of carotenoids	
	Mix[2] (Mybasol, Italy)	Increase the production of ascorbic acid, luteine (xanthophylls)	Bona *et al.* (2017)

2.24 MYCORRHIZAE AS A SOURCE OF RESTORATION OF DISTURBED ECOSYSTEMS

The association between roots and fungi seems an essential element of global ecosystem functioning. They help in the improvement of soil conditions, which are crucial to agriculture; this is especially pertinent in the context of climate change (Brundrett *et al.*, 1996). According to Brundrett (2009), approximately 92% of land plant families and 93% of flowering plant families showed mycorrhizal relationships. These associations are divided into seven categories based on their physical attributes and physiological interrelationships with symbionts. Among these associations, arbuscular mycorrhizal fungi are among the leading ones (Navarro-Fernández *et al.*, 2011). Arbuscular mycorrhizae, ecto-endo-mycorrhizae, ectomycorrhizae (ECM), and ericoid mycorrhizae are the four main forms of mycorrhizae. Approximately 5,000–6,000 various species of ascomycetes, basidiomycetes, and zygomycetes are linked with ECM (Brundrett *et al.*, 1996).

A disturbed soil ecosystem has low levels of infectious AMF; therefore, it is less probable that nursery seedlings near disturbed ecosystems will sufficiently become infected with AMF (Michelsen *et al.*, 1996). The introduction of appropriate fungal inocula can support the growth of late-successional trees. Late-successional tree species require AMF for proper growth and functioning. Therefore, they are known as obligatory mycotrophic species (Weremijewicz, and Janos, 2019). Mycorrhizal successional plant species are more dependent on AMF at early seedling growth as compared to lateral growth (Kiers *et al.*, 2000). Hence, inoculation of plant species with AMF could be an essential tool in the restoration of degraded ecosystems. AMF is not host specific and can be found in every soil (Brundrett and Abbott, 2002; Abbott *et al.*, 1992; Barea *et al.*, 2011; Al-Karaki, 2013).

Many researchers have stated that AMF inoculation is beneficial in specific conditions, such as in areas where AMF doesn't naturally exist, like minefields (Brundrett and Abbott, 2002). According to Koide and Mosse (2004), it is suitable to concentrate on regulating the AMF population in an area instead of pursuing AMF inoculation. Scientists have come to understand the value of microbial biotechnology for the reclamation and rehabilitation of damaged areas. Various research has determined plant–microbe relationship and their rhizospheric interaction are the key factors that enhance plant productivity and soil fertility. Initiation of these microbial activities is essential for the successful restoration of destroyed ecosystems. Mycorrhizal associations are among the leading determinants that ensure successful vegetation growth on reestablished sites (Danielson and Visser, 1989; Robertson, 2020; Khan, 2006). It is documented that the application of mycorrhizal fungi and Frankia strains enhances the growth, yield, and stability of conifers and alders on challenging materials generated by Alberta's oil sand industry (Fung and Macyk, 2000). For the transformation of damaged mining lands into a boreal plant community, re-inoculation of ectomycorrhizae, arbuscular mycorrhizae, ecto-endo-mycorrhizae, and ericoid mycorrhizae fungal associations should be considered. A long-term cooperative effort involving the oil sands business, university, and microbial biotechnology company is necessary to accomplish this goal (Brundrett *et al.*, 1996). Surface mining activities cause the degradation of ecosystems in various regions of the world, so there is a great need for the reclamation of these productive lands. The process of assisting the reclamation of degraded land is termed ecosystem restoration. SER (2004) documented that an ecosystem is called destroyed when the leaf litter layer on the forest floor is under layer 21 of an ecosystem. Disturbance of the soil ecosystem significantly affects the composition and functioning of mycorrhizal fungi (Allen *et al.*, 2005). Microsymbionts are essential not only for plants but for various important ecological functions based on mycorrhizal and related microbes. Mycorrhizal fungi are crucial for the productivity and dynamics of an environment. Individual plants benefit from microsymbionts, but the majority of crucial ecological functions also rely on various mycorrhizal fungi and related microbes. Mycorrhizal fungi can be extremely important to the dynamics and productivity of an ecosystem (Brundrett *et al.*, 1996). Inoculation of nursery seedlings with site-specific mycorrhizal fungi is an effective and environmentally friendly approach for the restoration of forests. For the reclamation of a disturbed ecosystem, there is a need to inoculate seedlings with stress-tolerant and host-specific microsymbionts. Sustainable wooded areas are produced by ecological techniques like utilization of mycorrhizal and actinorhizal plants.

2.25 THE MAJORITY OF THE GLOBAL POPULATION RELIES ON MEDICINAL PLANTS AS THEIR MAIN SOURCE OF LIFE-SAVING DRUGS

According to fossil records, arbuscular mycorrhizal associations appeared approximately 400–450 million years ago. Mycorrhizal association is possibly the most significant symbiosis on earth (Smith and Read, 2010). Mycorrhizal association is beneficial for both partners, as they exchange nutrients in a bidirectional pathway. Mycorrhizal fungi also enhance stress resistance against both biotic and abiotic stressors by supplying nutrients like phosphate and nitrogen.

Mycorrhizal fungi and their beneficial role in the assimilation of phosphate from soil have been studied and are presented in Figure 2.9. ERM scavenges phosphate (Pi) and is supplied to the plants through root cortical cells (Jakobsen *et al.*, 1992; Jakobsen *et al.*, 2001). Fungal hyphae enable them to incorporate into tiny spaces between soil particles and help them scavenge phosphorous, increasing the influx rate of phosphorous per surface unit (Jakobsen *et al.*, 1992; Li *et al.*, 2008). Hijikata *et al.* (2010) documented that mycorrhizal fungi can store phosphorous in polyphosphate form; therefore, the internal inorganic phosphorous concentration of fungi is lower and allows efficient transfer of P from extra-radical mycelium to the intra-radical mycelium. According to Ezawa *et al.* (2005) and Alvarez *et al.* (2012), the secretion of acid phosphatases enables the release of phosphorous from complex substances. Just like plants, mycorrhizal fungi have two systems for the uptake of phosphorous. The first system takes up inorganic phosphate from the soil through proton co-transport, and this system is known as a high-affinity system, as it works against an electrochemical gradient. The second system is a low-affinity system that works on the diffusion principle in which inorganic phosphorous penetrates the fungal plasma membrane (Bucher, 2007; Javot *et al.*, 2007; Thomson *et al.*, 1990).

Different studies have shown that mycorrhizal fungi can take up nitrogen in both forms, inorganic (NO_3 and $NH4^+$) and organic (amino acids), and incorporate the nitrogen into the host plant (Nasholm *et al.*, 1998; Govindarajulu *et al.*, 2005; Whiteside *et al.*, 2012). Mycorrhizal interaction between fungi and plant roots can enhance the nitrogen assimilation from soil. Studies showed that nitrogen in the form of ammonium is mainly absorbed by mycorrhizal fungi; however, various mycorrhizal fungi can absorb organic nitrogen compounds and nitrate (Table 2.2). Mycorrhizal

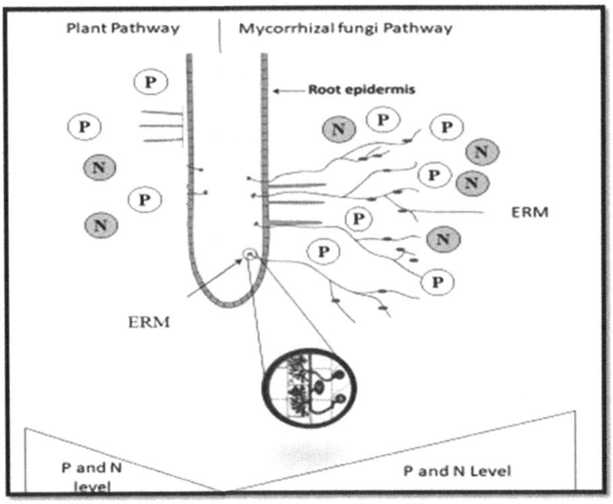

FIGURE 2.9 Phosphorous and nitrogen uptake pathway of the plant and fungus.

TABLE 2.2
Nitrogen Metabolic Pathways in Mycorrhiza

Pathways	Mycorrhizal Strains/Enzymes	References
Inorganic Nitrogen Uptake		
Uptake of ammonium	*Funneliformis mosseae, Rhizophagus irregularis, Acaulospora laevis, Gigaspora margarita*	(Jin *et al.*, 2005; Ames *et al.*, 1983; Frey and Schüepp, 1993; Johansen *et al.*, 1993; Hawkins *et al.*, 2000; Johansen *et al.*, 1996)
Ammonium (NH_4^+) transporter	*Rhizophagus irregularis*	(Pérez-Tienda *et al.*, 2011)
Uptake of nitrate	*Glomus fasciculatum, Rhizophagus irregularis, Funneliformis mosseae*	(Hawkins *et al.*, 2000; Tobar *et al.*, 1994; Bago *et al.*, 1996)
Nitrate (NO_3^-) transporter	*Rhizophagus irregularis*	(Gachomo *et al.*, 2009; Gachomo *et al.*, 2009; Tisserant *et al.*, 2012)
Nitrate (NO_3^-) permease	*Rhizophagus irregularis*	(Tisserant *et al.*, 2012)
Uptake of Organic Nitrogen		
Amino-acid transporter	*Acaulospora laevis, rhie, Gigaspora margarita, Funneliformis mosseae*	(Frey and Schüepp, 1993; Ames *et al.*, 1983)
Amino-acid permease	*Funneliformis mosseae*	(Cappellazzo *et al.*, 2008)
Alanine	*Rhizophagus irregularis*	(Gachomo *et al.*, 2009)
Peptide transporter	*Rhizophagus irregularis*	(Belmondo *et al.*, 2014)
Arginine	*Rhizophagus irregularis*	(Jin *et al.*, 2005; Gachomo *et al.*, 2009)
Glycine	*Funneliformis mosseae*	(Hawkins *et al.*, 2000; Hawkins *et al.*, 1999)
Cysteine	*Rhizophagus irregularis*	(Allen and Shachar-Hill, 2009)
Glutamine	*Funneliformis mosseae, Rhizophagus irregularis*	(Breuninger *et al.*, 2004)
Glutamate	*Funneliformis mosseae, Rhizophagus irregularis*	(Hawkins *et al.*, 2000)
Methionine	*Rhizophagus irregularis*	(Allen and Shachar-Hill, 2009)
Ornithine	*Rhizophagus irregularis*	(Gachomo *et al.*, 2009)
Urea	*Gigaspora margarita*	
Nitrate Reduction		
Nitrate reductase	*Glomus fasciculatum, Rhizophagus irregularis, Funneliformis mosseae, glma (Glomus macrocarpus), glsp (Glomus* species)	(Cliquet and Stewart, 1993; Subramanian and Charest, 1998; Kaldorf *et al.*, 1994; Fellbaum *et al.*, 2012)
Nitrate reductase (NADH)	*Funneliformis mosseae*	(Oliveer *et al.*, 1983; Cliquet and Stewart, 1993)
Nitrate reductase (NADPH)	*Rhizophagus irregularis*	(Kaldorf *et al.*, 1998)
Nitrite reductase	*Rhizophagus irregularis*	(Subramanian and Charest, 1998)
GDH Pathway		
GDH	*Claroideoglomus etunicatum*	(Nemec, 1981)
GDH (NADH)	*Funneliformis mosseae, Rhizophagus irregularis*	(MacDonald and Lewis, 1978)
GDH (NADPH)	*Glomus fasciculatum*	(Cliquet and Stewart, 1993; Smith *et al.*, 1985; Kaldorf *et al.*, 1994)
GS/GOGAT Pathway		
Glutamine synthetase	*Glomus fasciculatum, Funneliformis mosseae Rhizophagus irregularis*	(Cliquet and Stewart, 1993; Subramanian and Charest, 1998; Smith *et al.*, 1985; Kaldorf *et al.*, 1994; Govindarajulu *et al.*, 2005)
Glutamate synthase (NADH)	*Rhizophagus irregularis*	(Johansen *et al.*, 1996; Subramanian and Charest, 1998; Govindarajulu *et al.*, 2005)
Glutamate synthase (NADPH)	*Glomus fasciculatum, Rhizophagus irregularis*	(Cliquet and Stewart, 1993)

(*Continued*)

TABLE 2.2 (*Continued*)
Nitrogen Metabolic Pathways in Mycorrhiza

Pathways	Mycorrhizal Strains/Enzymes	References
Amino Acid Biosynthesis		
Transaminases	*Glomus fasciculatum*	(Cliquet and Stewart, 1993)
Asparagine synthase	*Rhizophagus irregularis*	(Gachomo *et al.*, 2009)
Arginine biosynthesis		
Carbamoyl-P synthase	*Rhizophagus irregularis*	(Gachomo *et al.*, 2009; Tian *et al.*, 2010; Fellbaum *et al.*, 2012)
Argininosuccinate synthase	*Rhizophagus irregularis*	
Argininosuccinate lyase	*Rhizophagus irregularis*	
Arginine Breakdown		
Arginase	*Rhizophagus irregularis*	(Gachomo *et al.*, 2009; Tian *et al.*, 2010; Fellbaum *et al.*, 2012)
Urease	*Rhol, Rhizophagus irregularis*	
Urease accessory protein	*Rhizophagus irregularis*	(Govindarajulu *et al.*, 2005)
Ornithine aminotransferase	*Rhizophagus irregularis*	(Gachomo *et al.*, 2009; Tian *et al.*, 2010; Fellbaum *et al.*, 2012; Govindarajulu *et al.*, 2005)
Polyamine Biosynthesis		
Ornithine decarboxylase	*Rhizophagus irregularis*	(Gachomo *et al.*, 2009; Tian *et al.*, 2010; Fellbaum *et al.*, 2012)
Nitrogen uptake from the interface	Host plants	
Plant ammonium transporter	*Medicago truncatula*	(Gomez *et al.*, 2009)
	Lotus japonicas	(Guether *et al.*, 2009)
	Glycine max	(Kobae *et al.*, 2010)
	Sorghum bicolor	(Koegel *et al.*, 2013)

fungi may act as an assimilator and pH regulator during ammonium assimilation by proton extrusion. In several situations, the growth of mycorrhiza hypha from fungi plays a crucial role in enhancing ammonium assimilation and nitrate uptake and competing with soil microbes for simple organic nitrogen (Nasholm *et al.*, 1998).

2.26 MYCORRHIZAE AS A SOURCE OF BIOPROTECTANTS AGAINST PATHOGENS

The presence of beneficial microbes in the root zone is among the key factors that determine the health of plants and the fertility level of soil. Studies have shown that mycorrhizal fungi are among the most essential microbial symbiosis, as under stress conditions they improve the assimilation of micronutrients, macronutrients, water relations, and productivity of the plant. AM association was reported to decrease the effect of pathogenic deterioration on the plant (Jeffries *et al.*, 2003). According to Gryndler (2000), mycorrhiza enhances soil's nutritional and physiological properties. Mycorrhizosphere and mycorrhizal fungi (AM) both interact with indigenous and exotic microbes and enhance the quality of soil (Figure 2.10). The effect of mycorrhizal fungi (AM) against the nematodes that cause lesions, galls, and root rots and against soil toxic fungal strains (*Phytophthora*, *Fusarium*, *Aphanomyces*, and *Verticillium*) was observed by Pinochet *et al.* (1996). Garcia-Garrido and Ocampo (1989) observed beneficial effects of mycorrhizal fungi against *Pseudomonas syringae* and *Erwinia carotovora*. AM fungi depend on the rhizosphere for their life cycle, and in return, they lessen the deteriorative effect of harmful microbes present in the soil (Linderman 1997).

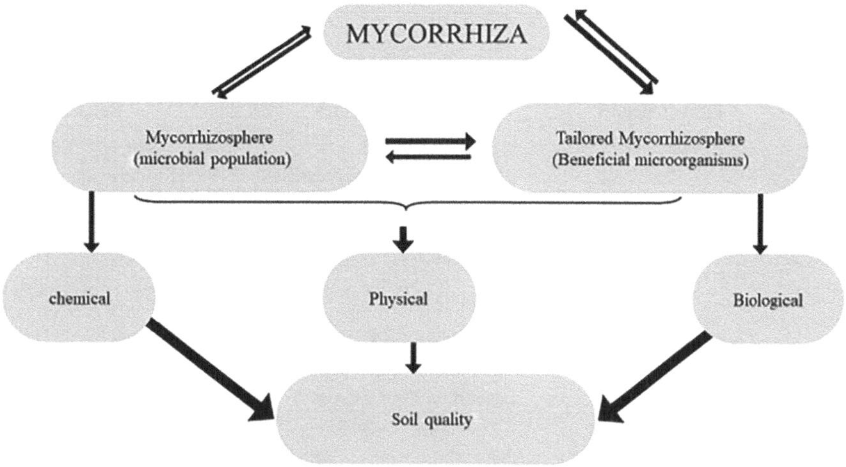

FIGURE 2.10 Mycorrhizal fungi interact with indigenous and non-indigenous microbes present in the root zone and improve the soil fertility.

2.27 MYCORRHIZAE AS A SOURCE OF TOLERANCE TO ABIOTIC STRESS

Abiotic stress is a term that is used to describe how non-living entities affect living organisms in a specific environment. Water scarcity, salinity, and extreme fluctuation in temperature are among the leading factors that cause problems in agriculture production, yield, and quality (Figure 2.11). Various strategies have been adopted to mitigate the negative effects of abiotic stresses on crop quality, production, and quality (Degani *et al.*, 2019). Globally, numerous important food crops are growing under less suitable climatic conditions due to water scarcity and land deterioration brought on by soil salinity (Lesk and Anderson, 2021). It is anticipated that extreme climatic changes may worsen the situation (Sheffield *et al.*, 2012). Studies have shown that plants regulate the expression of various defensive genes in response to stress. Regulation of expression of defensive genes differs from their responses to a single stress or on the specific environmental condition they are exposed to (Atkinson and Urwin, 2012).

Abiotic stress can be lessened by inoculation of plants with beneficial microbes. Mycorrhizal strains and plants exhibit a symbiotic relationship (Nath *et al.*, 2017). Fungal associations are linked with every plant, where they colonize partly or wholly in the interior or external regions of the

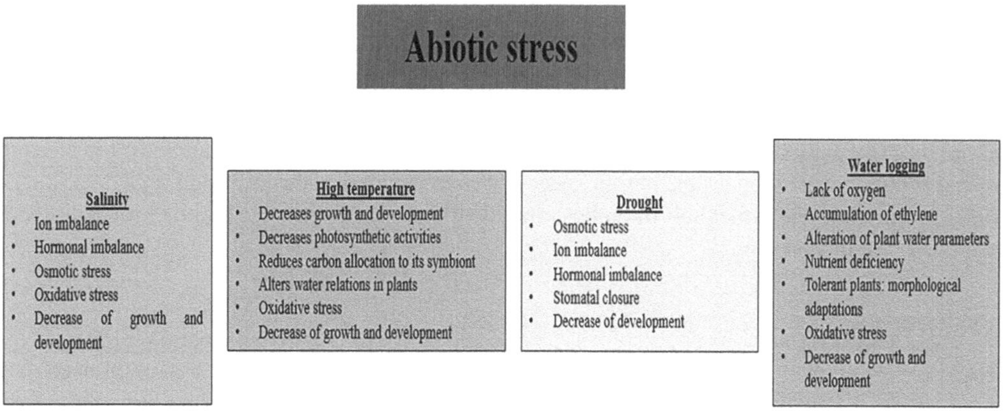

FIGURE 2.11 Effect of abiotic stressors on plants.

TABLE 2.3
Effect of Mycorrhizal Strains on Different Crop Yields

Abiotic Stressors	Fungal Strains	Host Plant Species	References
Drought	*Neotyphodium* sp.	*Festuca pratensis, F. arizonica*, perennial ryegrass	(Malinowski *et al.*, 1997; Barker *et al.*, 1997; Morse *et al.*, 2002
	Neotyphodium lolii	Perennial ryegrass	(Latch *et al.*, 1985; Ravel *et al.*, 1997)
	Neotyphodium coenophialum	Tall fescue	(Belesky *et al.*, 1989; De Battista *et al.*, 1990)
	Acremonium sp.	Tall fescue	(White *et al.*, 1992)
	Phialophora sp.	*F. pratensis*	(Malinowski *et al.*, 1997)
	Curvularia protuberate (Cp4666D)	*Dichanthelium lanuginosum, Leymus mollis, Oryza sativa, Lycopersicon esculentum*	(Rodriguez *et al.*, 2008)
	Curvularia protuberata (CpMH206)	*Dichanthelium lanuginosum*	(Rodriguez *et al.*, 2008)
	Curvularia protuberate (Cp4666D)	*Triticum aestivum*, watermelon	(Rodriguez *et al.*, 2008)
	Curvularia sp.	*Lycopersicon esculentum*	(Rodriguez and Redman, 2008)
Heat	*Curvularia protuberate*	*Lycopersicon esculentum, Dichanthelium lanuginosum*	(Rodriguez *et al.*, 2008)
Salinity	*Piriformospora indica*	*Hordeum vulgare*	(Waller *et al.*, 2005)
	Piriformospora indica	*Hordeum vulgare* cv. Ingrid	(Baltruschat *et al.*, 2008)
	Fusarium culmorum (FcRed1)	*L. esculentum*	(Rodriguez *et al.*, 2008)

host plant. Fungal strains can form a variety of relationships with different host species, such as commensalism, mutualistic, parasitic, and symbiotic. The mycorrhizal relationship with host plants depends upon the host genotype and abiotic factors. Mycorrhizal species have been used to overcome the harmful effects of abiotic stressors on crop yield (Table 2.3).

2.28 CONCLUSIONS AND FUTURE PERSPECTIVES

Mycorrhizal associations enhance the growth, yield, and production of plant species. Mycorrhizal fungi are linked to the roots of the plants and help them in the absorption of nutrients from the soil. Approximately 90% of vascular plants have vesicular–arbuscular mycorrhizae (VAM), whereas some tropical and economically important plant species have been found to have ectomycorrhizae. Researchers are focusing on the symbiotic relationships of fungi and plants. Mycorrhizal strains now are being used for various purposes, for example, as pesticides, to enhance the tolerance of the plant species against biotic and abiotic stresses, to enhance the production of secondary metabolites, and for the reclamation of destroyed lands. Now, various mycorrhizal strains are used for the commercial production of food crops, but the basic questions related to the persistence of these strains in soil, interaction with other soil microbes, and identification of effective strains for specific host species remain unanswered.

REFERENCES

Abbott, L. K., Robson, A. D. and Gazey, C. (1992). Selection of inoculant vesicular-arbuscular mycorrhizal fungi. In: *Methods in microbiology*. Academic Press, Elsevier, vol 24, 1–21.

Abu-Zeyad, R., Khan, A. G. and Khoo, C. (1999). Occurrence of arbuscular *Mycorrhiza* in *Castanospermum australe* A. Cunn. and *C. Fraser* and effects on growth and production of castanospermine. *Mycorrhiza*, 9, 111–117.

Al-Karaki, G. N. (2013). The role of *Mycorrhiza* in the reclamation of degraded lands in arid environments. In *Developments in soil classification, land use planning, and policy implications: Innovative thinking of soil inventory for land use planning and management of land resources*. Elsevier. 823–836.

Allen, J. W. and Shachar-Hill, Y. (2009). Sulfur transfer through an arbuscular *Mycorrhiza. Plant Physiology*, 149(1), 549–560.

Allen, M. F., Allen, E. B. and Gómez-Pompa, A. (2005). Effects of *Mycorrhizae* and non-target organisms on the restoration of a seasonal tropical forest in Quintano Roo, Mexico: Factors limiting tree establishment. *Restoration Ecology*, 13, 325–533.

Alvarez, M., Huygens, D., Diaz, L. M., Villanueva, C. A., Heyser, W. and Boeckx, P. (2012). The spatial distribution of acid phosphatase activity in ectomycorrhizal tissues depends on soil fertility and morphotype and relates to host plant phosphorus uptake. *Plant, Cell and Environment*, 35(1), 126–135.

Amani Machiani, M., Javanmard, A., Morshedloo, M. R., Janmohammadi, M. and Maggi, F. (2021). Funneliformis mosseae application improves the oil quantity and quality and eco-physiological characteristics of soybeans (*Glycine max* L.) under water stress conditions. *Journal of Soil Science and Plant Nutrition*, 21, 3076–3090.

Ames, R. N., Reid, C. P. P., Porter, L. K. and Cambardella, C. (1983). Hyphal uptake and transport of nitrogen from two 15N-labelled sources by Glomus mosseae, a vesicular---arbuscular *Mycorrhizal* fungus. *New Phytologist*, 95(3), 381–396.

Andrea, C., Lingua, G., Bardi, L., Masoero, G. and Berta, G. (2007). Influence of arbuscular *Mycorrhizal* fungi on growth and essential oil composition in Ocimum basilicum var. Genovese. *Caryologia*, 60(1–2), 106–110.

Araim, G., Saleem, A., Arnason, J. T. and Charest, C. (2009). Root colonization by an arbuscular *Mycorrhizal* (AM) fungus increases growth and secondary metabolism of purple coneflower, Echinacea purpurea (L.) moench. *Journal of Agriculture and Food Chemistry*, 57, 2255–2258.

Atkinson, N. J. and Urwin, P. E. (2012). The interaction of plant biotic and abiotic stresses: From genes to the field. *Journal of Experimental Botany*, 63(10), 3523–3543.

Avio, L., Sbrana, C., Giovannetti, M. and Frassinetti, S. (2017). Arbuscular *Mycorrhizal* fungi affect total phenolic content and antioxidant activity in leaves of oak leaf lettuce varieties. *Scientia Horticulturae*, 224, 265–271.

Bago, B., Vierheilig, H., Piché, Y. and Azcón-Aguilar, C. (1996). Nitrate depletion and pH changes induced by the extraradical mycelium of the arbuscular *Mycorrhizal* fungus Glomus intraradices grown in monoxenic culture. *New Phytologist*, 133(2), 273–280.

Baltruschat, H., Fodor, J., Harrach, B. D., Niemczyk, E., Barna, B., Gullner, G. and Skoczowski, A. (2008). Salt tolerance of barley induced by the root endophyte Piriformospora indica is associated with a strong increase in antioxidants. *New Phytologist*, 180(2), 501–510.

Barea, J., Palenzuela, J., Cornejo, P., Sánchez-Castro, I., Navarro-Fernández, C. and Lopéz-García, A. (2011). Ecological and functional roles of *Mycorrhizas* in semi-arid ecosystems of Southeast Spain. *Journal of Arid Environments*, 75(12), 1292–1301.

Barker, D. J., Hume, D. E. and Quigley, P. E. (1997). Negligible physiological responses to water deficit in endophyte-infected and uninfected perennial ryegrass. *Neotyphodium/Grass Interactions*, 137–139.

Barman, J., Samanta, A., Saha, B. and Datta, S. (2016). *Mycorrhiza*: The oldest association between plants and fungi. *Resonance*, 21, 1093–1104.

Baslam, M., Garmendia, I. and Goicoechea, N. (2011). Arbuscular *Mycorrhizal* fungi (AMF) improved the growth and nutritional quality of greenhouse-grown lettuce. *Journal of Agricultural and Food Chemistry*, 59(10), 5504–5515.

Battini, F., Bernardi, R., Turrini, A., Agnolucci, M. and Giovannetti, M. (2016). Rhizophagus intraradices or its associated bacteria affect gene expression of key enzymes involved in the rosmarinic acid biosynthetic pathway of basil. *Mycorrhiza*, 26, 699–707.

Behie, S. W. and Bidochka, M. J. (2014). Nutrient transfer in plant-fungal symbioses. *Trends in Plant Sciences*, 19, 734–740.

Belesky, D. P., Stringer, W. C. and Hill, N. S. (1989). Influence of endophyte and water regime upon tall fescue accessions. I. Growth characteristics. *Annals of Botany*, 63(5), 495–503.

Belmondo, S., Fiorilli, V., Pérez-Tienda, J., Ferrol, N., Marmeisse, R. and Lanfranco, L. (2014). A dipeptide transporter from the arbuscular *Mycorrhizal* fungus Rhizophagus irregularis is upregulated in the intraradical phase. *Frontiers in Plant Science*, 5, 436.

Bennett, A. E. and Classen, A. T. (2020). Climate change influences mycorrhizal fungal-plant interactions, but conclusions are limited by geographical study bias. *Ecology*, 101(4), e02978.

Blilou, I., Ocampo, J. A. and Garcia, G. J. (2000). Induction of LTP (lipid transfer protein) and Pal (phenylalanine ammonia-lyase) gene expression in rice roots colonized by the arbuscular *Mycorrhizal* fungus Glomus mosseae. *Journal of Experimental Botany*, 51, 1969–1977.

Bona, E., Cantamessa, S., Massa, N., Manassero, P., Marsano, F., Copetta, A., Lingua, G., D'Agostino, G., Gamalero, E. and Berta, G. (2017). Arbuscular *Mycorrhizal* fungi and plant growth-promoting pseudomonads improve yield, quality and nutritional value of tomato: A field study. *Mycorrhiza*, 27, 1–11.

Bonanomi, A., Oetiker, J. H., Guggenheim, R., Boller, T., Wiemken, A. and Vögeli-Lange, R. (2001). Arbuscular *Mycorrhiza* in mini-mycorrhizotrons: First contact of Medicago truncatula roots with Glomus intraradices induces chalcone synthase. *New Phytologist*, 150, 573–582.

Bonfante, P. and Anca, I. A. (2009). Plants, *Mycorrhizal* fungi, and bacteria: A network of interactions. *Annual Review of Microbiology*, 63, 363–383.

Bonito, G. M., Gryganskyi, A. P., Trappe, J. M. and Vilgalys, R. (2010). A global meta-analysis of Tuber ITS rDNA sequences: Species diversity, host associations and long-distance dispersal. *Molecular Ecology*, 19, 4994–5008.

Breuninger, M., Trujillo, C. G., Serrano, E., Fischer, R. and Requena, N. (2004). Different nitrogen sources modulate activity but not expression of glutamine synthetase in arbuscular *Mycorrhizal* fungi. *Fungal Genetics and Biology*, 41(5), 542–552.

Brundrett, M. C. (2009). *Mycorrhizal* associations and other means of nutrition of vascular plants: Understanding the global diversity of host plants by resolving conflicting information and developing reliable means of diagnosis. *Plant and Soil*, 320, 37–77.

Brundrett, M. C. and Abbott, L. K. (2002). Arbuscular *Mycorrhizas* in plant communities. *Microorganisms in Plant Conservation and Biodiversity*, 151–193.

Brundrett, M. C., Bougher, N., Dell, B., Grove, T. and Malajczuk, N. (1996). Working with *Mycorrhizas* in forestry and agriculture. *Canberra: Australian Centre for International Agricultural Research*, 32, 374.

Bucher, M. (2007). Functional biology of plant phosphate uptake at root and *Mycorrhiza* interfaces. *New Phytologist*, 173(1), 11–26.

Bücking, H., Liepold, E. and Ambilwade, P. (2012). The role of the mycorrhizal symbiosis in nutrient uptake of plants and the regulatory mechanisms underlying these transport processes. *Plant Science*, 4, 108–132.

Cappellazzo, G., Lanfranco, L., Fitz, M., Wipf, D. and Bonfante, P. (2008). Characterization of an amino acid permease from the endomycorrhizal fungus Glomus mosseae. *Plant Physiology*, 147(1), 429–437.

Caser, M., Demasi, S., Victorino, Í. M. M., Donno, D., Faccio, A., Lumini, E., Bianciotto, V. and Scariot, V. (2019). Arbuscular *Mycorrhizal* fungi modulate the crop performance and metabolic profile of saffron in soilless cultivation. *Agronomy*, 9, 232.

Castellanos Morales, V., Villegas, J., Wendelin, S., Vierheilig, H., Eder, R. and Cárdenas Navarro, R. (2010). Root colonization by the arbuscular *Mycorrhizal* fungus Glomus intraradices alters the quality of strawberry fruits (Fragaria× ananassa Duch.) at different nitrogen levels. *Journal of the Science of Food and Agriculture*, 90, 1774–1782.

Cecatto, A. P., Ruiz, F. M., Calvete, E. O., Martínez, J. and Palencia, P. (2016). *Mycorrhizal* inoculation affects the phytochemical content of strawberry fruits. *Acta Scientiarum Agronomy*, 38, 227–237.

Ceccarelli, N., Curadi, M., Martelloni, L., Sbrana, C., Picciarelli, P. and Giovannetti, M. (2010). *Mycorrhizal* colonization impacts on phenolic content and antioxidant properties of artichoke leaves and flower heads two years after field transplant. *Plant and Soil*, 335, 311–323.

Chaudhary, V. B., Holland, E. P., Charman-Anderson, S., Guzman, A., Bell-Dereske, L., Cheeke, T. E. and Helgason, T. (2022). What are mycorrhizal traits? *Trends in Ecology & Evolution*, 37(7), 573–581.

Chen, S., Zhao, H., Zou, C., Li, Y., Chen, Y., Wang, Z. and Ahammed, G. J. (2017). Combined inoculation with multiple arbuscular *Mycorrhizal* fungi improves growth, nutrient uptake, and photosynthesis in cucumber seedlings. *Frontiers in Microbiology*, 8, 2516.

Cliquet, J. B. and Stewart, G. R. (1993). Ammonia assimilation in Zea mays L. infected with a vesicular-arbuscular *Mycorrhizal* fungus Glomus fasciculatum. *Plant Physiology*, 101(3), 865–868.

Copetta, A., Lingua, G. and Berta, G. (2006). Effects of three AM fungi on growth, distribution of glandular hairs, and essential oil production in Ocimum basilicum L. var. Genovese. *Mycorrhiza*, 16, 485–494.

Danielson, R. M. and Visser, S. (1989). Host response to inoculation and behavior of induced and indigenous ectomycorrhizal fungi of jack pine grown on oil-sands tailings. *Canadian Journal of Forest Research*, 19, 1412–1421.

De Battista, J. P., Bouton, J. H., Bacon, C. W. and Siegel, M. R. (1990). Rhizome and herbage production of endophyte-removed tall fescue clones and populations. *Agronomy Journal*, 82(4), 651–654.

Degani, E., Leigh, S. G., Barber, H. M., Jones, H. E., Lukac, M., Sutton, P. and Potts, S. G. (2019). Crop rotations in a climate change scenario: short-term effects of crop diversity on resilience and ecosystem service provision under drought. *Agriculture, Ecosystems & Environment*, 285, 106625.

Dugassa, G. D., Von Alten, H. and Schönbeck, F. (1996). Effects of arbuscular *Mycorrhiza* (AM) on the health of Linum usitatissimum L. infected by fungal pathogens. *Plant and Soil*, 185, 173–182.

El Gabardi, S., Mouden, N., Chliyeh, M., Selmaoui, K., Touhami, A. O. and Douira, A. (2022). Mycorrhizae: Diversity and roles in plant ecosystems. *Moroccan Journal of Agricultural Sciences*, 3(4).

Ezawa, T., Hayatsu, M. and Saito, M. (2005). A new hypothesis on the strategy for acquisition of phosphorus in arbuscular *Mycorrhiza*: Up-regulation of secreted acid phosphatase gene in the host plant. *Molecular Plant-Microbe Interactions*, 18(10), 1046–1053.

Fellbaum, C. R., Gachomo, E. W., Beesetty, Y., Choudhari, S., Strahan, G. D., Pfeffer, P. E. and Bücking, H. (2012). Carbon availability triggers fungal nitrogen uptake and transport in arbuscular *Mycorrhizal* symbiosis. *Proceedings of the National Academy of Sciences*, 109(7), 2666–2671.

Fester, T.; Strack, D. and Hause, B. (2001). Reorganization of tobacco root plastids during arbuscule development. *Planta*, 213, 864–868.

Frey, B. and Schüepp, H. (1993). Acquisition of nitrogen by external hyphae of arbuscular *Mycorrhizal* fungi associated with Zea mays L. *New Phytologist*, 124(2), 221–230.

Fung, M. Y. and Macyk, T. M. (2000). Reclamation of oil sands mining areas. *Reclamation of Drastically Disturbed Lands*, 41, 755–774.

Gachomo, E., Allen, J. W., Pfeffer, P. E., Govindarajulu, M., Douds, D. D., Jin, H. and Bücking, H. (2009). Germinating spores of Glomus intraradices can use internal and exogenous nitrogen sources for de novo biosynthesis of amino acids. *New Phytologist*, 184(2), 399–411.

Garcia-Garrido, J. M. and Ocampo, J. A. (1989). Effect of VA *Mycorrhizal* infection of tomato on damage caused by Pseudomonas syringae. *Soil Biology and Biochemistry*, 21(1), 165–167.

Gardes, M. (2003). An orchid-fungus marriage-physical promiscuity, conflict and cheating. *New Phytology*, 154, 46.

Gomez, S. K., Javot, H., Deewatthanawong, P., Torres-Jerez, I., Tang, Y., Blancaflor, E. B. and Harrison, M. J. (2009). Medicago truncatula and Glomus intraradices gene expression in cortical cells harboring arbuscules in the arbuscular *Mycorrhizal* symbiosis. *BMC Plant Biology*, 9(1), 1–19.

Govindarajulu, M., Pfeffer, P. E., Jin, H., Abubaker, J., Douds, D. D., Allen, J. W. and Shachar-Hill, Y. (2005). Nitrogen transfer in the arbuscular *Mycorrhizal* symbiosis. *Nature*, 435(7043), 819–823.

Gryndler, M. (2000). Interactions of arbuscular *Mycorrhizal* fungi with other soil organisms. Arbuscular *Mycorrhizas. Physiology and Function*, 239–262.

Guether, M., Neuhäuser, B., Balestrini, R., Dynowski, M., Ludewig, U. and Bonfante, P. (2009). A *Mycorrhizal*-specific ammonium transporter from *Lotus japonicus* acquires nitrogen released by arbuscular *Mycorrhizal* fungi. *Plant Physiology*, 150(1), 73–83.

Handa, Y., Nishide, H., Takeda, N., Suzuki, Y., Kawaguchi, M. and Saito, K. (2015). RNA-seq transcriptional profiling of an arbuscular *Mycorrhiza* provides insights into regulated and coordinated gene expression in Lotus japonicus and Rhizophagus irregularis. *Plant Cell Physiology*, 56, 1490–1511.

Harrison, M. J. and Dixon, R. A. (1993). Isoflavonoid accumulation and expression of defense gene transcripts during the establishment of vesicular-arbuscular *Mycorrhizal* associations in roots of Medicago truncatula. *Molecular Plant-Microbe Interaction*, 6, 643–654.

Harrison, M. J. and Dixon, R. A. (1994). Spatial patterns of expression of flavonoid/isoflavonoid pathway genes during interactions between roots of Medicago truncatula and the *Mycorrhizal* fungus Glomus versiforme. *The Plant Journal*, 6, 9–20.

Hart, M., Ehret, D. L., Krumbein, A., Leung, C., Murch, S., Turi, C. and Franken, P. (2015). Inoculation with arbuscular *Mycorrhizal* fungi improves the nutritional value of tomatoes. *Mycorrhiza*, 25, 359–376.

Hawkins, H. J., Cramer, M. D. and George, E. (1999). Root respiratory quotient and nitrate uptake in hydroponically grown non-*mycorrhizal* and *Mycorrhizal* wheat. *Mycorrhiza*, 9, 57–60.

Hawkins, H. J., Johansen, A. and George, E. (2000). Uptake and transport of organic and inorganic nitrogen by arbuscular *Mycorrhizal* fungi. *Plant and Soil*, 226, 275–285.

Hijikata, N., Murase, M., Tani, C., Ohtomo, R., Osaki, M. and Ezawa, T. (2010). Polyphosphate has a central role in the rapid and massive accumulation of phosphorus in the extraradical mycelium of an arbuscular *Mycorrhizal* fungus. *The New Phytologist*, 186(2), 285–289.

Imhof, S., Massicote, H. B., Melville, L. H. and Petereson, R. L. (2013). Subterranean morphology and mycorrhizal structures. In: V. Merckx, ed. *Mycoheterotrophy*. New York, NY, USA: Springer International, 157–214.

Jakobsen, I., Abbott, L. K. and Robson, A. D. (1992). External hyphae of vesicular-arbuscular *Mycorrhizal* fungi associated with *Trifolium subterraneum* L. 1. Spread of hyphae and phosphorus inflow into roots. *New Phytologist*, 120(3), 371–380.

Jakobsen, I., Gazey, C. and Abbott, L. K. (2001). Phosphate transport by communities of arbuscular *Mycorrhizal* fungi in intact soil cores. *New Phytologist*, 95–103.

Javot, H., Pumplin, N. and Harrison, M. J. (2007). Phosphate in the arbuscular *Mycorrhizal* symbiosis: Transport properties and regulatory roles. *Plant, Cell and Environment*, 30(3), 310–322.

Jeffries, P., Gianinazzi, S., Perotto, S., Turnau, K. and Barea, J. (2003). The contribution of arbuscular *Mycorrhizal* fungi in sustainable maintenance of plant health and soil fertility. *Biology and Fertility of Soils*, 37(1), 1–16.

Jin, H., Pfeffer, P. E., Douds, D. D., Piotrowski, E., Lammers, P. J. and Shachar-Hill, Y. (2005). The uptake, metabolism, transport, and transfer of nitrogen in an arbuscular *Mycorrhizal* symbiosis. *New Phytologist*, 168(3), 687–696.

Johansen, A., Finlay, R. D. and Olsson, P. A. (1996). Nitrogen metabolism of external hyphae of the arbuscular *Mycorrhizal* fungus *Glornus intraradices*. *New Phytologist*, 133(4), 705–712.

Johansen, A., Jakobsen, I. and Jensen, E. S. (1993). Hyphal transport by a vesicular-arbuscular *Mycorrhizal* fungus of N applied to the soil as ammonium or nitrate. *Biology and Fertility of Soils*, 16, 66–70.

Jurkiewicz, A., Ryszka, P., Anielska, T., Waligorski, P., Białonska, D., Goralska, K., Tsimilli-Michael, M. and Turnau, K. (2010). Optimization of culture conditions of Arnica montana L.: Effects of *Mycorrhizal* fungi and competing plants. *Mycorrhiza*, 20, 293–306.

Kaldorf, M., Schmelzer, E. and Bothe, H. (1998). Expression of maize and fungal nitrate reductase genes in arbuscular *Mycorrhiza*. *Molecular Plant-Microbe Interactions*, 11(6), 439–448.

Kaldorf, M., Zimmer, W. and Bothe, H. (1994). Genetic evidence for the occurrence of assimilatory nitrate reductase in arbuscular *Mycorrhizal* and other fungi. *Mycorrhiza*, 5, 23–28.

Kapoor, R., Giri, B. and Mukerji, K. G. (2002). *Glomus macrocarpum*: A potential bioinoculant to improve essential oil quality and concentration in Dill (Anethum graveolens L.) and Carum (Trachyspermum ammi (Linn.) Sprague). *World Journal of Microbiology and Biotechnology*, 18, 459–463.

Khan, A. G. (2006). Mycorrhizoremediation – an enhanced form of phytoremediation. *Journal of Zhejiang University Science Bulletin*, 7, 503–514.

Khaosaad, T., Krenn, L., Medjakovic, S., Ranner, A., Lössl, A., Nell, M., Jungbauer, A. and Vierheilig, H. (2008). Effect of *Mycorrhization* on the isoflavone content and the phytoestrogen activity of red clover. *Journal of Plant Physiology*, 165, 1161–1167.

Kiers, J. L., Rombouts, F. M. and Nout, M. J. R. (2000). In vitro digestibility of Bacillus fermented soya bean. *International Journal of Food Microbiology*, 60(2–3), 163–169.

Kitamura, K., Tamai, Y., Yajima, T. and Miyamoto, T. (2023). Mycorrhizal morphology and symbiotic fungi on the root of *Pyrola renifolia*. *Asian Journal of Research in Botany*, 6(2), 148–157.

Klingner, A., Hundeshagen, B., Kernebeck, H. and Bothe, H. (1995). Localization of the yellow pigment formed in roots of gramineous plants colonized by arbuscular fungi. *Protoplasma*, 185, 50–57.

Kobae, Y., Tamura, Y., Takai, S., Banba, M. and Hata, S. (2010). Localized expression of arbuscular *Mycorrhiza*-inducible ammonium transporters in soybean. *Plant and Cell Physiology*, 51(9), 1411–1415.

Koegel, S., Ait Lahmidi, N., Arnould, C., Chatagnier, O., Walder, F., Ineichen, K. and Courty, P. E. (2013). The family of ammonium transporters (AMT) in *Sorghum bicolor*: Two AMT members are induced locally, but not systemically in roots colonized by arbuscular *Mycorrhizal* fungi. *New Phytologist*, 198(3), 853–865.

Kohout P. (2017). Biogeography of ericoid mycorrhiza. In: L. Tedersoo, ed. *Biogeography of mycorrhizal symbiosis*. Cham, Switzerland: Springer, 179– 193.

Koide, R. T. and Mosse, B. (2004). A history of research on arbuscular *Mycorrhiza*. *Mycorrhiza*, 14, 145–163.

Latch, G. C. M., Hunt, W. F. and Musgrave, D. R. (1985). Endophytic fungi affect the growth of perennial ryegrass. *New Zealand Journal of Agricultural Research*, 28(1), 165–168.

Latef, A. A. H. A., Hashem, A., Rasool, S., Abd_Allah, E. F., Alqarawi, A. A., Egamberdieva, D., Jan, S., Anjum, N. A. and Ahmad, P. (2016). Arbuscular *Mycorrhizal* symbiosis and abiotic stress in plants: A review. *Journal Plant Biology*, 59, 407–426.

Lee, J. and Scagel, C. F. (2009). Chicoric acid is found in basil (Ocimum basilicum L.) leaves. *Food Chemistry*, 115, 650–656.

Lesk, C. and Anderson, W. (2021). Decadal variability modulates trends in concurrent heat and drought over global croplands. *Environmental Research Letters*, 16(5), 055024.

Li, H., Smith, F. A., Dickson, S., Holloway, R. E. and Smith, S. E. (2008). Plant growth depressions in arbuscular *Mycorrhizal* symbioses: Not just caused by carbon drain? *New Phytologist*, 178(4), 852–862.

Li, Q., Yang, L., Xiang, D., Wan, Y., Wu, Q., Huang, W. and Zhao, G. (2020). The complete mitochondrial genomes of two model ectomycorrhizal fungi (Laccaria): Features, intron dynamics, and phylogenetic implications. *International Journal of Biological Macromolecules*, 145, 974–984.

Linderman, R. G. (1997). Vesicular-arbuscular *Mycorrhizal* (VAM) fungi. *Plant Relationships Part B: Part B*, 117–128.

Lingua, G., Bona, E., Manassero, P., Marsano, F., Todeschini, V., Cantamessa, S., Copetta, A., D'Agostino, G., Gamalero, E. and Berta, G. (2013). Arbuscular *Mycorrhizal* fungi and plant growth-promoting pseudomonads increase anthocyanin concentration in strawberry fruits (*Fragaria* × *ananassa* var. Selva) in conditions of reduced fertilization. *International Journal of Molecular Sciences*, 14, 16207–16225.

Liu, B., Dong, P., Zhang, X., Feng, Z., Wen, Z., Shi, L., Xia, Y., Chen, C., Shen, Z., Lian, C. and Chen, Y. (2022). Identification and characterization of eight metallothionein genes involved in heavy metal tolerance from the ectomycorrhizal fungus *Laccaria bicolor. Environmental Science and Pollution Research*, 1–13.

Liu, J., Maldonado-Mendoza, I., Lopez-Meyer, M., Cheung, F., Town, C. D. and Harrison, M. J. (2007). Arbuscular Mycorrhizal symbiosis is accompanied by local and systemic alterations in gene expression and an increase in disease resistance in the shoots. *The Plant Journal*, 50, 529–544.

Lohse, S., Schliemann, W., Ammer, C., Kopka, J., Strack, D. and Fester, T. (2005). Organization and metabolism of plastids and mitochondria in arbuscular *Mycorrhizal* roots of Medicago truncatula. *Plant Physiology*, 139, 329–340.

MacDonald, R. M. and Lewis, M. (1978). The occurrence of some acid phosphatases and dehydrogenases in the vesicular-arbuscular *Mycorrhizal* fungus glomusmosseae. *New Phytologist*, 80(1), 135–141.

Malinowski, D., Leuchtmann, A., Schmidt, D. and Nsberger, J. (1997). Growth and water status in meadow fescue (*Festuca pratensis*) is differently affected by its two natural endophytes. *Agronomy Journal*, 89, 673–678.

Maurich, T., Iorio, M., Chimenti, D. and Turchi, G. (2006). Erybraedin C and bitucarpin A, two structurally related pterocarpans purified from *Bituminaria bituminosa*, induced apoptosis in human colon adenocarcinoma cell lines MMR-and p53-proficient and-deficient in a dose-, time-, and structure-dependent fashion. *Chemico-Biological Interaction*, 159, 104–116.

Michelsen, A., Schmidt, I. K., Jonasson, S., Quarmby, C. and Sleep, D. (1996). Leaf 15 N abundance of subarctic plants provides field evidence that ericoid, ectomycorrhizal, and non-and arbuscular *Mycorrhizal* species access different sources of soil nitrogen. *Oecologia*, 105, 53–63.

Morse, L. J., Day, T. A. and Faeth, S. H. (2002). Effect of Neotyphodium endophyte infection on growth and leaf gas exchange of Arizona fescue under contrasting water availability regimes. *Environmental and Experimental Botany*, 48(3), 257–268.

Nahin, R. L., Pecha, M., Welmerink, D. B., Sink, K., DeKosky, S. T. and Fitzpatrick, A. L. (2009). Ginkgo Evaluation of Memory Study Investigators. Concomitant use of prescription drugs and dietary supplements in ambulatory elderly people. *Journal of the American Geriatrics Society*, 57(7), 1197–1205.

Nasholm, T., Ekblad, A., Nordin, A., Giesler, R., Högberg, M. and Högberg, P. (1998). Boreal forest plants take up organic nitrogen. *Nature*, 392(6679), 914–916.

Nath, M., Bhatt, D., Prasad, R. and Tuteja, N. (2017). Reactive oxygen species (ROS) metabolism and signaling in plant-*mycorrhizal* association under biotic and abiotic stress conditions. *Mycorrhiza-eco-physiology, Secondary Metabolites, Nanomaterials*, 223–232.

Navarro-Fernández, C. M., Aroca, R. and Barea, J. M. (2011). Influence of arbuscular *Mycorrhizal* fungi and water regime on the development of endemic Thymus species in dolomitic soils. *Applied Soil Ecology*, 48(1), 31–37.

Nazir, R., Rehman, S., Nisa, M., & ali Baba, U. (2019). Exploring bacterial diversity: From cell to sequence. In: *Freshwater microbiology*. Academic Press, Elsevier, 263–306.

Nemec, S. (1981). Histochemical characteristics of Glomus etunicatus infection of Citrus limon fibrous roots. *Canadian Journal of Botany*, 59(5), 609–617.

Oliveer, A. J., Smith, S. E., Nicholas, D. J. D., Wallace, W. and Smith, F. A. (1983). Activity Of nitrate reductase in trifoljumsubterraneum: Effects of *Mycorrhizal* infection and phosphate nutrition. *New Phytologist*, 94(1), 63–79.

Ori, F., Leonardi, M., Faccio, A., Sillo, F., Iotti, M., Pacioni, G. and Balestrini, R. (2020). Synthesis and ultrastructural observation of arbutoid mycorrhizae of black truffles (*Tuber melanosporum* and *T. aestivum*). *Mycorrhiza*, 30, 715–723.

Pacheco Flores de Valgaz, A., Barcos-Arias, M., Naranjo-Morán, J., Peña Tapia, D. and Moreira-Gómez, R. (2022). Ericaceous plants: A review for the bioprospecting of ericoid mycorrhizae from Ecuador. *Diversity*, 14(8), 648.

Parniske, M. (2008). Arbuscular *Mycorrhiza*: The mother of plant root endosymbioses. *Nature Reviews Microbiology*, 6(10), 763–775.

Paszkowski, U. (2006). A journey through signaling in arbuscular mycorrhizal symbioses 2006. *New Phytologist*, 172(1), 35–46.

Pérez-Brocal, V., Latorre, A. and Moya, A. (2011). Symbionts and pathogens: What is the difference? *Between Pathogenicity and Commensalism*, 215–243.

Pérez-Tienda, J., Testillano, P. S., Balestrini, R., Fiorilli, V., Azcón-Aguilar, C. and Ferrol, N. (2011). GintAMT2, a new member of the ammonium transporter family in the arbuscular *Mycorrhizal* fungus Glomus intraradices. *Fungal Genetics and Biology*, 48(11), 1044–1055.

Peterson, R. L., Massicotte, H. B. and Melville, L. H. (2004). *Mycorrhizas: Anatomy and Cell Biology*. Ottawa, ON, Canada: National Research Council Research Press.

Pineda, A., Zheng, S. J., van Loon, J. J., Pieterse, C. M. and Dicke, M. (2010). Helping plants to deal with insects: The role of beneficial soil-borne microbes. *Trends in Plant Science*, 15(9), 507–514.

Pinochet, J., Calvet, C., Camprubi, A. and Fernandez, C. (1996). Interactions between migratory endoparasitic nematodes and arbuscular mycorrhizal fungi in perennial crops: A review. *Plant and Soil*, 185, 183–190.

Pistelli, L., Ulivieri, V., Giovanelli, S., Avio, L., Giovannetti, M. and Pistelli, L. (2017). Arbuscular *Mycorrhizal* fungi alter the content and composition of secondary metabolites in Bituminaria bituminosa L. *Plant Biology*, 19, 926–933.

Piszczek, P., Kuszewska, K., Błaszkowski, J., Sochacka-Obruśnik, A., Stojakowska, A. and Zubek, S. (2019). Associations between root-inhabiting fungi and 40 species of medicinal plants with potential applications in the pharmaceutical and biotechnological industries. *Applied Soil Ecology*, 137, 69–77.

Ponce, M. A., Scervino, J. M., Erra-Balsells, R., Ocampo, J. A. and Godeas, A. M. (2004). Flavonoids from shoots and roots of Trifolium repens (white clover) are grown in the presence or absence of the arbuscular *Mycorrhizal* fungus Glomus intraradices. *Phytochemistry*, 65, 1925–1930.

Ran, Z., Ding, W., Cao, S., Fang, L., Zhou, J. and Zhang, Y. (2022). Arbuscular *Mycorrhizal* fungi: Effects on secondary metabolite accumulation of traditional Chinese medicines. *Plant Biology*, 24(6), 932–938.

Rasouli-Sadaghiani, M., Hassani, A., Barin, M., Rezaee Danesh, Y. and Sefidkon, F. (2010). Effects of AM fungi on growth, essential oil production, and nutrient uptake in basil. *Journal of Medicinal Plant Research*, 4, 2222–2228.

Ravel, C., Courty, C., Coudret, A. and Charmet, G. (1997). Beneficial effects of *Neotyphodium lolii* on the growth and the water status in perennial ryegrass cultivated under nitrogen deficiency or drought stress. *Agronomie*, 17(3), 173–181.

Raven, J. A. and Allen, J. F. (2003). Genomics and chloroplast evolution: What did cyanobacteria do for plants? *Genome Biology*, 4, 1–5.

Read, D. J., Duckett, J. G., Francis, R., Ligrone, R. and Russell, A. (2000). Symbiotic fungal associations in 'lower land plants. *Philosophical Transactions of the Royal Society of London. Series B: Biological Sciences*, 355(1398), 815–831.

Robertson, O. (2020). Fungal future: A review of mycelium biocomposites as an ecological alternative insulation material. *DS 101: Proceedings of NordDesign 2020*, Lyngby, Denmark, 12th–14th August 2020, 1–13.

Rodriguez, R. and Redman, R. (2008). More than 400 million years of evolution and some plants still can't make it on their own: Plant stress tolerance via fungal symbiosis. *Journal of Experimental Botany*, 59(5), 1109–1114.

Rodriguez, R. J., Henson, J., Van Volkenburgh, E., Hoy, M., Wright, L., Beckwith, F. and Redman, R. S. (2008). Stress tolerance in plants via habitat-adapted symbiosis. *The ISME Journal*, 2(4), 404–416.

Roy, R., Reinders, A., Ward, J. M. and McDonald, T. R. (2020). Understanding transport processes in lichen, Azolla – cyanobacteria, ectomycorrhiza, endomycorrhiza, and rhizobia – legume symbiotic interactions. *F1000Research*, 9.

Saleh, A. M., Abdel-Mawgoud, M., Hassan, A. R., Habeeb, T. H., Yehia, R. S. and AbdElgawad, H. (2020). Global metabolic changes induced by arbuscular *Mycorrhizal* fungi in oregano plants grown under ambient and elevated levels of atmospheric CO2. *Plant Physiology and Biochemistry*, 151, 255–263.

Schliemann, W., Ammer, C. and Strack, D. (2008). Metabolite profiling of *Mycorrhizal* roots of Medicago truncatula. *Phytochemistry*, 69, 112–146.

Schneider-Maunoury, L., Deveau, A., Moreno, M., Todesco, F., Belmondo, S., Murat, C., Courty, P. E., Jąkalski, M. and Selosse, M. A. (2020). Two ectomycorrhizal truffles, *Tuber melanosporum*, and *T. aestivum*, endophytically colonize roots of non-ectomycorrhizal plants in natural environments. *New Phytologist*, 225, 2542– 2556.

SER (Society for Ecological Restoration International Science and Policy Working Group). (2004). *The SER international primer on ecological restoration* (available from www.ser.org) accessed in July 2005. Tucson, AZ: Society for Ecological Restoration International.

Sheffield, J., Wood, E. F. and Roderick, M. L. (2012). Little change in global drought over the past 60 years. *Nature*, 491(7424), 435–438.

Shu, B., Li, W., Liu, L., Wei, Y. and Shi, S. (2016). Transcriptomes of arbuscular *Mycorrhizal* fungi and litchi host interaction after tree girdling. *Frontiers in Microbiology*, 7, 408.

Singh, A. (2007). Molecular basis of plant-symbiotic fungi interaction: An overview. *Scientific World*, 5(5), 115–131.

Singh, L. P., Gill, S. S. and Tuteja, N. (2011). Unraveling the role of fungal symbionts in plant abiotic stress tolerance. *Plant Signaling and Behavior*, 6(2), 175–191.

Singh, V., Naveenkumar, R. and Muthukumar, A. (2019). Arbuscular mycorrhizal fungi and their effectiveness against soil-borne diseases. *Management*, 183, 199.

Smith, S. E. and Read, D. J. (2008). *Mycorrhizal symbiosis*, 3rd edn. New York, NY: Academic Press.

Smith, S. E. and Read, D. J. (2010). *Mycorrhizal symbiosis*. Academic Press, Elsevier.

Smith, S. E., St John, B. J., Smith, F. A. and Nicholas, D. J. D. (1985). Activity of glutamine synthetase and glutamate dehydrogenase in Trifolium subterraneum L. and Allium cepa L: Effects of *Mycorrhizal* infection and phosphate nutrition. *New Phytologist*, 99(2), 211–227.

Subramanian, K. S. and Charest, C. (1998). Arbuscular *Mycorrhizae* and nitrogen assimilation in maize after drought and recovery. *Physiologia Plantarum*, 102(2), 285–296.

Tedersoo, L., Bahram, M. and Zobel, M. (2020). How mycorrhizal associations drive plant population and community biology. *Science*, 367(6480).

Thomson, B. D., Clarkson, D. T. and Brain, P. (1990). Kinetics of phosphorus uptake by the germ-tubes of the vesicular-arbuscular *Mycorrhizal* fungus, *Gigaspora margarita*. *New Phytologist*, 116(4), 647–653.

Tian, C., Kasiborski, B., Koul, R., Lammers, P. J., Bücking, H. and Shachar-Hill, Y. (2010). Regulation of the nitrogen transfer pathway in the arbuscular *Mycorrhizal* symbiosis: Gene characterization and the coordination of expression with nitrogen flux. *Plant Physiology*, 153(3), 1175–1187.

Tisserant, E., Kohler, A., Dozolme-Seddas, P., Balestrini, R., Benabdellah, K., Colard, A. and Martin, F. (2012). The transcriptome of the arbuscular *Mycorrhizal* fungus Glomus intraradices (DAOM 197198) reveals functional tradeoffs in an obligate symbiont. *New Phytologist*, 193(3), 755–769.

Tobar, R., Azcón, R. and Barea, J. M. (1994). Improved nitrogen uptake and transport from 15N-labelled nitrate by external hyphae of arbuscular *Mycorrhiza* under water-stressed conditions. *New Phytologist*, 126(1), 119–122.

Todeschini, V., AitLahmidi, N., Mazzucco, E., Marsano, F., Gosetti, F., Robotti, E., Bona, E., Massa, N., Bonneau, L. and Marengo, E. (2018). Impact of beneficial microorganisms on strawberry growth, fruit production, nutritional quality, and volatilome. *Frontiers in Plant Science*, 9, 1611.

Toussaint, J. P., Kraml, M., Nell, M., Smith, S. E., Smith, F. A., Steinkellner, S., Schmiderer, C., Vierheilig, H. and Novak, J. (2008). Effect of *Glomus mosseae* on concentrations of rosmarinic and caffeic acids and essential oil compounds in basil inoculated with *Fusarium oxysporum* f. sp. basilici. *Plant Pathology*, 57, 1109–1116.

Toussaint, J. P., Smith, F. A. and Smith, S. E. (2007). Arbuscular *Mycorrhizal* fungi can induce the production of phytochemicals in sweet basil irrespective of phosphorus nutrition. *Mycorrhiza*, 17, 291–297.

Turk, M. A., Assaf, T. A., Hameed, K. M. and Al-Tawaha, A. M. (2006). Significance of *Mycorrhizae*. *World Journal of Agricultural Sciences*, 2(1), 16–20.

Van Hunsel, F., Van de Koppel, S., Skalli, S., Kuemmerle, A., Teng, L., Wang, J. B. and Barnes, J. (2019). Analysis of hepatobiliary disorder reports associated with the use of herbal medicines in the global suspected ADR database Vigibase. *Frontiers in Pharmacology*, 10, 1326.

Verma, R., Kumar, D., Nagraik, R., Sharma, A., Tapwal, A. and Puri, S. (2021). *Mycorrhizal* inoculation impact on Acorus calamus L. – An ethnomedicinal plant of western Himalaya and its in silico studies for anti-inflammatory potential. *Journal of Ethnopharmacology*, 265, 113353.

Waller, F., Achatz, B., Baltruschat, H., Fodor, J., Becker, K., Fischer, M. and Kogel, K. H. (2005). The endophytic fungus Piriformospora indica reprograms barley to salt-stress tolerance, disease resistance, and higher yield. *Proceedings of the National Academy of Sciences*, 102(38), 13386–13391.

Wang, M., Schäfer, M., Li, D., Halitschke, R., Dong, C., McGale, E., Paetz, C., Song, Y. and Li, S. (2018). Blumenols as shoot markers of root symbiosis with arbuscular *Mycorrhizal* fungi. *Elife*, 7, e37093.

Ward, E. B., Duguid, M. C., Kuebbing, S. E., Lendemer, J. C. and Bradford, M. A. (2022). The functional role of ericoid mycorrhizal plants and fungi on carbon and nitrogen dynamics in forests. *New Phytologist*, 235(5), 1701–1718.

Weremijewicz, J. and Janos, D. P. (2019). Investigation of plant interactions across common *Mycorrhizal* networks using rotated cores. *Journal of Visualized Experiments*, (145), e59338.

White, R. H., Engelke, M. C., Morton, S. J., Johnson-Cicalese, J. M. and Ruemmele, B. A. (1992). Acremonium endophyte effects on tall fescue drought tolerance. *Crop Science*, 32(6), 1392–1396.

Whiteside, M. D., Garcia, M. O. and Treseder, K. K. (2012). Amino acid uptake in arbuscular *Mycorrhizal* plants. *PLoS ONE*, 7(10), e47643.

Yan, H., Freschet, G. T., Wang, H., Hogan, J. A., Li, S., Valverde-Barrantes, O. J. and Kou, L. (2022). Mycorrhizal symbiosis pathway and edaphic fertility frame root economics space among tree species. *New Phytologist*, 234(5), 1639–1653.

Zeng, Y., Guo, L. P., Chen, B. D., Hao, Z. P., Wang, J. Y., Huang, L. Q., Yang, G., Cui, X. M., Yang, L. and Wu, Z. X. (2013). Arbuscular *Mycorrhizal* symbiosis and active ingredients of medicinal plants: Current research status and prospectives. *Mycorrhiza*, 23, 253–265.

Zhang, X. Q., Bai, L., Sun, H. B., Yang, C. and Cai, B. Y. (2020). Transcriptomic and proteomic analysis revealed the effect of Funneliformis mosseae in soybean roots differential expression genes and proteins. *Journal of Proteome Research*, 19, 3631–3643.

Zou, Y. N., Wu, Q. S. and Kuča, K. (2021). Unraveling the role of arbuscular mycorrhizal fungi in mitigating the oxidative burst of plants under drought stress. *Plant Biology*, 23, 50–57.

3 Core Microbiome of *Salvia miltiorrhiza* Seed

A Rich Reservoir of Beneficial Microbes for Secondary Metabolism

Wajahat Ullah, Abdul Khaliq Jan, Dur Muhammad, Zubair Ahmad, Abdur Rauf, and Farman Ullah

3.1 INTRODUCTION

Medicinal plants are a vast and diverse group of plants that have been used for centuries in traditional medicine. They offer a source of natural remedies that are often safer and more effective than synthetic drugs [1, 2]. Many of these plants have been extensively studied for their therapeutic properties, and they continue to play a crucial role in the development of modern medicine. However, the use of medicinal plants is not limited to their therapeutic properties alone. Recent research has shown that the microbial communities associated with medicinal plants can significantly impact their secondary metabolism. The secondary metabolites produced by medicinal plants are responsible for their therapeutic properties, and the microbial communities associated with these plants can alter the chemical composition of these metabolites [3]. The study of medicinal plant–microbe interactions has gained significant attention in recent years due to the potential benefits that may be derived from these interactions. These benefits include the discovery of new natural products, the enhancement of existing natural products, and the development of more effective and sustainable therapeutic strategies [4].

This chapter will provide an overview of medicinal plants and their associated microbial communities. We will explore the core microbiome of medicinal plant seeds and their potential benefits in secondary metabolism. Additionally, we will discuss the key mechanisms involved in secondary metabolism and their potential implications for the discovery and development of new natural products and therapies. Ultimately, this chapter aims to highlight the importance of medicinal plant–microbe interactions as a crucial factor in the discovery and development of new natural products and therapies. Through a comprehensive review of existing literature and case studies, we hope to provide insights into the potential of these interactions and the techniques used to study them [3, 5].

3.2 OVERVIEW OF MEDICINAL PLANT SEEDS

Medicinal plant seeds are an integral part of humanity's long-standing relationship with nature and its healing properties. They hold a significant place in traditional medicine systems, and their potential in modern pharmaceuticals is increasingly being recognized. With a vast array of plant species across the globe, each harbouring unique seeds, the diversity, and richness of medicinal plant seeds are truly remarkable [6].

DOI: 10.1201/b23155-3

Throughout history, civilizations have relied on medicinal plant seeds for their therapeutic benefits. These seeds are packed with a wide range of chemical compounds and secondary metabolites that possess medicinal properties. The chemical composition of these seeds varies from plant to plant, giving rise to their distinctive therapeutic effects. For instance, fenugreek (*Trigonella foenum-graecum*, as shown in Figure 3.1a) seeds have been prized for their role in promoting digestion, managing blood sugar levels, and enhancing lactation. Black seed (*Nigella sativa*, as shown in Figure 3.1b) has been used for centuries in traditional medicine practices for its anti-inflammatory, antioxidant, and antimicrobial properties. Fennel (*Foeniculum vulgare*, as shown in Figure 3.1c) seeds are known for their carminative and digestive properties [7–9].

The physical attributes of medicinal plant seeds, such as their size, colour, and oil content, also contribute to their diverse applications. Some seeds are tiny and inconspicuous, while others are larger and more recognizable. The colour of the seeds can vary from pale to dark shades, reflecting the presence of specific compounds. Additionally, the oil content in certain seeds is of particular interest due to the presence of essential fatty acids, which have numerous health benefits [10, 11].

In recent years, the potential of medicinal plant seeds in modern pharmaceuticals has gained considerable attention. Pharmaceutical companies are now exploring the chemical constituents of these seeds to develop new drugs and treatments. The rich repository of bioactive compounds within the seeds holds promise for addressing various health issues. By harnessing the power of these natural compounds, researchers aim to create innovative therapies for conditions ranging from chronic diseases to infectious ailments [12, 13].

Seed-based products have also gained popularity in the health and wellness industry. The extraction of oils, powders, and extracts from medicinal plant seeds has paved the way for their integration into a wide range of products. Seed oils, such as black seed oil and fenugreek seed oil, are commonly used in skincare products for their nourishing and rejuvenating properties (as shown in Table 3.1). These oils are believed to promote healthy skin, reduce inflammation, and combat signs of aging. Additionally, seed powders and extracts are utilized as dietary supplements, adding

FIGURE 3.1 (a) *Trigonella foenum-graecum*, (b) *Nigella sativa*, (c) *Foeniculum vulgare*.

TABLE 3.1
Examples of Medicinal Plant Seeds and Their Therapeutic Properties

Medicinal Plant Seed	Therapeutic Properties
Black seed (*Nigella sativa*)	Anti-inflammatory, antioxidant, antimicrobial properties
Fenugreek seed (*Trigonella foenum-graecum*)	Digestive support, blood sugar regulation, lactation enhancement
Fennel seed (*Foeniculum vulgare*)	Carminative, digestive support

nutritional value and potential health benefits to various formulations. Food additives incorporating medicinal plant seeds contribute not only to flavour and aroma but also to the overall nutritional profile of food products [14–16].

The significance of medicinal plant seeds goes beyond their individual therapeutic properties. These seeds represent an abundant source of untapped potential for the discovery of new natural products and therapies. Researchers and industry professionals alike are motivated to uncover the hidden treasures within these seeds, hoping to unlock novel treatments for prevalent health issues. The exploration of medicinal plant seeds involves extensive scientific studies, including phytochemical analysis, pharmacological evaluations, and clinical trials. By bridging the gap between traditional knowledge and modern science, researchers aim to harness the vast healing potential of these seeds and transform them into safe and effective medicinal interventions [17–20].

Furthermore, the sustainable cultivation and conservation of medicinal plant species are crucial for the preservation of these invaluable resources. Efforts are being made to ensure the responsible sourcing of medicinal plant seeds, promoting ethical practices, and supporting local communities involved in their production. By fostering sustainable practices, it becomes possible to maintain the delicate balance between human needs and the preservation of biodiversity [21, 22].

Medicinal plant seeds are a fascinating and valuable aspect of traditional and modern medicine. Their diverse chemical compositions, unique properties, and therapeutic potential make them a rich source of natural remedies and a subject of ongoing scientific research. From traditional practices to pharmaceutical advancements, the significance of medicinal plant seeds in health care continues to evolve. By harnessing their power and preserving their ecological habitats, we can unlock the full potential of these remarkable seeds for the benefit of human health and well-being [23].

3.3 CORE MICROBIOME OF MEDICINAL PLANT SEEDS

The core microbiome refers to the set of microorganisms that are consistently found in a particular habitat or environment. In the case of medicinal plant seeds, the core microbiome refers to the microorganisms that are consistently found within the seeds. The study of the core microbiome of medicinal plant seeds is of great interest as it can significantly impact the secondary metabolism of the plants. Several studies have identified the microbial communities commonly found in medicinal plant seeds. Bacterial communities such as Bacillus, Pseudomonas, and Enterobacter are commonly found in the seeds of medicinal plants such as fenugreek, cumin, and fennel. Fungal communities such as Aspergillus, Penicillium, and Fusarium (as shown in Figure 3.2) are also commonly found in medicinal plant seeds [24–26].

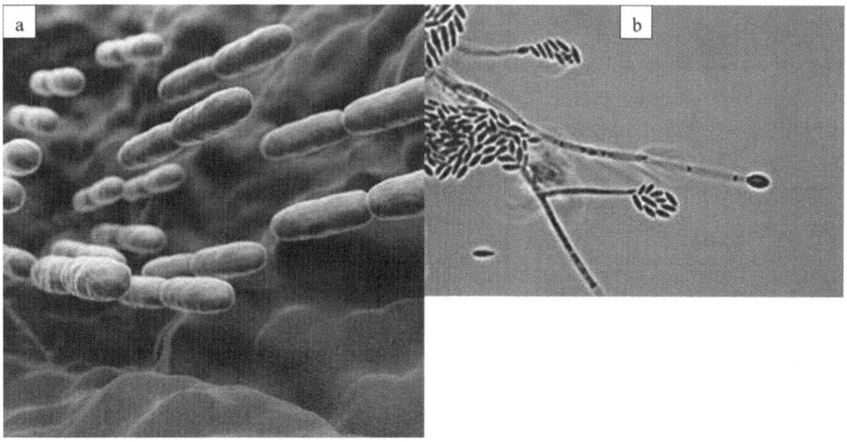

FIGURE 3.2 (a) *Enterobacter*, (b) *Fusarium*.

Understanding the microbiome of medicinal plant seeds can be challenging due to the complex nature of the seed microbiome and the limitations of traditional microbiological techniques. However, advances in next-generation sequencing technologies have provided new insights into the microbial communities associated with medicinal plant seeds. These methods have enabled researchers to identify and characterize the diversity of microorganisms present in seeds and to study their functional roles. Several techniques are used to study the microbiome of medicinal plant seeds, including culture-dependent and culture-independent methods. Culture-dependent methods involve the isolation and cultivation of microorganisms from seeds on various media, whereas culture-independent methods involve the direct extraction and analysis of microbial DNA from seeds. Both methods have their advantages and limitations, and researchers often use a combination of techniques to study the microbiome of medicinal plant seeds [27–30].

So, understanding the core microbiome of medicinal plant seeds is crucial for understanding the secondary metabolism of these plants and the potential for the development of new natural products and therapies [31].

The core microbiome of medicinal plant seeds is a topic of great interest and importance in the field of plant–microbe interactions and natural product discovery. The microbial communities associated with seeds play a significant role in seed germination, seedling establishment, and overall plant health. Moreover, these microorganisms can influence the secondary metabolism of medicinal plants, leading to the production of bioactive compounds with potential therapeutic applications [32, 33].

The core microbiome of medicinal plant seeds refers to the group of microorganisms that are consistently present in the seeds, regardless of environmental factors or geographical locations. These microorganisms can include bacteria, fungi, viruses, and other microbes. The study of the core microbiome involves identifying and characterizing these microorganisms to understand their diversity, abundance, and functional roles within the seed ecosystem [34].

In recent years, advances in next-generation sequencing technologies have revolutionized the field of microbial ecology. These techniques, such as high-throughput sequencing of the 16S rRNA gene for bacteria or the internal transcribed spacer (ITS) region for fungi, allow for the rapid and comprehensive analysis of microbial communities. Researchers can now obtain a detailed snapshot of the microbial diversity within medicinal plant seeds and gain insights into the potential functions of these microorganisms [35].

Culture-dependent methods have traditionally been used to study the microbiome of seeds. These methods involve isolating microorganisms from seeds and cultivating them on specific growth media. Culture-dependent techniques allow for the isolation and identification of individual microorganisms, which can then be further studied for their potential applications in medicine, agriculture, or biotechnology. However, these methods have limitations, as many microorganisms present in seeds are difficult to culture or may have specific growth requirements that are not easily replicated in the laboratory. Therefore, culture-dependent methods may underestimate the true microbial diversity within medicinal plant seeds [36, 37].

To overcome the limitations of culture-dependent techniques, culture-independent methods have gained popularity in studying the seed microbiome. These methods involve the direct extraction of microbial DNA from seeds, followed by high-throughput sequencing to identify the microbial taxa present. By bypassing the need for cultivation, culture-independent methods provide a more comprehensive view of the microbial diversity within seeds. Additionally, these techniques can reveal the presence of non-culturable or previously unknown microorganisms, expanding our understanding of the seed microbiome [38, 39].

In addition to DNA-based methods, other omics approaches, such as metagenomics, metatranscriptomics, and metabolomics, have been employed to study the core microbiome of medicinal plant seeds. Metagenomics allows for the characterization of the collective genomes of the microbial community, providing insights into the functional potential of the seed microbiome. Metatranscriptomics, on the other hand, focuses on the analysis of microbial gene expression, revealing the active functions and metabolic pathways within the seed microbiome. Metabolomics can complement these approaches by identifying the bioactive compounds produced by the microbial community and their potential interactions with the host plant [40, 41].

TABLE 3.2
Common Microbial Communities in Medicinal Plant Seeds

Medicinal Plant Seeds	Bacterial Communities	Fungal Communities
Fenugreek	Bacillus, Pseudomonas, Enterobacter	Aspergillus, Penicillium, Fusarium
Cumin	Bacillus, Pseudomonas, Enterobacter	Aspergillus, Penicillium, Fusarium
Fennel	Bacillus, Pseudomonas, Enterobacter	Aspergillus, Penicillium, Fusarium

The findings from studies on the core microbiome of medicinal plant seeds have significant implications for plant health, agricultural practices, and the development of new natural products and therapies. Microorganisms within seeds can influence plant growth and development through various mechanisms, such as nutrient acquisition, hormone production, and pathogen suppression. Understanding these interactions can help improve seed germination rates, enhance crop yields, and reduce the need for chemical fertilizers or pesticides in agriculture [42–44].

Furthermore, the microbial communities (Shown in Table 3.2) within medicinal plant seeds can contribute to the synthesis of bioactive compounds. Many medicinal plants produce secondary metabolites with therapeutic properties, such as antimicrobial, anti-inflammatory, or anticancer activities [45]. The presence of specific microorganisms in seeds can enhance the production of these bioactive compounds or even contribute new chemical entities to the plant's secondary metabolism. By studying the core microbiome (shown in Table 3.2) of medicinal plant seeds, researchers can identify key microbial players that are responsible for the synthesis of bioactive compounds. This knowledge can be leveraged to develop novel strategies for optimizing plant growth and enhancing the production of valuable natural products [46].

The core microbiome of medicinal plant seeds represents a fascinating and important research area. The microbial communities associated with seeds have profound effects on plant health and secondary metabolism. Advances in sequencing technologies and other omics approaches have enabled researchers to uncover the diversity and functional potential of microorganisms within seeds [47]. By understanding the core microbiome, we can harness the power of plant–microbe interactions to improve agricultural practices, discover new natural products, and develop innovative therapies. Continued research in this field promises to unveil the hidden potential of medicinal plant seeds and contribute to sustainable agriculture and human health [48].

3.4 POTENTIAL BENEFITS OF THE CORE MICROBIOME OF MEDICINAL PLANT SEEDS IN SECONDARY METABOLISM

Microbial communities associated with medicinal plant seeds have been shown to play a crucial role in shaping secondary metabolism, which ultimately affects the production of bioactive compounds and therapeutic properties of these plants. These microbial communities can contribute to the synthesis of new compounds, modify existing compounds, and increase the yield of bioactive molecules [49, 50].

Several case studies have demonstrated the impact of microbial communities on the secondary metabolism of medicinal plants [51]. The endophytic fungi isolated from the seeds of *Euphorbia pekinensis* were found to produce secondary metabolites with potent antitumour activity. Similarly, the endophytic bacterium *Bacillus subtilis* isolated from the seeds of *Ocimum basilicum* was shown to produce compounds with antimicrobial and antioxidant activities [52].

Seed-based products have also been shown to have significant potential as a source of new natural products and therapies. The seed of *Nigella sativa* (black cumin) has been traditionally used for its therapeutic properties and is currently being investigated for its potential anti-inflammatory, anticancer, and antimicrobial activities. The seed of *Trigonella foenum-graecum* (fenugreek) is another example of a medicinal plant seed with potential health benefits, including its potential to lower blood sugar levels and improve insulin sensitivity in individuals with type 2 diabetes. Studying the core microbiome of medicinal plant seeds can provide valuable insights into the mechanisms underlying the production of bioactive compounds and the potential for developing new natural products and therapies. Techniques such as metagenomics and metaproteomics have been used to study the microbial communities associated with medicinal plant seeds, allowing for a more comprehensive understanding of their role in secondary metabolism [53].

The potential benefits of the core microbiome of medicinal plant seeds in secondary metabolism are vast and represent a promising area of research for the development of new natural products and therapies [54].

The core microbiome of medicinal plant seeds, consisting of the microbial communities that establish and thrive in association with these seeds, has emerged as an intriguing area of research. These microbial communities, comprising bacteria, fungi, and other microorganisms, have been found to exert a significant influence on secondary metabolism in medicinal plants (as shown in Table 3.3). Secondary metabolism refers to the production of specialized compounds that are not essential for the plant's basic functions but play crucial roles in defence mechanisms and interactions with the environment [55].

The interaction between medicinal plant seeds and their associated microbes is believed to be mutually beneficial. The seeds provide a nutrient-rich environment for microbial colonization, while the microbes contribute to the overall health and fitness of the plants. These microbial communities can affect secondary metabolism in various ways, leading to the synthesis of novel compounds, alteration of existing compounds, and enhancement of bioactive molecule production [56].

One mechanism by which the core microbiome influences secondary metabolism is through the production of enzymes. Microbes associated with medicinal plant seeds possess a diverse array of enzymes capable of modifying and transforming plant metabolites. For example, they may participate in processes such as glycosylation, methylation, or hydroxylation, which can modify the chemical structure of bioactive compounds. These modifications can enhance the bioavailability, stability, or pharmacological activity of the compounds, thereby influencing their therapeutic potential [57].

Furthermore, the core microbiome of medicinal plant seeds can contribute to the production of new bioactive compounds [58]. Microbes can produce a wide range of secondary metabolites, including antibiotics, antitumour agents, antioxidants, and immunomodulatory compounds [59]. Through their interactions with the plant host, these microbes may induce the expression of specific genes involved in secondary metabolism, leading to the synthesis of previously unknown compounds with potentially beneficial properties [60].

Studies have highlighted the potential of the core microbiome to enhance the yield of bioactive molecules in medicinal plants. For instance, the presence of specific microbial species has been correlated with increased production of valuable compounds in plants. Researchers have identified certain bacterial strains associated with medicinal plant seeds that can promote the synthesis of secondary metabolites. These findings open up avenues for optimizing cultivation practices by harnessing the beneficial effects of these microbial communities, ultimately leading to higher yields of bioactive compounds [61].

Metagenomics and metaproteomics have become powerful tools in studying the core microbiome of medicinal plant seeds. Metagenomics involves the direct sequencing of microbial DNA from environmental samples, enabling the identification and characterization of the microbial community

TABLE 3.3
Examples of Microbial Communities Associated with Medicinal Plant Seeds and Their Effects on Secondary Metabolism

Medicinal Plant Seed	Microbial Community	Effects on Secondary Metabolism
Euphorbia pekinensis	Endophytic fungi	Production of secondary metabolites with antitumour activity
Ocimum basilicum	Endophytic bacterium (*Bacillus subtilis*)	Production of compounds with antimicrobial and antioxidant activities

TABLE 3.4
Techniques for Studying the Core Microbiome of Medicinal Plant Seeds

Technique	Description
Metagenomics	Analysis of microbial community composition, genetic diversity, and functional potential through DNA sequencing.
Metaproteomics	Investigation of functional activities, metabolic potential, and protein expression within the seed microbiome through proteomic analysis.
Metatranscriptomics	Study of gene expression and transcriptional activity of the seed microbiome through RNA sequencing.
Cultivation-based	Isolation and characterization of microorganisms from seeds using various growth media and conditions.
Fluorescence in situ hybridization (FISH)	Visualization of specific microorganisms within seeds using fluorescent probes targeting specific taxa.

composition. Metaproteomics, on the other hand, focuses on the analysis of the proteins expressed by the microbial community, providing insights into the functional activities and metabolic potential of the microbes (as shown in Table 3.4). These techniques have allowed researchers to unravel the complex interactions between medicinal plant seeds and their associated microbial communities and elucidate their role in shaping secondary metabolism [62].

Understanding the potential benefits of the core microbiome of medicinal plant seeds in secondary metabolism has significant implications for drug discovery and the development of new natural products. Traditional medicine systems have long relied on plant-based remedies, and the exploration of microbial contributions to secondary metabolism adds a new dimension to this field. By harnessing the interactions between medicinal plant seeds and their associated microbes, researchers can potentially identify novel compounds, optimize production methods, and improve the therapeutic efficacy of existing herbal remedies [63].

The integration of modern scientific techniques with traditional knowledge can further enhance our understanding of the core microbiome and its impact on secondary metabolism. Collaborative efforts between scientists, ethnobotanists, and traditional healers can facilitate the identification of plant–microbe interactions that have been harnessed by indigenous communities for centuries. This interdisciplinary approach allows for a holistic understanding of medicinal plants and their microbial partners, opening up new possibilities for sustainable drug discovery and the preservation of traditional medicinal practices [64].

The core microbiome of medicinal plant seeds represents a fascinating area of research with immense potential for the development of new natural products and therapies. The microbial communities associated with these seeds have been found to influence secondary metabolism, leading to the production of bioactive compounds with therapeutic properties. By studying the core microbiome using advanced techniques, researchers can gain valuable insights into the mechanisms underlying secondary metabolism and potentially unlock the discovery of novel compounds. The integration of traditional knowledge with modern scientific approaches paves the way for the sustainable utilization of medicinal plants and their associated microbial communities, ultimately benefiting human health and well-being [65].

3.5 KEY MECHANISMS INVOLVED IN SECONDARY METABOLISM

Secondary metabolism refers to the biosynthesis of compounds that are not essential for the survival of an organism but are important for its ecological function, such as defence against predators or attraction of pollinators. These compounds, also known as specialized metabolites or secondary metabolites, are diverse in structure and function and are often biologically active, making them valuable for human use. The biosynthesis of secondary metabolites involves a complex series of enzymatic reactions and pathways, which vary depending on the organism and the compound being produced. However, some key mechanisms are involved in many types of secondary metabolism. One of these is the shikimate pathway, which produces aromatic amino acids that serve as precursors for many specialized metabolites, including alkaloids, flavonoids, and terpenoids [66].

Other key enzymes involved in secondary metabolism include polyketide synthases (PKSs) and non-ribosomal peptide synthetases (NRPSs), which are responsible for the biosynthesis of a wide range of natural products, including antibiotics, anticancer agents, and immunosuppressants. These enzymes are modular and can be combined in different ways to generate diverse chemical structures [67].

Microbial communities associated with medicinal plant seeds can significantly impact secondary metabolism by modulating these key mechanisms and pathways. For example, certain microorganisms can produce enzymes that modify or activate specific biosynthetic pathways, leading to the production of novel compounds with enhanced therapeutic potential. Additionally, microbial communities can interact with the host plant to promote the production of certain compounds or to activate specific pathways [68].

Understanding these mechanisms and the role of microbial communities in modulating them is crucial for the development of new natural products and therapies. By studying the interactions between medicinal plant seeds and their associated microbes, researchers can identify new compounds with therapeutic potential and optimize production methods to increase yield and efficacy [69].

Secondary metabolism plays a vital role in the survival and adaptation of organisms, including plants, fungi, and bacteria. The biosynthesis of secondary metabolites involves a complex network of enzymatic reactions and pathways, which contribute to the production of diverse compounds with various functions and activities. These compounds can have significant implications for human health, as they often possess therapeutic properties and have been used in traditional medicine for centuries [70].

One key mechanism involved in secondary metabolism is the shikimate pathway. The shikimate pathway is a series of enzymatic reactions that leads to the synthesis of aromatic amino acids, such as phenylalanine, tyrosine, and tryptophan. These amino acids serve as precursors for the production of a wide range of secondary metabolites, including alkaloids, flavonoids, lignans, and terpenoids. The shikimate pathway is conserved in many organisms, including plants, bacteria, and fungi, highlighting its importance in secondary metabolism across different kingdoms of life [71].

Polyketide synthases and non-ribosomal peptide synthetases are another set of key enzymes involved in secondary metabolism. These enzymes are responsible for the biosynthesis of polyketides and non-ribosomal peptides, respectively. Polyketides are a diverse class of compounds with various biological activities, including antibiotics, antifungals, anticancer agents, and immunosuppressants. Non-ribosomal peptides, on the other hand, are involved in the synthesis of compounds such as siderophores, which are essential for iron acquisition in bacteria, as well as cyclic peptides with antimicrobial and anticancer properties. PKSs and NRPSs are modular enzymes, meaning they consist of different domains that can be combined in various ways to produce a wide range of chemical structures, leading to the diversity of secondary metabolites observed in nature [72].

TABLE 3.5
Mechanisms and Descriptions of Secondary Metabolite Biosynthesis and Regulation

Mechanism	Description
Shikimate pathway	Biosynthesis of aromatic amino acids (phenylalanine, tyrosine, tryptophan), precursors for many secondary metabolites.
Polyketide synthases	Enzymes were responsible for the biosynthesis of polyketides, a diverse class of compounds with various biological activities.
Non-ribosomal peptide synthetases	Enzymes are involved in the biosynthesis of non-ribosomal peptides, including cyclic peptides and siderophores.
Regulation	Tight control of secondary metabolism at transcriptional, post-transcriptional, and post-translational levels is influenced by various factors.

Regulation is another key aspect of secondary metabolism (as shown in Table 3.5). The production of secondary metabolites is tightly regulated at the transcriptional, post-transcriptional, and post-translational levels. Various factors, such as environmental cues, nutrient availability, and signalling molecules, can influence the expression of genes involved in secondary metabolism. Transcription factors and regulatory proteins play crucial roles in controlling the expression of biosynthetic genes, ensuring that secondary metabolites are produced under appropriate conditions [73].

In recent years, the role of microbial communities in shaping secondary metabolism has gained considerable attention. Microbes associated with medicinal plant seeds can impact secondary metabolism through various mechanisms. For instance, certain microorganisms can produce enzymes that modify or activate specific biosynthetic pathways, leading to the production of novel compounds. These enzymes may catalyse reactions such as oxidation, reduction, glycosylation, or methylation, altering the chemical structure and properties of secondary metabolites. Microbes can also influence secondary metabolism through cross-talk with the host plant. This interaction can involve the exchange of signalling molecules or the activation of specific defence or stress response pathways, ultimately influencing the production of secondary metabolites [74].

Studying the interactions between medicinal plant seeds and their associated microbial communities can provide valuable insights into the modulation of secondary metabolism. Advanced techniques such as metagenomics, metatranscriptomics, and metabolomics enable researchers to characterize the microbial composition, gene expression profiles, and metabolite profiles of these communities. By comparing the microbiome of healthy plants with those experiencing perturbations or stress, researchers can identify key microbial players and their contributions to secondary metabolism [75].

The understanding of these mechanisms and the role of microbial communities in secondary metabolism has significant implications for the development of new natural products and therapies. The exploration of the core microbiome associated with medicinal plant seeds offers a unique opportunity to uncover novel compounds and optimize their production. By harnessing the interactions between seeds and their associated microbes, researchers can potentially enhance the production of desired metabolites, improve the yield and quality of natural products, and discover new therapeutic compounds with enhanced efficacy [76].

Furthermore, the integration of traditional knowledge and modern scientific approaches is crucial for maximizing the potential of secondary metabolism in medicinal plants. Traditional medicinal practices often involve the use of plant materials in combination with other natural substances, such as microbial fermentation products, to enhance therapeutic effects. By bridging the gap between traditional knowledge and scientific research, researchers can identify synergistic interactions between medicinal plant compounds and microbial metabolites, leading to the development of effective and safe natural products [77].

Secondary metabolism is a complex process involving various mechanisms and pathways that contribute to the production of diverse bioactive compounds. The shikimate pathway, PKSs, NRPSs, and regulatory mechanisms play key roles in shaping secondary metabolism. Microbial communities associated with medicinal plant seeds can influence secondary metabolism through enzymatic modifications, cross-talk with the host plant, and the production of signalling molecules. Understanding these interactions and mechanisms is essential for unlocking the full potential of medicinal plant seeds in the development of new natural products and therapies. By harnessing the power of secondary metabolism and the contributions of microbial communities, researchers can pave the way for innovative approaches to health care and promote the sustainable utilization of medicinal plants [78].

3.6 MICROBIAL INTERACTIONS WITHIN THE CORE MICROBIOME

The human microbiome is a giant and diverse network of microorganisms that live in the human body, along with microorganisms, fungi, viruses, and different microbes. These microbes colonize numerous body sites, including the skin, oral cavity, gastrointestinal tract, and reproductive organs, and form complex ecosystems that play an important role in human health and disease. Within the microbiome, there exists a subset of microorganisms referred to as the core microbiome. The core microbiome includes microorganisms that are consistently present across people and have a big impact on host physiology and immune features [79].

Microbial interactions inside the middle microbiome are dynamic and complex, regarding diverse mechanisms that impact the composition and balance of the microbial network. These interactions can be classified into several categories, such as conversation, cooperation, competition, and niche partitioning [80]. Communication among microorganisms inside the core microbiome is critical for maintaining homeostasis and coordinating microbial sports. Microbes communicate through the secretion of signalling molecules, such as quorum-sensing molecules, which permit them to feel and respond to changes in their surroundings. Quorum sensing allows microorganisms to coordinate gene expression, modify biofilm formation, and modulate virulence elements [81].

Cooperation is another important factor in microbial interactions inside the middle microbiome. Microbes can interact in mutualistic interactions in which distinctive species paintings collectively for mutual advantage. As an example, some microorganisms within the intestinal microbiome produce essential vitamins that can be utilized by other microbes, promoting their growth and survival. Cooperative interactions contribute to the overall balance and characteristics of the middle microbiome. Opposition within the core microbiome arises because of restrained assets and area. Microbes compete for nutrients, attachment sites, and niche functions. This competition can shape the structure and diversity of the microbial network. Opposition also drives the development of antimicrobial techniques consisting of the manufacturing of antimicrobial peptides and bacteriocins, which assist microorganisms in establishing dominance over others [82].

Niche partitioning is a phenomenon determined in the middle microbiome, where microorganisms occupy unique ecological niches primarily based on their physiological capabilities and metabolic options. This partitioning permits the efficient utilization of assets and decreases direct competition between microbial species. As an example, different species of microorganisms within the gut microbiome can also specialize in breaking down specific nutritional components, ensuring the efficient digestion and absorption of nutrients [83].

Understanding microbial interactions in the gut microbiome is vital for unraveling the complexities of host-microbe relationships and their impact on human health and disease. Disruption of these interactions can result in dysbiosis, in which the microbial community will become imbalanced, resulting in numerous pathological conditions. Dysbiosis has been associated with severe sicknesses, such as inflammatory bowel sickness, obesity, diabetes, or even intellectual fitness problems [84].

Advances in sequencing technologies, metagenomics, and metabolomics have revolutionized our capability to have a look at microbial interactions in the centre microbiome. These techniques permit comprehensive profiling of microbial groups and their purposeful capability, imparting important insights into the complicated dynamics of microbial interactions.

The information gained from studying microbial interactions within the gut microbiome opens up new avenues for therapeutic interventions aimed at modulating the microbiome to promote health and prevent or treat microbiome-related diseases. Strategies such as probiotics, prebiotics, and fecal microbiota transplantation are being explored to restore microbial balance and enhance beneficial microbial interactions in the gut microbiome.

In conclusion, microbial interactions within the core microbiome are essential for maintaining host health and contribute to the stability and resilience of the microbial network. Communication, cooperation, competition, and niche partitioning are key mechanisms driving these interactions. Understanding these interactions provides insights into the complex dynamics of the core microbiome and offers potential for therapeutic strategies to improve human health. Furthermore, research in this field will undoubtedly enhance our understanding of the intricate web of microbial interactions and their implications for human well-being. [85].

3.7 FACTORS INFLUENCING THE COMPOSITION OF THE CORE MICROBIOME

The composition of the core microbiome, which refers to the subset of microorganisms consistently present across individuals, is influenced by a multitude of factors. These factors can broadly be categorized into host-related factors, environmental factors, and microbial factors. Understanding the intricate interplay between these factors is crucial for understanding the dynamics of the core microbiome and its impact on human health [86].

3.7.1 Host-Related Factors

3.7.1.1 Genetic Predisposition

The genetic makeup of the host plays a significant role in shaping the composition of the core microbiome. Host genes influence immune responses, epithelial barrier function, and mucus production, which in turn affect the microbial community structure.

3.7.1.2 Age

The core microbiome evolves, and the age of the host influences its composition. Infants, adults, and the elderly have distinct microbial profiles, reflecting developmental changes, immune maturation, and alterations in lifestyle and diet.

3.7.1.3 Diet

Dietary habits have a profound impact on the composition of the core microbiome. Different nutrients, such as fibre, complex carbohydrates, and certain fats, serve as substrates for specific microbial groups, influencing their abundance and activity.

3.7.1.4 Lifestyle and Hygiene

Environmental factors, such as personal hygiene practices, exposure to pets, and living conditions, can influence the core microbiome. For example, individuals living in rural areas or those with frequent contact with natural environments may have distinct microbial profiles compared to urban dwellers.

3.7.2 Environmental Factors

3.7.2.1 Geographic Location

Regional differences in environmental conditions, climate, and diet contribute to variations in the core microbiome. Factors such as temperature, humidity, and altitude shape the microbial community structure in different geographic locations.

3.7.2.2 Seasonality

Seasonal changes in diet, exposure to environmental factors, and fluctuations in host physiology can impact the core microbiome. The availability of certain foods during specific seasons can influence the abundance and diversity of microbial populations.

3.7.2.3 Antibiotic Usage

Antibiotics can profoundly alter the composition of the core microbiome by selectively eliminating susceptible microbial populations. Prolonged or repeated antibiotic treatments can lead to long-lasting disruptions in the microbial community structure.

3.7.3 Microbial Factors

3.7.3.1 Microbial Succession

The establishment and development of the core microbiome involve a process of microbial succession. Initially, the colonization of microorganisms is driven by environmental factors and interactions with the host. Over time, a stable and diverse microbial community is established.

3.7.3.2 Inter-Species Interactions

Microbes within the core microbiome engage in intricate interactions such as cooperation, competition, and metabolic cross-feeding. These interactions can shape the composition and functional potential of the microbial community.

3.7.3.3 Horizontal Gene Transfer

Microbes can acquire new genetic material through horizontal gene transfer, allowing them to adapt and thrive in specific environments. This process influences the diversity and functional capabilities of the core microbiome.

Bacteriophages, or viruses that infect bacteria, play a crucial role in modulating the composition of the core microbiome. Viral predation can selectively target certain microbial populations, leading to shifts in community structure.

The interplay between these host-related, environmental, and microbial factors is complex and highly dynamic. Additionally, individual variations in these factors contribute to the unique composition of the core microbiome in each individual. Understanding these factors and their interactions is vital for deciphering the functional implications of the core microbiome and developing targeted interventions to promote a healthy microbial community [87].

Emerging research suggests that alterations in the composition of the core microbiome, known as dysbiosis, may be associated with various diseases, including metabolic disorders, gastrointestinal conditions, autoimmune diseases, and mental health disorders. Consequently, efforts are underway to modulate the core microbiome using strategies such as probiotics, prebiotics, dietary interventions, and fecal microbiota transplantation to restore microbial balance and promote host health.

In conclusion, the composition of the core microbiome is influenced by a multitude of factors, including host-related factors, environmental factors, and microbial factors. These factors interact in a complex manner, shaping the microbial community structure and function. Understanding the dynamics of the core microbiome and the factors that influence it holds immense potential for improving human health and preventing or treating microbiome-related disorders. Further research in this field will contribute to our knowledge of the core microbiome and its role in human well-being [88].

3.8 METAGENOMICS AND METAPROTEOMICS APPROACHES

Metagenomics and metaproteomics are powerful techniques used to study the genetic and protein content of complex microbial communities, such as those found in the human microbiome and various environmental samples. These approaches provide valuable insights into the composition, function, and interactions of microorganisms within these communities. This section explores the principles, methodologies, and applications of metagenomics and metaproteomics, highlighting their significance in advancing our understanding of microbial ecology and their potential in various fields, including medicine, environmental science, and biotechnology [89].

3.8.1 Metagenomics

Metagenomics is a field of study that focuses on analysing the collective genomic material present in a microbial community without the need for cultivation. It involves the extraction of DNA directly from environmental samples, followed by high-throughput sequencing to generate large datasets of genetic information. Metagenomic analysis enables the identification and characterization of microbial species, the reconstruction of microbial genomes, and the exploration of microbial gene functions and metabolic pathways.

3.8.1.1 Sample Collection and DNA Extraction

Metagenomic studies begin with the collection of environmental samples, such as soil, water, or human feces. These samples are processed to extract total DNA, which includes genomic DNA from all microorganisms present in the community.

3.8.1.2 Sequencing and Data Analysis

The extracted DNA is then subjected to sequencing using technologies like next-generation sequencing (NGS). The resulting sequence data is processed through bioinformatic pipelines to remove artifacts, perform quality control, and reconstruct individual microbial genomes, known as metagenome assembly. Comparative analysis, functional annotation, and taxonomic classification are performed to gain insights into the microbial community structure, gene content, and potential functional capabilities.

3.8.1.3 Applications of Metagenomics

Metagenomics has widespread applications, including:

- Identifying novel microbial species and genes
- Characterizing microbial community dynamics in different environments
- Exploring the functional potential of microbial communities
- Discovering novel enzymes and biomolecules for biotechnological applications
- Studying microbial contributions to health and disease

3.8.2 Metaproteomics

Metaproteomics focuses on characterizing the proteins expressed by microbial communities in their natural environment. It involves the extraction and analysis of proteins directly from environmental samples, providing information on the functional activities of the microbial community. Metaproteomics complements metagenomics by providing insights into the actual protein expression and post-translational modifications, helping to understand the functional diversity and activities of microbial populations.

3.8.2.1 Sample Collection and Protein Extraction

Environmental samples are collected and processed to extract proteins. Various extraction methods are employed to solubilize and enrich proteins while removing interfering substances.

3.8.2.2 Protein Separation and Identification

The extracted proteins are separated using techniques such as gel electrophoresis or liquid chromatography, and the resulting protein fractions are analysed using mass spectrometry (MS) for identification and quantification. Database search algorithms are used to match the acquired MS data against protein databases for identification.

3.8.2.3 Functional Analysis and Data

Functional analysis of the identified proteins involves determining their biological functions, metabolic pathways, and interactions within the microbial community. Integration with metagenomic data allows correlation between gene content and protein expression, providing a more comprehensive understanding of microbial community dynamics and functions.

3.8.2.4 Applications of Metaproteomics

Metaproteomics has several applications, including:

- Characterizing functional activities of microbial communities in different environments
- Identifying microbial responses to environmental changes or perturbations
- Elucidating host-microbiome interactions
- Discovering novel biomarkers for disease diagnosis or environmental monitoring
- Evaluating the efficacy of microbiome-based therapies

Both metagenomics and metaproteomics offer valuable insights into the structure and function of microbial communities. Their combined use provides a more comprehensive understanding of the complex interactions and activities within these communities. These approaches have revolutionized our understanding of microbial ecology, paving the way for discoveries and applications in diverse fields. Continued advancements in sequencing technologies, bioinformatics, and data analysis will further enhance the potential of metagenomics and metaproteomics in unravelling the mysteries of microbial communities [90, 91].

3.9 BIOTECHNOLOGICAL APPLICATIONS OF THE CORE MICROBIOME

The core microbiome, which represents the subset of microorganisms consistently present in a particular environment or host, has gained significant attention in the field of biotechnology. The diverse and dynamic microbial communities within the core microbiome possess a vast array of functional capabilities that can be harnessed for various biotechnological applications. This section explores some of the key biotechnological applications of the core microbiome, highlighting their potential in agriculture, environmental remediation, industrial processes, and human health [92].

3.9.1 Agricultural Applications

The core microbiome has great potential in agriculture to improve crop productivity, nutrient uptake, and plant health. Some biotechnological applications include:

- **Biofertilizers:** Certain core microbiome members can enhance nutrient availability and uptake by plants, improving soil fertility and reducing the need for chemical fertilizers.
- **Biocontrol Agents:** Beneficial microbes within the core microbiome can act as biocontrol agents, protecting plants from pathogens and pests.
- **Plant Growth Promotion:** Core microbiome members can stimulate plant growth through mechanisms such as hormone production, nutrient cycling, and stress tolerance.

3.9.2 Environmental Remediation

The core microbiome holds promise for environmental remediation strategies by contributing to the breakdown of pollutants and restoration of ecosystems. Some applications include:

- **Bioremediation:** Certain core microbiome organisms possess the ability to degrade pollutants, such as hydrocarbons, pesticides, and heavy metals, thereby aiding in the cleanup of contaminated sites.
- **Wastewater Treatment:** Core microbiome members play a vital role in the treatment of wastewater by breaking down organic matter and removing pollutants.

3.9.3 Industrial Processes

The core microbiome can be harnessed for various industrial processes, offering sustainable and efficient alternatives to conventional methods. Some applications include:

- **Bioprocessing and Biocatalysis**: Microbes within the core microbiome produce enzymes and metabolites that can be utilized in industrial processes, such as the production of biofuels, pharmaceuticals, and biopolymers.
- **Food and Beverage Production:** The core microbiome contributes to the fermentation and flavour development of food and beverage products, such as cheese, wine, and yogurt.

3.9.4 Human Health

The core microbiome has significant implications for human health, and biotechnological applications in this domain are rapidly emerging. Some areas of focus include:

- **Probiotics and Therapeutics:** Core microbiome organisms can be used as probiotics to promote gut health and enhance immune function. Additionally, they hold potential as therapeutic agents for the treatment of various diseases, including gastrointestinal disorders, metabolic disorders, and immune-related conditions.
- **Personalized Medicine:** Understanding the individual-specific core microbiome composition can aid in the development of personalized medical interventions, including tailored dietary recommendations and targeted therapies.

3.9.5 Bioprospecting and Discovery

The core microbiome represents a vast source of untapped genetic and functional diversity. Bioprospecting efforts involve exploring the core microbiome for the discovery of novel enzymes, bioactive compounds, and metabolic pathways. These discoveries have the potential to drive advancements in various fields, including drug development, enzyme engineering, and biotechnology.

The biotechnological applications of the core microbiome hold immense potential in addressing global challenges related to agriculture, environmental sustainability, industrial processes, and human health. Continued research and technological advancements will further unravel the functional capabilities of the core microbiome and pave the way for innovative biotechnological solutions [93, 94].

3.10 CONSERVATION AND PRESERVATION OF MEDICINAL PLANT SEEDS AND THEIR MICROBIOME

Medicinal plants have been used for hundreds of years as a supply of herbal treatments and recuperation compounds. However, many medicinal plant species face the hazard of extinction due to habitat loss, over-harvesting, and environmental changes. To ensure the long-term survival of those treasured plant species, the conservation and preservation of medicinal plant seeds and their related microbiome are crucial. This section explores the significance of maintaining medicinal plant seeds and their microbiome, the challenging situations involved, and the techniques employed for their protection [95].

3.10.1 Significance of Medicinal Plant Seeds

Medicinal plant seeds play an important role in the maintenance and propagation of plant species. They serve as a repository of genetic diversity and contain essential compounds potentially responsible for their therapeutics. With the resource of preserving medicinal plant seeds, we're able to preserve the genetic integrity of these plants and guard their potential for future research, sustainable cultivation, and the improvement of new medicinal products.

3.10.2 Microbiome Related to Medicinal Plant Seeds

The microbiome related to medicinal plant seeds refers to the diverse community of microorganisms, together with bacteria, fungi, and viruses, that stay on and in the seed. These microorganisms play an essential role in plant fitness, growth, and protection mechanisms. They contribute to nutrient acquisition, disorder resistance, and the synthesis of bioactive compounds inside the plant. Retaining the microbiome related to medicinal plant seeds is crucial for retaining the plant's regular health and its capability to offer therapeutically active compounds.

3.10.3 Factors Affecting Medicinal Plant Seeds

Conserving and safeguarding medicinal plant seeds and their microbiome presents numerous challenging situations, including:

- **Constrained Seed Availability:** Many medicinal plant species have constrained seed availability, making it hard to acquire and store enough quantities for conservation purposes.
- **Seed Dormancy and Viability:** A few medicinal plant seeds have dormancy mechanisms, requiring specific treatments to break dormancy and promote germination. Moreover, seed viability can be short lived, necessitating careful upkeep techniques.
- **Genetic Variants:** Medicinal plant populations regularly exhibit genetic variation, requiring complete sampling strategies to capture the total genetic variety within and among populations.
- **Microbial Range and Dynamics:** The microbial composition of medicinal plant seeds can vary significantly, making it hard to install standardized safety protocols that can effectively capture and maintain the associated microbiome.

3.10.4 Strategies for Medicinal Plant Seeds

Numerous strategies are employed for the conservation and safety of medicinal plant seeds and their associated microbiome:

- **Seed Banks:** Setting up seed banks or gene banks dedicated to keeping medicinal plant seeds is a critical step in their long-term conservation. These facilities use specialized storage techniques consisting of cryopreservation and desiccation to maintain seed viability and genetic integrity.
- **In Vitro Way of Life:** In vitro tissue culture techniques, including somatic embryogenesis and micropropagation, can be employed to multiply and conserve selected genotypes of medicinal plants under controlled conditions.
- **Cryopreservation:** Cryopreservation involves the ultra-low-temperature storage of seeds or plant tissues in liquid nitrogen, ensuring the long-term preservation of genetic material.
- **Microbiome Characterization:** Accurate characterization of the microbiome associated with medicinal plant seeds is crucial for understanding their composition and functional dynamics. High-throughput sequencing strategies, including amplicon sequencing and metagenomics, are utilized to investigate microbial diversity and community dynamics.
- **Microbial Conservation Strategies:** Preservation of the associated microbiome can be achieved through cryopreservation, lyophilization (freeze-drying), or culture-based methods, wherein specific microbial strains are isolated and conserved for future application.

3.10.5 Advantages of Medicinal Plant Seed Conservation and Microbiome Protection

The conservation and upkeep of medicinal plant seeds and their associated microbiome provide several benefits:

- **Conservation of Genetic Range:** Maintaining the genetic diversity of medicinal plant species ensures their adaptability to changing environmental conditions and provides a valuable resource for future research and development.
- **Sustainable Medicinal Plant Manufacturing:** Conserved seeds may be used for sustainable cultivation and propagation, reducing dependence on wild harvesting and supporting the production of high-quality medicinal plants.
- **Discovery of Novel Compounds:** Preserved seeds and their associated microbiome serve as a valuable resource for the discovery of novel bioactive compounds and potential drug candidates.
- **Ecological Recovery:** Conserving medicinal plant seeds contributes to ecosystem restoration and rehabilitation, promoting biodiversity conservation and supporting ecosystem services.

Conserving medicinal plant seeds and their associated microbiome is vital for maintaining biodiversity, preserving traditional medicine practices, and harnessing their therapeutic potential. Effective seed collection, storage, and microbiome protection, combined with collaborative research and sustainable use, will safeguard the availability and sustainability of medicinal plants and their precious genetic and microbial resources for future generations [96–98].

3.11 CLINICAL APPLICATIONS AND TRANSLATIONAL RESEARCH

Clinical applications and translational research play a pivotal role in bridging the gap between scientific discoveries and their implementation in clinical practice. Translational research aims to translate basic scientific findings into practical applications that benefit patient care, improve diagnostics, and enhance treatment outcomes. This section explores the importance of clinical applications and translational research, highlights key areas of focus, and discusses the impact of these efforts on health care [99].

3.11.1 Importance of Clinical Applications and Translational Research

Clinical applications and translational research are crucial for several reasons:

- **Advancing Patient Care:** Translating scientific discoveries into clinical applications allows for the development of innovative diagnostic tools, therapies, and preventive strategies, ultimately improving patient care and outcomes.
- **Bench-to-Bedside Translation:** Translational research facilitates the efficient transfer of knowledge from the laboratory bench to the patient's bedside, ensuring that scientific advancements are effectively applied in clinical settings.
- **Personalized Medicine:** Clinical applications and translational research contribute to the development of personalized medicine approaches, tailoring treatments and interventions to individual patients based on their unique characteristics.
- **Closing the Gap:** These efforts address the gap between scientific knowledge and its practical implementation, ensuring that promising discoveries are effectively utilized in health care settings [100].

3.11.2 Key Areas of Focus in Clinical Applications and Translational Research

Clinical applications and translational research encompass a wide range of areas, including:

- **Diagnostics:** Developing improved diagnostic tools, such as biomarkers, imaging techniques, and molecular assays, to aid in early disease detection, accurate diagnosis, and monitoring of treatment response [101].
- **Therapeutics:** Translating novel therapeutic approaches, including targeted therapies, immunotherapies, gene therapies, and regenerative medicine, from preclinical studies to clinical trials and subsequent implementation in patient care.
- **Prevention and Public Health:** Implementing preventive strategies based on scientific evidence, such as vaccination programs, lifestyle modifications, and public health interventions, to reduce the burden of disease and improve population health.
- **Precision Medicine:** Integrating genomic, proteomic, and other omics data into clinical decision-making processes to personalize treatment plans and optimize therapeutic outcomes [102].
- **Health Technology:** Developing and implementing innovative health technologies, such as wearable devices, telemedicine platforms, and digital health solutions, to improve patient monitoring, remote health care delivery, and disease management.

3.11.3 Impact on Health Care Clinical Applications

Translational research has a profound impact on health care:

- **Improved Patient Outcomes:** By translating scientific discoveries into clinical practice, health care providers can access the latest diagnostic tools, treatments, and interventions, leading to improved patient outcomes, reduced morbidity, and enhanced quality of life.
- **Precision and Personalized Medicine:** Translational research enables the development of tailored treatment approaches, allowing health care professionals to customize therapies based on individual patient characteristics, leading to more effective and targeted interventions.
- **Evidence-Based Practice:** Clinical applications and translational research provide the evidence base necessary to guide clinical decision-making, ensuring that health care practices are grounded in scientific knowledge and research findings.
- **Collaboration and Interdisciplinary Approaches:** These efforts foster collaboration between researchers, clinicians, industry partners, and policymakers, facilitating interdisciplinary approaches that drive innovation and accelerate the translation of research findings into clinical applications [103].

3.11.4 Challenges and Future Directions

Despite the significant advancements in clinical applications and translational research, several challenges persist:

- **Complex Regulatory Landscape:** Navigating regulatory requirements for the translation of scientific discoveries into clinical practice can be complex and time consuming, requiring careful adherence to ethical, legal, and safety considerations.
- **Funding and Resources:** Adequate funding and resources are essential to support translational research efforts, including clinical trials, infrastructure development, and multidisciplinary collaborations.
- **Data Integration and Sharing:** Effective integration and sharing of research and clinical data are essential for the success of translational research, enabling the generation of actionable insights and facilitating collaborative efforts.

- **Implementation Science:** Incorporating translational research findings into routine clinical practice requires a focus on implementation science, addressing barriers, and optimizing strategies for successful adoption and scalability.

To overcome these challenges, future directions in clinical applications and translational research include increased collaboration between academia, industry, and health care institutions; enhanced data sharing and integration; continued investment in research infrastructure; and the promotion of multidisciplinary approaches that bring together experts from various fields.

Clinical applications and translational research are integral to the advancement of health care by effectively translating scientific discoveries into practical applications that benefit patients. These efforts drive innovation, improve diagnostics and therapeutics, and promote personalized and evidence-based medicine. Continued focus on translational research, collaboration, and overcoming existing challenges will ensure the successful implementation of scientific advancements for the betterment of patient care and population health [104, 105].

3.12 CONCLUSIONS AND FUTURE PERSPECTIVES

In conclusion, this chapter highlights the important role of microbial communities in shaping the chemical diversity and therapeutic potential of medicinal plants, particularly through their impact on secondary metabolism. Through a review of existing literature, we have demonstrated the potential for utilizing seed-based products as a source of new natural products and therapies, and the need for further exploration of medicinal plant–microbe interactions. The core microbiome of medicinal plant seeds is diverse and unique, with the potential to significantly impact the secondary metabolism of the plants. Understanding the mechanisms and pathways involved in secondary metabolism is crucial for unlocking the full potential of medicinal plants and their associated microbes. There are several challenges and opportunities for future research in this field. One of the major challenges is the limited understanding of the complex interactions between medicinal plants and their microbial communities, which requires the development of new and innovative research techniques. Additionally, there is a need to explore the potential of seed-based products in the development of new natural products and therapies [106, 107].

In recent years, there has been a growing recognition of the importance of medicinal plants and their associated microbial communities in the field of natural product discovery. The exploration of the core microbiome of medicinal plant seeds has provided valuable insights into the mechanisms underlying secondary metabolism and the potential for developing new natural products and therapies. However, there is still much to learn and explore in this emerging field of research [108].

One of the major challenges in studying medicinal plant–microbe interactions is the complexity and diversity of microbial communities. The core microbiome of medicinal plant seeds is composed of a diverse array of bacteria, fungi, and other microorganisms, each with its own unique genetic and metabolic potential. Understanding the composition, dynamics, and functional roles of these microbial communities is crucial for harnessing their potential in secondary metabolism. High-throughput sequencing technologies, such as metagenomics and metatranscriptomics, have provided powerful tools for characterizing microbial communities and their functional potential. Integrating these technologies with advanced bioinformatics and data analysis methods will allow for a more comprehensive understanding of the complex interactions between medicinal plants and their associated microbes [109].

Another important aspect for future research is the exploration of seed-based products as a source of new natural products and therapies. Seed oils, powders, and extracts have gained popularity in various industries, including pharmaceuticals, cosmetics, and food. However, there is still much untapped potential in these seed-based products, and further research is needed to unlock their full therapeutic benefits. This includes the identification and characterization of bioactive compounds present in seed-based products, elucidation of their mechanisms of action, and optimization of production methods to ensure consistent quality and efficacy [110].

In addition to the challenges, there are also exciting opportunities for future research in this field. Advances in omics technologies, such as metabolomics and proteomics, hold promise for identifying and characterizing novel bioactive compounds from medicinal plant seeds. These technologies allow for the comprehensive profiling of metabolites and proteins present in seeds, providing valuable insights into the chemical diversity and potential therapeutic properties. Furthermore, interdisciplinary collaborations between microbiologists, botanists, chemists, and pharmacologists are crucial for integrating diverse expertise and approaches to fully understand and harness the potential of medicinal plant–microbe interactions [111, 112].

Furthermore, sustainable cultivation and conservation practices are essential for the long-term viability of medicinal plant species and their associated microbial communities. Overharvesting, habitat destruction, and climate change pose significant threats to medicinal plant biodiversity. To ensure the availability of these valuable resources for future generations, it is important to promote sustainable cultivation methods, protect natural habitats, and support local communities involved in the cultivation and harvesting of medicinal plant seeds [113].

In conclusion, the study of the core microbiome of medicinal plant seeds and its impact on secondary metabolism holds great promise for the discovery of new natural products and therapies. By understanding the mechanisms and pathways involved in secondary metabolism, researchers can unlock the full potential of medicinal plants and their associated microbial communities. However, there are still many challenges and opportunities for future research in this field. By addressing these challenges and embracing interdisciplinary collaborations, researchers can pave the way for the development of innovative and sustainable approaches to health care based on the rich resources of medicinal plant seeds and their microbial partners [114].

REFERENCES

1. Choudhury, A., P. A. Singh, N. Bajwa, S. Dash, and P. Bisht. "Pharmacovigilance of herbal medicines: Concerns and prospects." *Journal of Ethnopharmacology* 309 (2023): 116383. https://doi.org/10.1016/j.jep.2023.116383.
2. Maroyi, Alfred. "An ethnobotanical survey of medicinal plants used by the people in Nhema communal area, Zimbabwe." *Journal of Ethnopharmacology* 136, no. 2 (June 2011): 347–354. www.ethnopharmacologia.org/prelude2016/pdf/biblio-hm-45-maroyi.pdf, https://doi.org/10.1016/j.jep.2011.05.003. Accessed 13 Sept. 2020.
3. Elshafie, Hazem S., *et al.* "A comprehensive review on the biological, agricultural and pharmaceutical properties of secondary metabolites based-plant origin." *International Journal of Molecular Sciences* 24, no. 4 (7 Feb. 2023): 3266–3266. https://doi.org/10.3390/ijms24043266. Accessed 20 May 2023.
4. Ashim Das Astapati, and Soumitra Nath. The complex interplay between plant-microbe and virus interactions in sustainable agriculture: Harnessing phytomicrobiomes for enhanced soil health, designer plants, resource use efficiency, and food security. *Crop Design* (1 May 2023): 100028–100028. https://doi.org/10.1016/j.cropd.2023.100028. Accessed 20 May 2023.
5. Niranjan, V., et al. (2023). Applications of microbial consortia and microbiome interactions for augmenting sustainable agrobiology. In R. D. Singh & A. K. Sharma (Eds.), Sustainable Agrobiology: Recent Advances and Future Perspectives (pp. 275–316). Springer Nature. doi: 10.1007/978-981-19-9570-5_13
6. Pradeep, P. M., and Yadahally N. Sreerama. "Soluble and bound phenolics of two different millet genera and their milled fractions: Comparative evaluation of antioxidant properties and inhibitory effects on starch hydrolysing enzyme activities." *Journal of Functional Foods* 35 (Aug. 2017): 682–693. https://doi.org/10.1016/j.jff.2017.06.033. Accessed 26 Nov. 2020.
7. Das, Niloy. "A review on efficacy of spices and herbs as per Ayurveda and their role as potent antioxidant and antimicrobial agents." Journal of Ayurveda and Integrative Medicine, vol. 11, no. 4, 2020, pp. 100924, doi: 10.1016/j.jaim.2020.100924.Çelik, Hülya, and Filiz Akın. "Some medicinal aromatic plants and their pharmacological effects." *Journal of Pharmaceutical Negative Results* (2023): 1165–1175.
8. Çelik, Hülya, and Filiz Akın. "Some medicinal aromatic plants and their pharmacological effects." *Journal of Pharmaceutical Negative Results* (2023): 1165–1175.
9. Santos, Joshua H., and Mark Lloyd G. Dapar. "Application of essential oils in the pharmaceutical industry." *Essential Oils: Sources, Production and Applications* (2023): 177.

10. Visković, Jelena, Valtcho D. Zheljazkov, Vladimir Sikora, Jay Noller, Dragana Latković, Cynthia M. Ocamb, and Anamarija Koren. "Industrial hemp (Cannabis sativa L.) agronomy and utilization: A review." *Agronomy* 13, no. 3 (2023): 931.
11. Malík, Matěj, and Pavel Tlustoš. "Nootropic herbs, shrubs, and trees as potential cognitive enhancers." *Plants* 12, no. 6 (2023): 1364.
12. Arora, Charu, Dakeshwar Kumar Verma, Jeenat Aslam, and Pramod Kumar Mahish, eds. *Phytochemicals in Medicinal Plants: Biodiversity, Bioactivity and Drug Discovery*. Walter de Gruyter GmbH & Co KG, 2023.
13. Haros, Claudia M., María Reguera, Norma Sammán, and Octavio Paredes-López, eds. *Latin-American Seeds: Agronomic, Processing, and Health Aspects*. CRC Press, 2023.
14. Gitea, Manuel Alexandru, Simona Gabriela Bungau, Daniela Gitea, Bianca Manuela Pasca, Anamaria Lavinia Purza, and Andrei-Flavius Radu. "Evaluation of the phytochemistry – Therapeutic activity relationship for grape seeds oil." *Life*13, no. 1 (2023): 178.
15. Hussain, Ashiq, Tusneem Kausar, Mian Anjum Murtaza, Muhammad Abdullah Jamil, Muhammad Azhar Iqbal, Muhammad Abid Majeed, Ayesha Rafique, Khansa Iftikhar, Saima Noreen Muhammad Yousaf Quddoos, and Arshad Mehmood. "6. Production, characterization, food application and biological study of powder of pumpkin (Cucurbita maxima) parts (peel, flesh and seeds)." *Pure and Applied Biology (PAB)* 12, no. 1 (2023): 48–60.
16. Dikme, Tuğba Gül. "Use of Medicinal and Aromatic Plants in Food." Journal of Food Science and Technology, vol. 59, no. 5, 2022, pp. 2511–2525.
17. Chaurasia, S., and A. Pandey. "Phytochemistry and Pharmacology of Genus Artocarpus: A Review on Current Status of Knowledge." *Russian Journal of Bioorganic Chemistry* (2023): 1–34.
18. Mosaddad, Seyed Ali, Ahmed Hussain, and Hamid Tebyaniyan. "Green alternatives as antimicrobial agents in mitigating periodontal diseases: A narrative review." *Microorganisms* 11, no. 5 (2023): 1269.
19. Butler, Jason A., Genesee Herzberg, and Richard Louis Miller, eds. *Integral Psychedelic Therapy: The Non-ordinary Art of Psycho-spiritual Healing*. Taylor & Francis, 2023.
20. Kalachaveedu, Mangathayaru, Reshma Senthil, Sowndarya Azhagiyamanavalan, Ramnarayanan Ravi, Harikrishnan Meenakshisundaram, and Arunasalam Dharmarajan. "Traditional medicine herbs as natural product matrices in cancer chemoprevention: A trans pharmacological perspective (scoping review)." *Phytotherapy Research* 37, no. 4 (2023): 1539–1573.
21. Ricci, Marina, Annalisa Di Roma, Alessandra Scarcelli, and Michele Fiorentino. "Bio-sustainable materials for tailor-made well-being: A case study of behavioral packaging for cosmetics self-production." In *Developments in Design Research and Practice II: Best Papers from the 11th Senses and Sensibility 2021: Designing Next Genera (c) tions*, pp. 361–381. Springer Nature Switzerland, 2023.
22. Alberdi Aresti, Goiuri, Edurne Magro Montero, Mari José Aranguren Querejeta, and Mirene Begiristain Zubillaga. "A seed towards a sustainable food system in healthcare institutions: The case of the Basque Country." Sustainability, vol. 15, no. 11, 2023, pp. 9541.
23. Zareef, Huma, Muhammad Tayyab Gul, Rahmatullah Qureshi, Hanan Aati, and Mehmooda Munazir. "Application of ethnobotanical indices to document the use of plants in traditional medicines in Rawalpindi district, Punjab-Pakistan." *Ethnobotany Research and Applications* 25 (2023): 1–29.
24. Sun, Xiang, Or Sharon, and Amir Sharon. "Distinct features based on partitioning of the endophytic fungi of cereals and other grasses." *Microbiology Spectrum* (2023): e00611–23.
25. Rafieian, Fatemeh, Reza Amani, Atefe Rezaei, Aslı Can Karaça, and Seid Mahdi Jafari. "Exploring fennel (*Foeniculum vulgare*): Composition, functional properties, potential health benefits, and safety." *Critical Reviews in Food Science and Nutrition* (2023): 1–18.
26. Ghanbary, M. A. Tajick. "Bioactive Potential and GC-MS Fingerprinting of Extracts from Endophytic Fungi Associated with Seeds of Some Medicinal Plants." Journal of Fungi, vol. 9, no. 8, 2023, pp. 857.
27. Rai, Manoj, et al. "Understanding the microbiome of medicinal plant seeds can be challenging due to the complex nature of the seed microbiome and the limitations of traditional microbiological techniques." Journal of Applied Microbiology, vol. 134, no. 1, 2023, pp. 14–26.
28. Kumari, Poonam, Karuna Shanker, and Akanksha Singh. "Insight into Andrographis paniculata associated bacterial endo microbiome and assessment of culturable bacterial endophytes for enhancement of industrially important andrographolide content." *Industrial Crops and Products* 200 (2023): 116840.
29. Kumari, Poonam, Karuna Shanker, and Akanksha Singh. "Insight into Andrographis paniculata associated bacterial endo microbiome and assessment of culturable bacterial endophytes for enhancement of industrially important andrographolide content." *Industrial Crops and Products* 200 (2023): 116840.

30. Sondo, Moussa, Issa Wonni, Agnieszka Klonowska, Kadidia Koïta, and Lionel Moulin. "Quantification of diversity sampling bias resulting from rice root bacterial isolation on popular and nitrogen-free culture media using 16S amplicon barcoding." *PLOS One* 18, no. 4 (2023): e0279049.
31. Sharma, Garima, Surbhi Agarwal, Kavita Verma, Rashmi Bhardwaj, and Vartika Mathur. "Therapeutic compounds from medicinal plant endophytes: Molecular and metabolic adaptations." *Journal of Applied Microbiology* (2023): lxad074.
32. Abdelsattar, Amal M., Ashraf Elsayed, Mohamed A. El-Esawi, and Yasmin M. Heikal. "Enhancing *Stevia rebaudiana* growth and yield through exploring beneficial plant-microbe interactions and their impact on the underlying mechanisms and crop sustainability." *Plant Physiology and Biochemistry* (2023): 107673.
33. Martins, Samuel J., Josephine M. Pasche, Hiago A. Silva, Gijs G. Selten, Noah Savastano, Lucas Abreu, Harsh Bais, et al. "The use of synthetic microbial communities (SynComs) to improve plant health." Phytopathology, vol. 113, no. 4, 2023, pp. 743–754.
34. Zhao, Liang, Sean Walkowiak, and Wannakuwattewaduge Gerard Dilantha Fernando. "Artificial intelligence: A promising tool in exploring the phytomicrobiome in managing disease and promoting plant health." *Plants* 12, no. 9 (2023): 1852.
35. Chaudhari, Hiral G., Shobha Prajapati, Zuhour Hussein Wardah, Gopal Raol, Vimalkumar Prajapati, Rajesh Patel, Ali A. Shati, Mohammad Y. Alfaifi, Serag Eldin I. Elbehairi, and R. Z. Sayyed. "Decoding the microbial universe with metagenomics: A brief insight." *Frontiers in Genetics* 14 (2023).
36. Anguita-Maeso, Manuel, Juan A. Navas-Cortés, and Blanca B. Landa. "Insights into the methodological, biotic and abiotic factors influencing the characterization of xylem-inhabiting microbial communities of olive trees." *Plants* 12, no. 4 (2023): 912.
37. Santiago-Rodriguez, Tasha M., Brice Le François, Jean M. Macklaim, Evgueni Doukhanine, and Emily B. Hollister. "The skin microbiome: Current techniques, challenges, and future directions." *Microorganisms* 11, no. 5 (2023): 1222.
38. Rajendran, L., D. Durgadevi, R. Kavitha, T. Archana, S. Harish, V. Sendhilvel, T. Raguchander, and G. Karthikeyan. "Deciphering the role of growth-promoting bacterial endophytes in harmonizing plant health." In *Microbial Symbionts and Plant Health: Trends and Applications for Changing Climate*, pp. 265–300. Springer Nature Singapore, 2023.
39. Mukherjee, Arghya, Emilio D'Ugo, Roberto Giuseppetti, Fabio Magurano, and Paul D. Cotter. "Metagenomic approaches for understanding microbial communities in contaminated environments: Bioinformatic tools, case studies, and future outlook." In *Metagenomics to Bioremediation*, pp. 103–156. Academic Press, 2023.
40. Gupta, Hitakshi, Jahanvi Ganotra, Nikita Pathania, Tirth Bhargavbhai Patel, Nisha Choudhary, Reia Rani, Damini Supolia, and Deepak Kumar. "Unearthing the modern trends and concepts of rhizosphere microbiome in relation to plant productivity." *Plant Microbiome for Plant Productivity and Sustainable Agriculture* (2023): 19–54.
41. Valles-Colomer, Mireia, Cristina Menni, Sarah E. Berry, Ana M. Valdes, Tim D. Spector, and Nicola Segata. "Cardiometabolic health, diet, and the gut microbiome: A meta-omics perspective." *Nature Medicine* (2023): 1–11.
42. Abdelsattar, Amal M., Ashraf Elsayed, Mohamed A. El-Esawi, and Yasmin M. Heikal. "Enhancing Stevia rebaudiana growth and yield through exploring beneficial plant-microbe interactions and their impact on the underlying mechanisms and crop sustainability." *Plant Physiology and Biochemistry* (2023): 107673.
43. de Andrade, Luana Alves, Carlos Henrique Barbosa Santos, Edvan Teciano Frezarin, Luziane Ramos Sales, and Everlon Cid Rigobelo. "Plant growth-promoting rhizobacteria for sustainable agricultural production." *Microorganisms* 11, no. 4 (2023): 1088.
44. Wahab, Abdul, Asma Munir, Muhammad Hamzah Saleem, Mukhtar Iderawumi AbdulRaheem, Humera Aziz, Manar Fawzi Bani Mfarrej, and Gholamreza Abdi. "Interactions of metal-based engineered nanoparticles with plants: An overview of the state of current knowledge, research progress, and prospects." *Journal of Plant Growth Regulation* (2023): 1–21.
45. Soni, R., N. Singh, G. Singh, and S. Raj. "Significance of Plants in Vedic Astrology, Their Socio-Religious Beliefs, Conservational and Therapeutic Aspects." Journal of Ayurveda and Integrative Medicine, vol. 12, no. 2, 2021, pp. 153–162.
46. Shen, Shuangqian, Chuansong Zhan, Chenkun Yang, Alisdair R. Fernie, and Jie Luo. "Metabolomics-centered mining of plant metabolic diversity and function: Past decade and future perspectives." *Molecular Plant* 16, no. 1 (2023): 43–63.
47. Kumari, Poonam, Karuna Shanker, and Akanksha Singh. "Insight into Andrographis paniculata associated bacterial endo microbiome and assessment of culturable bacterial endophytes for enhancement of industrially important andrographolide content." *Industrial Crops and Products* 200 (2023): 116840.

48. Zhao, Liang, Sean Walkowiak, and Wannakuwattewaduge Gerard Dilantha Fernando. "Artificial intelligence: A promising tool in exploring the phytomicrobiome in managing disease and promoting plant health." *Plants* 12, no. 9 (2023): 1852.
49. Kumari, Poonam, Karuna Shanker, and Akanksha Singh. "Insight into Andrographis paniculata associated bacterial endo microbiome and assessment of culturable bacterial endophytes for enhancement of industrially important andrographolide content." *Industrial Crops and Products* 200 (2023): 116840.
50. Bala, Saroj, Diksha Garg, Kandi Sridhar, Baskaran Stephen Inbaraj, Ranjan Singh, Srinivasulu Kamma, Manikant Tripathi, and Minaxi Sharma. "Transformation of agro-waste into value-added bioproducts and bioactive compounds: Micro/nanoformulations and application in the agri-food-pharma sector." *Bioengineering* 10, no. 2 (2023): 152.
51. Nafees, Sana, Huda Nafees, S. Nizamudeen, and Rauoof Ahmad Rather. "Pharmacological profile of active phytometabolites from traditional medicinal plants." In *Phytohormones and Stress Responsive Secondary Metabolites*, pp. 75–88. Academic Press, 2023.
52. Teraiya, Sagar, Dhaval Nirmal, and Preetam Joshi. "Potential scope and prospects of plant growth-promoting microbes (PGPMs) in micropropagation technology." In *Plant-Microbe Interaction-Recent Advances in Molecular and Biochemical Approaches*, pp. 249–277. Academic Press, 2023.
53. Abotbina, Walid, S. M. Sapuan, M. T. H. Sultan, M. F. M. Alkbir, and R. A. Ilyas. "Effect of black seed fiber, on the physical, thermal, mechanical, morphological, and biodegradation properties of cornstarch-based biocomposites." *Fibers and Polymers* 24, no. 2 (2023): 681–692.
54. Bachheti, Rakesh Kumar, and Archana Bachheti, eds. *Secondary Metabolites from Medicinal Plants: Nanoparticles Synthesis and their Applications*. CRC Press, 2023.
55. Ahmed, Ayesha, Pengfei He, Yueqiu He, Brajesh K. Singh, Yixin Wu, Shahzad Munir, and Pengbo He. "Biocontrol of plant pathogens in omics era – with special focus on endophytic bacilli." *Critical Reviews in Biotechnology* (2023): 1–19.
56. Jha, Dhruva Kumar. "Problems and prospects of utilization of bacterial endophytes for the management of plant diseases." *Indian Phytopathology* 76, no. 1 (2023): 3–20.
57. Toppo, Prabha, Lahasang Lamu Kagatay, Ankita Gurung, Priyanka Singla, Rakhi Chakraborty, Swarnendu Roy, and Piyush Mathur. "Endophytic fungi mediates production of bioactive secondary metabolites via modulation of genes involved in key metabolic pathways and their contribution in different biotechnological sectors." *3 Biotech* 13, no. 6 (2023): 191.
58. Sharma, Garima, Surbhi Agarwal, Kavita Verma, Rashmi Bhardwaj, and Vartika Mathur. "Therapeutic compounds from medicinal plant endophytes: Molecular and metabolic adaptations." *Journal of Applied Microbiology* (2023): lxad074.
59. Anand, Kumar, Vijay Kumar, Indu Bhushan Prasher, Megha Sethi, Harsh Raj, Himanshu Ranjan, Subhash Chand, and Gaurav Kumar Pandey. "Bioactive Molecules from Fungal Endophytes and Their Applications in Pharmaceutical Industries: Challenges and Future Scope." Journal of Basic Microbiology, vol. 63, no. 4, 2023, pp. 257–273.
60. Toppo, Prabha, Lahasang Lamu Kagatay, Ankita Gurung, Priyanka Singla, Rakhi Chakraborty, Swarnendu Roy, and Piyush Mathur. "Endophytic fungi mediates production of bioactive secondary metabolites via modulation of genes involved in key metabolic pathways and their contribution in different biotechnological sectors." *3 Biotech* 13, no. 6 (2023): 191.
61. Castronovo, Lara Mitia, Alberto Vassallo, Alessio Mengoni, Elisangela Miceli, Patrizia Bogani, Fabio Firenzuoli, Renato Fani, and Valentina Maggini. "Medicinal plants and their bacterial microbiota: A review on antimicrobial compounds production for plant and human health." *Pathogens* 10, no. 2 (2021): 106.
62. Chaudhari, Hiral G., Shobha Prajapati, Zuhour Hussein Wardah, Gopal Raol, Vimalkumar Prajapati, Rajesh Patel, Ali A. Shati, Mohammad Y. Alfaifi, Serag Eldin I. Elbehairi, and R. Z. Sayyed. "Decoding the microbial universe with metagenomics: A brief insight." *Frontiers in Genetics* 14 (2023).
63. Ouyang, Yue, Qiqing Cheng, Chunsong Cheng, Ziyu Tang, Yufeng Huang, Eyu Tan, Shaofeng Ma, Xinheng Lin, Ying Xie, and Hua Zhou. "Effects of Plant-Associated Microbiota on Cultivation and Quality of Chinese Herbal Medicines." Chinese Herbal Medicines, vol. 15, no. 3, 2023, pp. 537–554.
64. Leventhal, M., Grant, D., et al. "Disease and Healing in Ancient Societies: Dental Calculus Residues and Skeletal Pathology Data Indicate Age and Sex-Biased Medicinal Practices Among Native Californians." Journal of Archaeological Science: Reports, vol. 48, 2023, pp. 103657.
65. Teuscher, Eberhard, and Ulrike Lindequist. *Natural Poisons and Venoms: Plant Toxins: Terpenes and Steroids*. Walter de Gruyter GmbH & Co KG, 2023.
66. Slavković, Filip, and Abdelhafid Bendahmane. "Floral phytochemistry: Impact of volatile organic compounds and nectar secondary metabolites on pollinator behavior and health." *Chemistry & Biodiversity* (2023): e202201139.

67. Lunde, Bradley M., Claire Moore, and Gabriele Varani. "RNA-binding proteins: Modular design for efficient function." *Nature Reviews Molecular Cell Biology* 8, no. 6 (2007): 479–490.
68. Toppo, Prabha, Lahasang Lamu Kagatay, Ankita Gurung, Priyanka Singla, Rakhi Chakraborty, Swarnendu Roy, and Piyush Mathur. "Endophytic fungi mediates production of bioactive secondary metabolites via modulation of genes involved in key metabolic pathways and their contribution in different biotechnological sectors." *3 Biotech* 13, no. 6 (2023): 191.
69. Fernandes, Aa, P. M. Rodrigues, Ma Pintado, and F. K. Tavaria. "A systematic review of natural products for skin applications: Targeting inflammation, wound healing, and photo-aging." *Phytomedicine* (2023): 154824.
70. Kumar, Sumit, Tulasi Korra, Rajneesh Thakur, R. Arutselvan, Abhijeet Shankar Kashyap, Yasser Nehela, Victor Chaplygin, Tatiana Minkina, and Chetan Keswani. "Role of Plant Secondary Metabolites in Defence and Transcriptional Regulation in Response to Biotic Stress." *Plant Stress* (2023): 100154.
71. Chagaleti, Bharath Kumar, MB Rahul Reddy, Venkatesan Saravanan, P. Senthil Kumar, and M. K. Kathiravan. "An overview of mechanism and chemical inhibitors of shikimate kinase." *Journal of Biomolecular Structure and Dynamics* (2023): 1–17.
72. Nielsen, Mikkel Rank. "Cloning, Reconstruction, and Heterologous Expression of Secondary Metabolite Gene Clusters from Fusarium." Frontiers in Microbiology, vol. 10, 2019, article 1539.
73. Gusain, Suman, Shubham Joshi, and Rohit Joshi. "Sensing, signaling, and regulatory mechanism of cold-stress tolerance in plants." *Plant Physiology and Biochemistry* (2023): 107646.
74. Berg, Gabriele, and Kornelia Smalla. "Plant species and soil type cooperatively shape the structure and function of microbial communities in the rhizosphere." *FEMS Microbiology Ecology* 68, no. 1 (2009): 1–13.
75. Zuñiga, Cristal, Livia Zaramela, and Karsten Zengler. "Elucidation of complexity and prediction of interactions in microbial communities." *Microbial Biotechnology* 10, no. 6 (2017): 1500–1522.
76. Cernava, Tomislav, Armin Erlacher, Jung Soh, Christoph W. Sensen, Martin Grube, and Gabriele Berg. "Enterobacteriaceae dominate the core microbiome and contribute to the resistome of arugula (Eruca sativa Mill.)." *Microbiome* 7, no. 1 (2019): 1–12.
77. Maridass, Muthia, and A. John De Britto. "Origins of plant-derived medicines." *Ethnobotanical Leaflets* 2008, no. 1 (2008): 44.
78. Giordano, Daniela, Daniela Coppola, Roberta Russo, Renata Denaro, Laura Giuliano, Federico M. Lauro, Guido di Prisco, and Cinzia Verde. "Marine microbial secondary metabolites: Pathways, evolution, and physiological roles." *Advances in Microbial Physiology* 66 (2015): 357–428.
79. Proctor, Diana M., Rebecca A. Drummond, Michail S. Lionakis, and Julia A. Segre. "One population, multiple lifestyles: Commensalism and pathogenesis in the human mycobiome." *Cell Host & Microbe* 31, no. 4 (2023): 539–553.
80. Faust, Karoline, J. Fah Sathirapongsasuti, Jacques Izard, Nicola Segata, Dirk Gevers, Jeroen Raes, and Curtis Huttenhower. "Microbial co-occurrence relationships in the human microbiome." *PLoS Computational Biology* 8, no. 7 (2012): e1002606.
81. Xiao, Yingping, Huicong Zou, Jingjing Li, Tongxing Song, Wentao Lv, Wen Wang, Zhenyu Wang, and Shiyu Tao. "Impact of quorum sensing signaling molecules in gram-negative bacteria on host cells: Current understanding and future perspectives." *Gut Microbes* 14, no. 1 (2022): 2039048.
82. Faust, Karoline, and Jeroen Raes. "Microbial interactions: From networks to models." *Nature Reviews Microbiology* 10, no. 8 (2012): 538–550.
83. Brochet, Silvia, Andrew Quinn, Ruben A. T. Mars, Nicolas Neuschwander, Uwe Sauer, and Philipp Engel. "Niche partitioning facilitates coexistence of closely related honey bee gut bacteria." *Elife* 10 (2021): e68583.
84. Cullen, Chad M., Kawalpreet K. Aneja, Sinem Beyhan, Clara E. Cho, Stephen Woloszynek, Matteo Convertino, Sophie J. McCoy *et al.* "Emerging priorities for microbiome research." *Frontiers in Microbiology* 11 (2020): 136.
85. Dalal, Nishu, Rekha Jalandra, Minakshi Sharma, Hridayesh Prakash, Govind K. Makharia, Pratima R. Solanki, Rajeev Singh, and Anil Kumar. "Omics technologies for improved diagnosis and treatment of colorectal cancer: Technical advancement and major perspectives." *Biomedicine & Pharmacotherapy* 131 (2020): 110648.
86. Levy, Maayan, Aleksandra A. Kolodziejczyk, Christoph A. Thaiss, and Eran Elinav. "Dysbiosis and the immune system." *Nature Reviews Immunology* 17, no. 4 (2017): 219–232.
87. Zhi, S. (2016). Host/niche adaptation and specificity in Escherichia coli. Infection, Genetics and Evolution, 43, 277–286. doi: 10.1016/j.meegid.2016.05.024

88. Carding, Simon, Kristin Verbeke, Daniel T. Vipond, Bernard M. Corfe, and Lauren J. Owen. "Dysbiosis of the gut microbiota in disease." *Microbial Ecology in Health and Disease* 26, no. 1 (2015): 26191.
89. Schneider, Thomas, and Kathrin Riedel. "Environmental proteomics: Analysis of structure and function of microbial communities." *Proteomics* 10, no. 4 (2010): 785–798.
90. Kumar, Vineet, Indu Shekhar Thakur, Ajay Kumar Singh, and Maulin P. Shah. "Application of metagenomics in remediation of contaminated sites and environmental restoration." In *Emerging technologies in environmental bioremediation*, pp. 197–232. Elsevier, 2020.
91. Madupu, Ramana, Sebastian Szpakowski, and Karen E. Nelson. "Microbiome in human health and disease." *Science Progress* 96, no. 2 (2013): 153–170.
92. Berg, Gabriele, Daria Rybakova, Doreen Fischer, Tomislav Cernava, Marie-Christine Champomier Vergès, Trevor Charles, Xiaoyulong Chen, *et al.* "Microbiome definition re-visited: Old concepts and new challenges." *Microbiome* 8 (2020): 1–22.
93. Igiehon, Nicholas Ozede, and Olubukola Oluranti Babalola. "Rhizosphere microbiome modulators: Contributions of nitrogen-fixing bacteria towards sustainable agriculture." *International Journal of Environmental Research and Public Health* 15, no. 4 (2018): 574.
94. Rehman, Hafiz Mamoon, James William Cooper, Hon-Ming Lam, and Seung Hwan Yang. "Legume biofortification is an underexploited strategy for combatting hidden hunger." *Plant, Cell & Environment* 42, no. 1 (2019): 52–70.
95. Munir, Shahzad, Ayesha Ahmed, Yongmei Li, Pengbo He, Brajesh K. Singh, Pengfei He, Xingyu Li, Suhail Asad, Yixin Wu, and Yueqiu He. "The Hidden Treasures of citrus: Finding Huanglongbing cure where it was lost." *Critical Reviews in Biotechnology* 42, no. 4 (2022): 634–649.
96. Khan, Md Mahmudul Hasan, Mohd Y. Rafii, Shairul Izan Ramlee, Mashitah Jusoh, and Md Al-Mamun. "Bambara groundnut (Vigna subterranea L. Verdc): A crop for the new millennium, its genetic diversity, and improvements to mitigate future food and nutritional challenges." *Sustainability* 13, no. 10 (2021): 5530.
97. Sidibe, M. *Baobab, Adansonia Digitata L.* Vol. 4. Crops for the Future, 2002.
98. Soomro, Shan-e-Hyder, Xiaotao Shi, Jiali Guo, Senfan Ke, Caihong Hu, Muhammad Asad, Shaista Jalbani, Haider M. Zwain, Parvaiz Khan, and Muhammad Waseem Boota. "Are global influences of cascade dams affecting river water temperature and fish ecology?" *Applied Water Science* 13, no. 4 (2023): 106.
99. Zerhouni, E. A. "Translational research: Moving discovery to practice." *Clinical Pharmacology & Therapeutics* 81, no. 1 (2007): 126–128.
100. Khodyakov, Dmitry, Lisa Mikesell, Ron Schraiber, Marika Booth, and Elizabeth Bromley. "On using ethical principles of community-engaged research in translational science." *Translational Research* 171 (2016): 52–62.
101. Naesens, Maarten, and Minnie M. Sarwal. "Molecular diagnostics in transplantation." *Nature Reviews Nephrology* 6, no. 10 (2010): 614–628.
102. Kumari, Smita, Rohan Gupta, Rashmi K. Ambasta, and Pravir Kumar. "Multiple therapeutic approaches of glioblastoma multiforme: From terminal to therapy." *Biochimica et Biophysica Acta (BBA)-Reviews on Cancer* (2023): 188913.
103. Nommsen-Rivers, Laurie, Maureen M. Black, Parul Christian, Sharon Groh-Wargo, M. Jane Heinig, Kiersten Israel-Ballard, Julie Obbagy, *et al.* "An equitable, community-engaged translational framework for science in human lactation and infant feeding – a report from 'Breastmilk Ecology: Genesis of Infant Nutrition (BEGIN)' Working Group 5." *The American Journal of Clinical Nutrition* 117 (2023): S87-S105.
104. Mohanty, Sachi Nandan, J. V. R. Ravindra, G. Surya Narayana, Chinmaya Ranjan Pattnaik, and Y. Mohamed Sirajudeen, eds. *Drone Technology: Future Trends and Practical Applications*. John Wiley & Sons, 2023.
105. Crombie, Kevin M., Tom G. Adams, Joseph E. Dunsmoor, Benjamin N. Greenwood, Jasper A. Smits, Charles B. Nemeroff, and Josh M. Cisler. "Aerobic exercise in the treatment of PTSD: An examination of preclinical and clinical laboratory findings, potential mechanisms, clinical implications, and future directions." *Journal of Anxiety Disorders* (2023): 102680.
106. Calcott, Mark J., David F. Ackerley, Allison Knight, Robert A. Keyzers, and Jeremy G. Owen. "Secondary metabolism in the lichen symbiosis." *Chemical Society Reviews* 47, no. 5 (2018): 1730–1760.
107. Radjasa, Ocky K., Yvette M. Vaske, Gabriel Navarro, Hélène C. Vervoort, Karen Tenney, Roger G. Linington, and Phillip Crews. "Highlights of marine invertebrate-derived biosynthetic products: Their biomedical potential and possible production by microbial associates." *Bioorganic & Medicinal Chemistry* 19, no. 22 (2011): 6658–6674.

108. Kumari, Poonam, Karuna Shanker, and Akanksha Singh. "Insight into Andrographis paniculata associated bacterial endo microbiome and assessment of culturable bacterial endophytes for enhancement of industrially important andrographolide content." *Industrial Crops and Products* 200 (2023): 116840.
109. Sohrabi, Reza, Bradley C. Paasch, Julian A. Liber, and Sheng Yang He. "Phyllosphere microbiome." *Annual Review of Plant Biology* 74 (2023).
110. Kaur, Gurinder, and Ronald Kander. "The sustainability of industrial hemp: A literature review of its economic, environmental, and social sustainability." *Sustainability* 15, no. 8 (2023): 6457.
111. Chen, Yue, Ying Ge, Xianlin Han, Ling Hao, Tao Huan, Liang Li, Lingjun Li, *et al.* "Meeting report on the 2nd Chinese American Society for Mass Spectrometry (CASMS) conference advancing biological and pharmaceutical mass spectrometry." *Molecular & Cellular Proteomics* (2023): 100559.
112. Dikobe, Tshegofatso, Kedibone Masenya, and Madira C. Manganyi. "Molecular technologies ending with 'omics': The driving force toward sustainable plant production and protection." *F1000Research* 12 (2023): 480.
113. Zehra, Andleeb, Mukesh Meena, Dhanaji M. Jadhav, and Prashant Swapnil. "Regulatory mechanisms for the conservation of endangered plant species, Chlorophytum tuberosum – Potential medicinal plant species." *Sustainability*15, no. 8 (2023): 6406.
114. Abdelsattar, Amal M., Ashraf Elsayed, Mohamed A. El-Esawi, and Yasmin M. Heikal. "Enhancing Stevia rebaudiana growth and yield through exploring beneficial plant-microbe interactions and their impact on the underlying mechanisms and crop sustainability." *Plant Physiology and Biochemistry* (2023): 107673.

4 Exploring the Impact of Endophytes on Secondary Metabolite Production in Medicinal Plants under Abiotic Stress

Zul Kamal, Muhammad Esa, Hasin Sahar, Asghar Ali, Uzma Sher, and Abdul Khaliq Jan

4.1 BACKGROUND

Medicinal plants have been utilized in traditional medicine since ancient times due to their well-known medicinal properties. Secondary metabolites are a group of chemical compounds found in these plants that contribute to their therapeutic properties. Many factors, including genetic and environmental factors, affect the amount of secondary metabolites that are produced in medicinal plants. Abiotic stressors can significantly affect a plant's growth, development, and levels of accumulation of secondary metabolites. These environmental conditions include drought, salt, excessively high or low temperatures, and exposure to heavy metals [1]. The study of endophytes and how they influence the production of secondary metabolites in therapeutic plants under abiotic stress has attracted a lot of attention in recent years. The term 'endophytes' refers to microorganisms that dwell harmlessly inside the tissues of plants. They are microscopic organisms that are found in plant seeds, roots, stems, and leaves. They include various microorganisms, such as fungi and bacteria. The symbiotic relationship between endophytes and their host plants has many advantages, including improved nutrient intake, better development, and stress tolerance [2].

As part of the evolutionary process, plants alter their physiology and biochemistry to adapt and survive in the presence of abiotic stress. These alterations may influence the metabolic pathways responsible for producing secondary metabolites. Through their interactions with their host plant and impact on the synthesis of secondary metabolites, endophytes play an essential part in these adaptations. They function as biotic mediators and stimulate the plant's defence mechanisms to encourage the production of secondary metabolites [3]. Endophytes are capable of producing bioactive chemicals and secondary metabolites, including antibiotics, antifungal agents, and compounds that boost plant growth and antioxidant defences, lowering oxidative stress caused by abiotic stresses [1]. Additionally, endophytes alter the signalling pathways by which the genes in charge of producing secondary metabolites are expressed. The reactivation of specific transcription factors and metabolic enzymes by endophytes results in an increase in the manufacture of secondary metabolites. These interactions between endophytes and hosts are highly specialized and rely on the endophyte species and the particular medicinal plant involved [4].

Finally, the assessment of the role of endophytes in the synthesis of secondary metabolites in medicinal plants under abiotic stresses is a fascinating area of research. The pharmaceutical and agricultural sectors stand to benefit significantly from knowing the mechanisms by which endophytes increase stress tolerance and secondary metabolite synthesis. By taking advantage of endophytes, we

DOI: 10.1201/b23155-4

can develop novel techniques to increase the production of highly potent secondary metabolites in medicinal plants, enhancing their general quality and therapeutic characteristics in the near future.

4.2 ENDOPHYTES AND THEIR CHARACTERISTICS

Due to their abundant biotechnological potential, endophytes survive inside the healthy tissues of plants without showing any harmful symptoms. It has been proven that almost every plant contains these endophytes. Lots of studies have shown that endophytes produce chemicals by helping their host plant's secondary metabolites to be synthesized [1]. De Barry first used the term 'endophytes' to describe the microbial community living inside plants in 1866 [5]. The presence of these endophytic microorganisms varies depending on the type of plant they inhabit, but they have mainly been found in both aerial and underground parts of plants [2].

4.2.1 Characteristics of Endophytes

Microorganisms identified as endophytes live in plant tissues without harming the host. Endophytes enter plants without causing visible signs of disease or inhibiting growth. This beneficial relationship improves plant development and stress durability [6]. There are many kinds and subtypes of endophytes. Endophytes are found in different plant tissues and can take many different forms depending on the environment and location [3]. Some endophytes choose particular plant organs or tissues due to the presence of nutrients, their physical characteristics, or chemical clues [7]. Endophytes can be transferred through seeds from parent plants to offspring, maintaining ongoing survival and benefits in the following generation [8]. Endophytes produce bioactive substances that help plants resist stress, such as antibiotics and molecules that promote growth. Since endophytic fungi show a range of capabilities, including antibacterial, antidiabetic, anticancer, antitumour, antifungal, antimalarial, antioxidant, antiviral, and immunomodulatory activity, endophytic microorganisms can serve as a good alternative source for pharmaceutically significant drugs [4].

Novel drugs can be developed by investigating the bioactive compounds derived from endophytes found in medicinal plants. This process has the particular advantage of making it possible to extract substances derived from plants while causing the least amount of damage to the plant ecosystem. Many biologically active compounds have been successfully extracted from endophytes, including taxol, vincristine, camptothecin, piperine, azadirachtin, and many more [9, 10]. Endophytes can survive in a variety of ecosystems, including harsh ones like polluted areas and deserts. This adaptability is advantageous for ecological restoration and bioremediation research studies [11]. Recent studies are investigating their potential to increase crop yield, create biocontrol agents, and find new bioactive compounds with therapeutic uses [12–15]. Endophytes have now been found to be a key source for the production of secondary metabolites and to provide important information on the biosynthetic routes for these naturally occurring compounds [16, 17].

4.3 SECONDARY METABOLITES

Secondary metabolites are an extensive class of organic chemicals produced by a wide range of species, such as fungi, bacteria, fungi, and plants. Secondary metabolites play specific roles such as predator defence, communication, resource competition, or pollinator attraction, in contrast to primary metabolites, which are essential for fundamental cellular processes and growth [18–20]. Secondary metabolites are very important in many different fields (Table 4.1). Due to unique structural and functional diversity, secondary metabolites have been divided into different classes such as alkaloids, terpenoids, phenolic compounds, flavonoids, steroids, and polyketoids [21–26].

4.3.1 Alkaloids

A large and diverse number of secondary metabolites with nitrogenous ring structures are known as alkaloids. They usually occur in both plants and some animals, including insects and

TABLE 4.1
Applications of Secondary Metabolites and Their Descriptions

Class	Description	Applications
Alkaloids	Nitrogenous ring structures with diverse chemical structures and properties	**Medicinal uses** Morphine (painkiller), quinine (antimalarial), vincristine (anticancer), ephedrine (decongestant) **Stimulants** Caffeine (stimulant), nicotine (stimulant) **Recreational drugs** Cocaine (stimulant), cannabinoids (e.g., THC)
Terpenoids	Derived from isoprene units and exhibit a wide range of structures and functions	**Essential oils** Aromatherapy, perfumes, flavourings **Pharmaceuticals** Taxol (anticancer), artemisinin (antimalarial) **Industrial uses** Biofuels, bioplastics, insecticides
Phenolics	Contains a phenol ring and possesses diverse structures and biological activities	**Antioxidants** Flavonoids, resveratrol (health benefits) **Natural dyes** Anthocyanins (food, cosmetics, textiles) **Preservatives** Tannins (food, beverages) **Medicinal uses** Curcumin (anti-inflammatory), ellagic acid (antitumour)
Flavonoids	Subclass of phenolics with flavone backbone and diverse subclasses and functions	**Health benefits** Antioxidant, anti-inflammatory, anticancer **Dietary supplements** Quercetin, catechins **Cosmetics** Skincare, anti-aging treatments **Natural food colorants** Quercetin, luteolin
Steroids	Lipids with a specific four-ring structure and diverse roles in hormone regulation	**Pharmaceuticals** Corticosteroids (anti-inflammatory), sex hormones (hormone replacement therapy) **Performance-enhancing drugs** Anabolic steroids **Veterinary medicine** Anti-inflammatory, immunosuppressants
Polyketides	Derived from acetyl or malonyl-CoA units and exhibit diverse biological activities	**Antibiotics** Erythromycin, tetracycline **Antifungals** Amphotericin B, nystatin **Anticancer drugs** Doxorubicin, paclitaxel **Other applications** Immunosuppressants, antiparasitic agents

amphibians, and are well known for their significant pharmacological properties, being used as both therapeutic and even recreational drugs. Examples include quinine, morphine, caffeine, and nicotine [21–26].

4.3.2 Terpenoids

Terpenoids, also known as isoprenoids, which are formed from isoprene units, appear to be common products in nature. They are essential for many biological functions, including hormone control, respiration, and photosynthesis. Terpenoids, which are used in essential oils, perfumes, and flavours, are the source of the attractive flavours and smells found in a variety of plants. Examples of the vast range of terpenoids include menthol, carotenoids, and cannabinoids [21–26].

4.3.3 Phenolics

A class of secondary metabolites known as phenolics is defined by the presence of a phenol ring. They are present in plant tissues and perform a variety of tasks, such as UV protection, pathogen defence, and antioxidant activity. Phenolic compounds have gained interest in recent studies because they may have positive effects on health and could potentially prevent chronic diseases. The most common include lignans, tannins, and flavonoids [21–26].

4.3.4 Flavonoids

Flowers, fruits, and leaves all have interesting variations due to flavonoids, a subclass of phenolic compounds. Flavonoids hold potential health benefits due to their antioxidant and anti-inflammatory abilities. They are commonly found in fruits, vegetables, tea, and cocoa [21–26].

4.3.5 Steroids

Steroids are a type of lipids, essential for cell membrane structure and function, identified by their characteristic four-ringed structure. Additionally, they act as building blocks for vital hormones like corticosteroids and sex hormones. Both the plant and animal kingdoms include steroid molecules, which have important physiological roles [21–26].

4.3.6 Polyketides

A wide group of secondary metabolites called 'polyketides' are produced when acetyl or malonyl-CoA units condense. These substances show a wide range of biological actions, such as antifungal, antibacterial, and anticancer potential. Various clinically significant drugs, such as antibiotics and chemotherapeutic agents, have their origins in the polyketides class of secondary metabolites [21–26].

Due to their pharmacological activities, secondary metabolites, in particular alkaloids, terpenoids, and polyketides, have been the subject of extensive studies. These substances possess remarkable antioxidant, anti-inflammatory, anti-cancer, and anti-microbial effects. Secondary metabolites are responsible for the development of many of today's pharmaceuticals, such as antibiotics, painkillers, and anticancer drugs. Secondary metabolites have a wide range of uses in the medicinal, cosmetic, and taste and fragrance industries. They help in the manufacturing of natural colours, food additives, fragrances, and cosmetics. Particularly, terpenoids and phenolics are well liked for their variety of qualities [27–31].

4.4 SECONDARY METABOLITES IN ENDOPHYTES

Because of their unique metabolic capabilities, endophytes can produce a wide variety of secondary metabolites. These compounds are commonly synthesized from basic metabolites via complex metabolic pathways. The endophyte's genetic makeup, the host plant's environment, and interactions with the host all have an impact on the synthesis of these metabolites. In general, they increase the host plant's ability to compete with other microorganisms, protect the host plant against diseases, and fend off attacks from herbivores. These secondary metabolites are necessary for the ecological interactions between endophytes and their host plants [32–34].

4.4.1 Characterization of Secondary Metabolites

First, a raw extract of the endophyte culture's secondary metabolites is taken to define the endophytes' by-products. Several chromatographic methods, including high-performance liquid chromatography (HPLC), gas chromatography (GC), and thin-layer chromatography (TLC), are used to separate the various compounds present in the crude extract. The desired secondary metabolites are simple to separate and purify using these techniques. Isolated compounds must be subjected to spectroscopic procedures like nuclear magnetic resonance (NMR) and mass spectrometry (MS) to ascertain their chemical structures [35–38].

4.4.2 Medicinal Value and Significance of Endophyte Secondary Metabolites

In plant–microbe interactions, endophyte secondary metabolites play key ecological roles. By preventing nourishment or limiting the development of herbivorous insects, they help protect the host plant against herbivores' attacks. Additionally, by affecting the nutrient cycle and supporting plant adaptation to external stressors, these metabolites contribute to ecosystems' overall biodiversity and ecological balance. Moreover, recognizing their ecological functions also improves our understanding of plant–microbe interactions and ecosystem dynamics [39–41]. Endophyte secondary metabolites possess important therapeutic abilities. Many activities, such as antibacterial, anticancer, antioxidant, anti-inflammatory, and immunomodulatory properties, are demonstrated by these substances. For example, endophytic fungi have produced bioactive substances with possible anticancer effects, including taxol and camptothecin [42–44]. Additionally, secondary metabolites from endophytes are becoming important sources for the discovery and development of new drugs. They are strong candidates for the discovery of new drugs to treat a variety of diseases due to their unique chemical structures and bioactive qualities. Researchers are exploring the therapeutic potential of secondary metabolites derived from endophytes for the treatment of malaria, diabetes, Alzheimer's disease, and cardiovascular problems. A wide range of research is being done on the synthesis, characterization, therapeutic potentials, and importance of the secondary metabolites produced by endophytes (Tables 4.2, 4.3). Finding new drugs and bioactive substances can be facilitated by understanding the metabolic capabilities of endophytes, identifying their secondary metabolites, and investigating their therapeutic characteristics [45–48].

4.5 ABIOTIC STRESSES

Abiotic stressors are non-living environmental elements that may be harmful to plant development, growth, and production. Extreme temperatures, flooding, drought, and salinity are examples of stresses that can be natural. In contrast, they can also be brought on by human activities such as pollution or chemical exposure. Agriculture faces major complications from abiotic stressors, which can result in decreased crop quality and severe yield losses. Temperature stress, drought, flooding, salinity, light stress, chemical and pollutant stress, and nutritional imbalances are examples of abiotic stressors [49, 50].

A common ecological stress that inhibits plant growth and development is the lack of water. Lack of water impairs photosynthesis, nutrient absorption, and biological functions of plants. Plants under drought stress begin to wilt, have fewer leaves, and may even die [51]. Moreover, hypoxia or an oxygen shortage can result from too much water being applied to plant roots. Flooding stress deteriorates roots, slows nutrient uptake, and promotes the development of potentially dangerous anaerobic microbes in plants [52]. Similarly, water imbalances and ion toxicity result from the

TABLE 4.2
Applications and Significance of Endophyte Secondary Metabolites

Application	Activity	Description
Medicinal value	**Antimicrobial**	Strong antibacterial efficacy against a variety of diseases, including drug-resistant strains, has been shown by endophyte secondary metabolites.
	Anticancer	Using some endophyte-derived compounds with promising anticancer effects may make it possible to develop new cancer treatments.
	Antioxidant	Secondary metabolites found in endophytes have antioxidant action, which can help prevent oxidative stress-related illnesses.
	Anti-inflammatory	Some endophyte-derived substances have anti-inflammatory properties, making them useful in the treatment of inflammatory illnesses.
Immunomodulatory		Endophyte metabolites have the potential to treat immunological-related diseases by modulating the immune response.
Drug discovery and development	**Novel drug sources**	Endophytes serve as a rich source for finding novel pharmaceuticals and bioactive substances with unique chemical structures and a range of biological activities.
	Targeting drug-resistant pathogens	Endophyte secondary metabolites may be able to attack bacteria that are resistant to drugs, addressing the growing issue of antibiotic resistance.
Therapeutic potential		Compounds derived from endophytes show potential for the development of therapeutics for illnesses like malaria, diabetes, Alzheimer's, and cardiovascular issues.
Ecological significance	**Plant protection**	By discouraging eating or inhibiting the growth of herbivorous insects, secondary metabolites made by endophytes aid in the defence of host plants against herbivores.
	Biodiversity and ecosystem balance	Endophyte metabolites influence nutrient cycling and aid plant response to environmental stresses, which contribute to the overall biodiversity and ecological balance of ecosystems.
Ecological interactions		Endophyte secondary metabolites are important in plant–microbe interactions and have an impact on how different organisms interact and coexist in ecosystems

disruption of the osmotic equilibrium within plant cells caused by high salt concentrations in water for irrigation or soil. Reduced water intake, nutritional deficiency, and tissue damage in plants are all effects of salinity stress [53]. Furthermore, high temperatures and low temperatures both have negative effects on plants. Heat stress harms cells, inhibits photosynthesis, and disrupts the metabolic functions of plants. Cold stress results in dehydration, the development of ice crystals, and reduced membrane integrity of plants [54]. Essential nutrient imbalances may limit plant growth or slow development, or signs of nutrient toxicity are caused by deficiencies or excesses of nutrients, which impair physiological functions, chlorophyll production, and enzyme activity of plants [55, 56]. Additionally, plants are also stressed by insufficient or excessive light levels. Insufficient light slows development and limits photosynthesis, while too much light, especially in hot climates, causes oxidative stress and harms the photosynthetic machinery of plants [57]. Last, plants may become poisonous when exposed to contaminants such as heavy metals, herbicides, or air pollution. These compounds interfere with cellular functions, impair photosynthesis, and harm or kill plant tissues, ultimately resulting in plant death [58]. Plants can respond to abiotic stresses in a number of physiological, biochemical, and molecular ways. Changes in gene expression and the synthesis of protective chemicals like secondary metabolites (antioxidants and stress-related proteins), as well as adjustments to root structure and water-use efficiency, are all examples of these adaptations [59].

TABLE 4.3
Endophyte Sources and Their Applications

Endophyte	Source	Applications
Penicillium sp.	Fungal endophyte	**Antimicrobial** Produces compounds with potent antimicrobial activity against various pathogens, including drug-resistant strains.
Aspergillus sp.	Fungal endophyte	**Antimicrobial** Exhibits strong antimicrobial properties, making it valuable in combating infectious diseases.
Taxomyces andreanae	Fungal endophyte	**Anticancer** Produces taxol, a well-known anticancer compound used in chemotherapy.
Colletotrichum sp.	Fungal endophyte	**Antioxidant** Produces secondary metabolites with strong antioxidant properties, protecting against oxidative stress-related diseases.
Xylaria sp.	Fungal endophyte	**Anti-inflammatory** Produces compounds with anti-inflammatory effects, potentially useful in treating inflammatory disorders.
Bacillus subtilis	Bacterial endophyte	**Immunomodulatory** Produces metabolites that modulate the immune response, offering potential applications in immune-related disorders.
Phoma sp.	Fungal endophyte	**Drug discovery** An abundance of bioactive chemicals with the potential for developing novel drugs and therapies.
Alternaria sp.	Fungal endophyte	**Drug discovery** Produces bioactive compounds with potential applications in the development of novel therapeutics.
Streptomyces sp.	Bacterial endophyte	**Targeting drug-resistant pathogens** Produces antibiotics effective against drug-resistant pathogens, addressing the challenge of antibiotic resistance.
Endophyllum osteospermi	Fungal endophyte	**Therapeutic potential** Produces compounds with potential antimalarial activity, offering new avenues for malaria treatment.
Fusarium oxysporum	Fungal endophyte	**Plant protection** Produces secondary metabolites that protect host plants against herbivory, deterring feeding or inhibiting the growth of herbivorous insects.
Pseudomonas fluorescens	Bacterial endophyte	**Biodiversity and ecosystem balance** Enhances nutrient cycling and ecological balance by promoting plant growth and enhancing soil fertility.
Phomopsis sp.	Fungal endophyte	**Ecological interactions** Promotes plant health and growth, increases resiliency to environmental stressors, and enhances ecosystem dynamics overall.

4.6 SECONDARY METABOLITE ACCUMULATION DURING ABIOTIC STRESSES

As previously mentioned, abiotic factors, including high temperatures, drought, and salinity, as well as the ongoing consequences of climate change, have a significant impact on the synthesis and composition of secondary metabolites in plants [60]. However, plants cope with these issues and produce more heat shock proteins and antioxidants to defend against cellular damage in response to high-temperature stress. Higher concentrations of phenolic substances, flavonoids, and terpenoids, which have antioxidant characteristics and offer protection, are frequently the result of this response [61]. In addition, changes in secondary metabolite profiles based on drought stress encourage plants to produce osmoprotectants such as proline, glycine betaine, and soluble carbohydrates [62, 63].

Furthermore, the osmotic balance in plants is disrupted by salinity stress, which encourages the formation of suitable solutes including alkaloids [62, 63]. Additionally, climate change, which is characterized by increased temperatures and changing rainfall patterns, has an impact on the generation of secondary metabolites. The production of secondary metabolites is impacted by altered rainfall patterns and increased temperatures [62, 63]. Understanding the variations in secondary metabolite production under diverse abiotic stresses and climate change circumstances is essential for predicting and controlling the effects of environmental changes on plant productivity, ecosystem dynamics, and human well-being [64]. By examining these variations, we can learn important things about how plants adapt, and we can make conservation and sustainable agricultural techniques to meet the challenges imposed by shifting environmental conditions.

4.7 CLIMATE CHANGE AND ITS IMPACT ON MEDICINAL PLANTS

'Global climate change' refers to long-term changes in temperature, precipitation patterns, and other environmental elements brought on by human activities, primarily the production of greenhouse gases into the atmosphere. The effects of these alterations on many different ecosystems and species, including medicinal plants, are huge and important to consider (Table 4.4) [65]. While medicinal plants face difficulties due to climate change, there is an advantage. Some plant species may grow in new climatic conditions and may have therapeutic characteristics that have only recently been found. Such unusual plant species have potential, which can be explored and used to synthesize new drug molecules and lead compounds [66, 67].

The timing of biological events including flowering, fruiting, and seed production is referred to as 'phenology'. The phenological patterns of medicinal plants can be disturbed by climate change. Timing mismatches with pollinators, seed dispersers, and other ecological interactions can result from variations in temperature and precipitation. Such disturbances may affect the ability of medicinal plants to reproduce and their long-term survival [68]. Traditional healers and many indigenous groups base their health care systems on the use of medicinal plants. Traditional knowledge may be lost, and cultural practices could suffer as a result of changes in the distribution and availability of medicinal plants brought on by climate change. These communities' well-being and sense of cultural identity may be greatly affected by this loss [69]. In addition, the geographical distribution of different species of medicinal plants may differ as a result of changes in rainfall and temperature patterns caused by global climate change. For some plants, some places might stop being appropriate, but new places might become highly suitable. Natural habitats may be lost as a result, forcing plants to adapt to new environmental factors [70]. Similarly, the physiological functions of medicinal plants can be affected

TABLE 4.4
Possible Impacts of Global Climate Change on Medicinal Plants

Impact	Description
Altered distribution	Changes in temperature and precipitation patterns may affect the medicinal plants' geographic propagation.
Habitat loss	Some areas may stop being suitable habitats, while new areas might start to be.
Phenological changes	Natural habitats for medicinal plants may become extinct as a result of climate change.
Altered plant physiology	The quality and quantity of bioactive chemicals may be affected by water stress, lower growth rates, and changes in plant metabolism brought on by rising temperatures and drought conditions.
Increased vulnerability to pests and diseases	Climate change may make it easier for pests and diseases to spread, making medicinal plants easier to harm and less productive.
Loss of traditional knowledge and cultural practices	Indigenous communities and healers may be affected by climate change-related changes in the distribution and availability of medicinal plants, which could result in the loss of traditional knowledge and disrupt cultural traditions.

by climate change. Increased water stress, slower growth rates, and alterations in plant metabolism are the effects of rising temperatures and increasing drought conditions. These physiological alterations may affect the number and quality of bioactive molecules produced by medicinal plants, which may disturb the therapeutic efficacy of those substances [71, 72]. The cycle of pests, pathogens, and infections that influence medicinal plants can shift as a result of global climate change. Increased moisture, shifting precipitation patterns, and rising temperatures can all lead to conditions that are more favourable to the growth of pests and the spread of plant diseases. This may make medicinal plants less resistant to harm, which would lower their availability and medicinal output [73].

4.8 VARIATION IN SECONDARY METABOLITES DURING DIFFERENT ABIOTIC STRESSES AND CLIMATE CHANGE

Abiotic stressors have a significant impact on the composition and synthesis of secondary metabolites in plants. Plants produce more heat shock proteins and antioxidants in response to high-temperature stress to prevent cellular damage. Higher quantities of phenolic chemicals, flavonoids, and terpenoids, which have antioxidant characteristics and offer protection, are frequently the result of this reaction [60, 74, 75]. Changes in secondary metabolites brought on by drought stress cause plants to produce osmoprotectants such as proline, glycine betaine, and soluble carbohydrates. Additionally, specialized secondary metabolites are produced to alter stomatal closure to regulate water conservation, such as the chemical abscisic acid (ABA) [76]. The osmotic balance in plants is disturbed by salinity stress, which leads to a buildup of suitable solutes like polyols, proline, and alkaloids. These secondary metabolites are essential for preserving the water balance within cells and balancing out the adverse effects of too much salt. In response to salt stress, certain plants also create particular secondary metabolites that support ion homeostasis and offer antioxidant defence mechanisms [63]. Secondary metabolite production has more challenges as a result of climate change, which is characterized by rising temperatures, changing precipitation patterns, and increased atmospheric CO_2 levels. Secondary metabolite synthesis may be impacted by primary metabolism when CO_2 levels are elevated. The composition and synthesis of secondary metabolites are also impacted by rising temperatures and shifting precipitation patterns, which may have an impact on how plants interact with herbivores, microorganisms, and other organisms as well as the nutritional value of plant-based products [77, 78].

4.9 CONCLUSIONS

In a fascinating field of research, endophytes are fundamental to the accumulation of secondary metabolites in medicinal plants under abiotic stress. Greater stress tolerance and the production of secondary metabolites are just a few of the advantages provided by endophytes, or microorganisms that reside inside plant tissues, in their symbiotic relationships with their host plants. They increase the production of secondary metabolites, turn on the plant's defence mechanisms, and regulate the rate of plant development.

Endophytes may trigger an accumulation of secondary metabolites in several ways. They produce phytohormones, enzymes that improve nutrient availability and promote the synthesis of secondary metabolites, as well as bioactive compounds that assist stress tolerance. Additionally, they regulate genes responsible for producing secondary metabolites. There are specific interactions between each species of endophyte and the medicinal plant.

By assessing the activity of endophytes in response to abiotic stress, it is possible to stimulate the synthesis of beneficial compounds in medicinal plants, improve their therapeutic potential, and enhance their general quality. Additionally, it provides chances for eco-friendly agricultural, pharmaceutical production, and drug discovery. Researchers can investigate new methods for increasing the production of secondary metabolites by identifying and investigating the potential of endophytes, which could result in improvements in natural product-based treatments, lead compounds, novel drug discovery, and health care benefits. Research into the complex interactions between plants and microbes will benefit biotechnology, agriculture, medicinal chemistry, drug discovery, and biomedical science.

REFERENCES

1. Fontana, D.C., *et al.*, Endophytic fungi: Biological control and induced resistance to phytopathogens and abiotic stresses. *Pathogens*, 2021. **10**(5): p. 570.
2. Anand, U., *et al.*, Current scenario and prospects of endophytic microbes: Promising candidates for abiotic and biotic stress management for agricultural and environmental sustainability. *Microbial Ecology*, 2023: p. 1–32.
3. Samreen, T., *et al.*, Seed-associated bacterial and fungal endophytes: Diversity, life cycle, transmission, and application potential. *Applied Soil Ecology*, 2021. **168**: p. 104191.
4. Guerrero-Zúñiga, A.L., *et al.*, Functional diversity of plant endophytes and their role in assisted phytoremediation, in *Bioremediation of industrial waste for environmental safety: Volume II: Biological agents and methods for industrial waste management*, 2020: p. 237–255.
5. Adeleke, B.S. and O.O. Babalola, Biotechnological overview of agriculturally important endophytic fungi. *Horticulture, Environment, and Biotechnology*, 2021. **62**: p. 507–520.
6. Kaushal, M., G. Mahuku, and R. Swennen, Metagenomic insights of the root colonizing microbiome associated with symptomatic and non-symptomatic bananas in Fusarium wilt infected fields. *Plants*, 2020. **9**(2): p. 263.
7. Jin, H., *et al.*, Phylogenic diversity and tissue specificity of fungal endophytes associated with the pharmaceutical plant, Stellera chamaejasme L. revealed by a cultivation-independent approach. *Antonie Van Leeuwenhoek*, 2015. **108**: p. 835–850.
8. Truyens, S., *et al.*, Bacterial seed endophytes: Genera, vertical transmission and interaction with plants. *Environmental Microbiology Reports*, 2015. **7**(1): p. 40–50.
9. Manganyi, M.C. and C.N. Ateba, Untapped potentials of endophytic fungi: A review of novel bioactive compounds with biological applications. *Microorganisms*, 2020. **8**(12): p. 1934.
10. Lacerda, Í.C.D.S., J.C. Polonio, and H.C. Golias, Endophytic fungi as a source of antiviral compounds – a review. *Chemistry & Biodiversity*, 2022. **19**(6): p. e202100971.
11. Li, F., *et al.*, Distinct endophytes are used by diverse plants for adaptation to karst regions. *Scientific Reports*, 2019. **9**(1): p. 1–9.
12. Schlegel, M., V. Queloz, and T.N. Sieber, The endophytic mycobiome of European ash and sycamore maple leaves – geographic patterns, host specificity and influence of ash dieback. *Frontiers in Microbiology*, 2018. **9**: p. 2345.
13. Dastogeer, K.M., *et al.*, Host specificity of endophytic mycobiota of wild Nicotiana plants from arid regions of Northern Australia. *Microbial Ecology*, 2018. **75**: p. 74–87.
14. Mohammad Golam Dastogeer, K., *et al.*, Host specificity of endophytic fungi from stem tissue of nature farming tomato (Solanum lycopersicum Mill.) in Japan. *Agronomy*, 2020. **10**(7): p. 1019.
15. Apigo, A. and R. Oono, Plant abundance, but not plant evolutionary history, shapes patterns of host specificity in foliar fungal endophytes. *Ecosphere*, 2022. **13**(1): p. e03879.
16. Slama, H.B., *et al.*, Potentials of endophytic fungi in the biosynthesis of versatile secondary metabolites and enzymes. *Forests*, 2021. **12**(12): p. 1784.
17. Lavanya, N., *et al.*, Endophytic fungal isolation from Blumea axillaris: Identification and biological activity of secondary metabolites. *Notulae Scientia Biologicae*, 2021. **13**(2): p. 10953–10953.
18. Demain, A.L. and A. Fang, The natural functions of secondary metabolites, in *History of modern biotechnology I*, 2000: p. 1–39.
19. Costa, T., *et al.*, Secondary metabolites. 2012.
20. Pagare, S., *et al.*, Secondary metabolites of plants and their role: Overview. *Current Trends in Biotechnology and Pharmacy*, 2015. **9**(3): p. 293–304.
21. Aswani, R., T. Vipina Vinod, and J. Ashitha, Benefits of plant – endophyte interaction for sustainable agriculture, in *Microbial endophytes: Functional biology and applications*. Kumar, A., Radhakrishnan, E., Eds, 2020: p. 35–55.
22. Tiwari, R. and C. Rana, Plant secondary metabolites: A review. *International Journal of Engineering Research and General Science*, 2015. **3**(5): p. 661–670.
23. Peters, K., *et al.*, Chemical diversity and classification of secondary metabolites in nine bryophyte species. *Metabolites*, 2019. **9**(10): p. 222.
24. Ahmed, E., *et al.*, Secondary metabolites and their multidimensional perspective in plant life. *Journal of Pharmacognosy and Phytochemistry*, 2017. **6**(2): p. 205–214.

25. Velu, G., V. Palanichamy, and A.P. Rajan, Phytochemical and pharmacological importance of plant secondary metabolites in modern medicine, In *Bioorganic phase in natural food: An overview*. 2018: p. 135–156.
26. Komaki, H., T. Tamura, and Y. Igarashi, Classification and secondary metabolite-biosynthetic gene clusters of marine streptomyces strains including a lobophorin-and divergolide-producer. *Hydrobiology*, 2023. **2**(1): p. 151–161.
27. Devi, R., *et al.*, Fungal secondary metabolites and their biotechnological applications for human health, in *New and future developments in microbial biotechnology and bioengineering*. 2020, Elsevier. p. 147–161.
28. Keswani, C., *et al.*, Unraveling the efficient applications of secondary metabolites of various Trichoderma spp. *Applied Microbiology and Biotechnology*, 2014. **98**: p. 533–544.
29. Yadav, A.N., *et al.*, Metabolic engineering to synthetic biology of secondary metabolites production, in *New and future developments in microbial biotechnology and bioengineering*. 2019, Elsevier. p. 279–320.
30. Isah, T., *et al.*, Secondary metabolism of pharmaceuticals in the plant in vitro cultures: Strategies, approaches, and limitations to achieving higher yield. *Plant Cell, Tissue and Organ Culture (PCTOC)*, 2018. **132**: p. 239–265.
31. Seca, A.M. and D.C. Pinto, *Biological potential and medical use of secondary metabolites*. 2019, MDPI. p. 66.
32. Rustamova, N., *et al.*, Novel secondary metabolites from endophytic fungi: synthesis and biological properties. *Phytochemistry Reviews*, 2020. **19**: p. 425–448.
33. Kumara, P.M., *et al.*, Endophytes and plant secondary metabolite synthesis: Molecular and evolutionary perspective. *Advances in Endophytic Research*, 2014: p. 177–190.
34. Meena, H., S. Hnamte, and B. Siddhardha, Secondary metabolites from endophytic fungi: Chemical diversity and application, in *Advances in endophytic fungal research: Present status and future challenges*. 2019: p. 145–169.
35. Liu, J. and G. Liu, Analysis of secondary metabolites from plant endophytic fungi, in *Plant pathogenic fungi and oomycetes: Methods and protocols*. 2018: p. 25–38.
36. Yan, J., *et al.*, Characterization of secondary metabolites of an endophytic fungus from Curcuma wenyujin. *Current Microbiology*, 2014. **69**: p. 740–744.
37. Chigozie, V.U., *et al.*, Bioactivities and HPLC analysis of secondary metabolites of a morphologically identified endophytic Aspergillus fungus isolated from Mangifera indica. *Natural Product Research*, 2022. **36**(22): p. 5884–5888.
38. Prasher, I.B. and R.K. Dhanda, GC-MS analysis of secondary metabolites of Endophytic Nigrospora sphaerica isolated from Parthenium hysterophorus. *International Journal of Pharmaceutical Sciences Review and Research*, 2017. **44**(1): p. 217–223.
39. Sridhar, K.R., Diversity, ecology, and significance of fungal endophytes, in *Endophytes and secondary metabolites*. 2019: p. 61–100.
40. Wani, Z.A., *et al.*, Plant-endophyte symbiosis, an ecological perspective. *Applied Microbiology and Biotechnology*, 2015. **99**: p. 2955–2965.
41. Rana, K.L., *et al.*, Endophytic fungi: biodiversity, ecological significance, and potential industrial applications, in *Recent advancement in white biotechnology through fungi: Volume 1: Diversity and enzymes perspectives*. 2019: p. 1–62.
42. Gouda, S., *et al.*, Endophytes: A treasure house of bioactive compounds of medicinal importance. *Frontiers in Microbiology*, 2016. **7**: p. 1538.
43. Zimowska, B., *et al.*, Bioactive products from endophytic fungi of sages (Salvia spp.). *Agriculture*, 2020. **10**(11): p. 543.
44. Joseph, B. and R.M. Priya, Bioactive compounds from endophytes and their potential in. *American Journal of Biochemistry and Molecular Biology*, 2011. **1**(3): p. 291–309.
45. Kamat, S., *et al.*, Marine endophytes from the Indian coasts: The untapped sources of sustainable anticancer drug discovery. *Sustainable Chemistry and Pharmacy*, 2022. **27**: p. 100675.
46. Staniek, A., H.J. Woerdenbag, and O. Kayser, Endophytes: Exploiting biodiversity for the improvement of natural product-based drug discovery. *Journal of Plant Interactions*, 2008. **3**(2): p. 75–93.
47. Demers, D.H., *et al.*, Exploitation of mangrove endophytic fungi for infectious disease drug discovery. *Marine Drugs*, 2018. **16**(10): p. 376.
48. Tiwari, P. and H. Bae, Endophytic fungi: Key insights, emerging prospects, and challenges in natural product drug discovery. *Microorganisms*, 2022. **10**(2): p. 360.
49. Zhang, Y., *et al.*, Interactions between endophytes and plants: Beneficial effect of endophytes to ameliorate biotic and abiotic stresses in plants. *Journal of Plant Biology*, 2019. **62**: p. 1–13.
50. Lata, R., *et al.*, Induction of abiotic stress tolerance in plants by endophytic microbes. *Letters in Applied Microbiology*, 2018. **66**(4): p. 268–276.

51. Azad, K. and S. Kaminskyj, A fungal endophyte strategy for mitigating the effect of salt and drought stress on plant growth. *Symbiosis*, 2016. **68**: p. 73–78.
52. Munir, N., *et al.*, Mechanisms and strategies of plant microbiome interactions to mitigate abiotic stresses. *Agronomy*, 2022. **12**(9): p. 2069.
53. Verma, H., *et al.*, The potential application of endophytes in management of stress from drought and salinity in crop plants. *Microorganisms*, 2021. **9**(8): p. 1729.
54. Bilal, S., *et al.*, Synergistic association of endophytic fungi enhances Glycine max L. resilience to combined abiotic stresses: Heavy metals, high temperature, and drought stress. *Industrial Crops and Products*, 2020. **143**: p. 111931.
55. Belesky, D.P. and C.P. West, Abiotic stresses and endophyte effects. *Tall Fescue for the Twenty-first Century*, 2009. **53**: p. 49–64.
56. Fouda, A., *et al.*, The interaction between plants and bacterial endophytes under salinity stress, in *Endophytes and secondary metabolites*. 2019, Springer. p. 591–607.
57. Godara, H. and W. Ramakrishna, Endophytes as nature's gift to plants to combat abiotic stresses. *Letters in Applied Microbiology*, 2023. **76**(2): p. ovac067.
58. Verma, A., *et al.*, Fungal endophytes to combat biotic and abiotic stresses for climate-smart and sustainable agriculture. *Frontiers in Plant Science*, 2022. **13**: p. 953836.
59. Khan, A.L., *et al.*, Endophytic fungi: A resource for gibberellins and crop abiotic stress resistance. *Critical Reviews in Biotechnology*, 2015. **35**(1): p. 62–74.
60. Akula, R. and G.A. Ravishankar, influence of abiotic stress signals on secondary metabolites in plants. *Plant Signaling & Behavior*, 2011. **6**(11): p. 1720–1731.
61. Ogbe, A.A., J.F. Finnie, and J. Van Staden, The role of endophytes in secondary metabolites accumulation in medicinal plants under abiotic stress. *South African Journal of Botany*, 2020. **134**: p. 126–134.
62. Yeshi, K., *et al.*, Plant secondary metabolites produced in response to abiotic stresses have potential application in pharmaceutical product development. *Molecules*, 2022. **27**(1): p. 313.
63. Ashraf, M.A., *et al.*, Environmental stress and secondary metabolites in plants: An overview, in *Plant metabolites and regulation under environmental stress*. 2018: p. 153–167.
64. Moore, B.D., *et al.*, Explaining intraspecific diversity in plant secondary metabolites in an ecological context. *New Phytologist*, 2014. **201**(3): p. 733–750.
65. Applequist, W.L., *et al.*, Scientists' warning on climate change and medicinal plants. *Planta Medica*, 2020. **86**(01): p. 10–18.
66. Gupta, A., *et al.*, Medicinal plants under climate change: impacts on pharmaceutical properties of plants, in *Climate change and agricultural ecosystems*. 2019, Elsevier. p. 181–209.
67. Zhao, Q., *et al.*, Modeling impacts of climate change on the geographic distribution of medicinal plant Fritillaria cirrhosa D. Don. *Plant Biosystems-An International Journal Dealing with all Aspects of Plant Biology*, 2018. **152**(3): p. 349–355.
68. Harish, B., *et al.*, Impact of climate change on medicinal plants – a review. *Ancient Science of Life*, 2012. **32**(Suppl 1): p. S23.
69. Son, H.N., A. Kingsbury, and H.T. Hoa, Indigenous knowledge and the enhancement of community resilience to climate change in the northern mountainous region of Vietnam. *Agroecology and Sustainable Food Systems*, 2021. **45**(4): p. 499–522.
70. Groner, V.P., *et al.*, Climate change, land cover change, and overharvesting threaten a widely used medicinal plant in South Africa. *Ecological Applications*, 2022. **32**(4): p. e2545.
71. Patni, B., *et al.*, Alarming influence of climate change and compromising quality of medicinal plants. *Plant Physiology Reports*, 2021: p. 1–10.
72. Kumari, A., *et al.*, Climate change and its impact on crops: A comprehensive investigation for sustainable agriculture. *Agronomy*, 2022. **12**(12): p. 3008.
73. Datta, S., Impact of climate change in Indian horticulture-a review. *International Journal of Science, Environment, and Technology*, 2013. **2**(4): p. 661–671.
74. Bont, Z., *et al.*, Heritable variation in root secondary metabolites is associated with recent climate. *Journal of Ecology*, 2020. **108**(6): p. 2611–2624.
75. Chen, F., *et al.*, Genomics of secondary metabolism in Populus: interactions with biotic and abiotic environments. *Critical Reviews in Plant Science*, 2009. **28**(5): p. 375–392.
76. Zahir, A., *et al.*, Synergistic effects of drought stress and photoperiods on phenology and secondary metabolism of Silybum marianum. *Applied Biochemistry and Biotechnology*, 2014. **174**: p. 693–707.
77. Misra, B.B. and S. Chen, Advances in understanding CO 2 responsive plant metabolomes in the era of climate change. *Metabolomics*, 2015. **11**: p. 1478–1491.
78. Skendžić, S., *et al.*, The impact of climate change on agricultural insect pests. *Insects*, 2021. **12**(5): p. 440.

5 Progress in Understanding the Positive Impact of Microbiota Connected to Medicinal Plants and Herbs

Ratna Prabha, DP Singh, Amrender Kumar, and Kumar M

5.1 INTRODUCTION

Medicinal plants represent an extensive bioresource that holds immense promise for applications in contemporary medicine and agriculture. Medicinal plants (MP) find utilization within the enduring customs of traditional medicines in numerous nations. As time progresses, particularly in the realm of ethnobotanical and ethnopharmacological investigations, formulations derived from MPs have progressively been assimilated into conventional medicine. In the contemporary era, these MPs serve as a reservoir of chemical constituents, either employed in their raw form (e.g., atropine, morphine, etc.) or as intermediates for the synthesis of chemotherapeutic and pharmaceutical compounds (e.g., acetylsalicylic acid, paclitaxel, etc.) (Castronovo *et al.*, 2021). Previous studies over medicinal plants have predominantly focused on the analysis of bioactive phytochemicals. However, the current paradigm is experiencing a notable transition owing to the acknowledgment that a considerable proportion of phytotherapeutic compounds are generated by symbiotic microorganisms or as a result of intricate host–microbe interactions. Nevertheless, our understanding of their microbiome, encompassing the complex microbial communities associated with these plants, remains predominantly unexplored (Köberl *et al.*, 2013).

In various plant-associated habitats, microorganisms thrive in abundant numbers, with particular emphasis on the nutrient-rich rhizosphere – a region of soil influenced by plant roots. This dynamic microenvironment harbours a diverse array of microbes that play a pivotal role in plant nutrition, overall health, and quality (Hiltner, 1904; Hartmann *et al.*, 2008; Berg, 2009; Mendes *et al.*, 2011, 2013). Remarkably, the rhizosphere can accommodate an impressive population density of up to 1011 microbial cells per gram of root, encompassing a vast assemblage of more than 30,000 distinct prokaryotic species (Berendsen *et al.*, 2012). The rhizospheric microorganisms, originating from a vast pool of soil microbes, are drawn to the rhizosphere by the plant root secretions and other rhizodeposits (Compant *et al.*, 2010). Their colonization is influenced by the varying composition of these exudates (Bais *et al.*, 2006; Doornbos *et al.*, 2012). It is noteworthy that each plant species harbours a distinct rhizosphere microbiome, which is contingent upon the existing soil community (Smalla *et al.*, 2001; Köberl *et al.*, 2013).

Plants harbour diverse bacterial communities both externally on their surfaces and internally within their organs. Among these microbial communities, those associated with the roots and leaf surfaces have been extensively studied and are collectively referred as plant microbiome or microbiota. These microbial consortia play crucial roles in providing various services to the host plant, including protection against pathogens and assisting in nutrient acquisition from the soil to support plant growth. Additionally, they actively contribute to the synthesis of growth-promoting hormones, thereby influencing plant development and physiology (Bulgarelli *et al.*, 2013).

DOI: 10.1201/b23155-5

Furthermore, certain members of the endophytic microbiota have demonstrated the ability to synthesize bioactive compounds within medicinal plants (Conti *et al.*, 2016). The plant microbiome represents a novel attribute that enhances the adaptability of plants to their surroundings. Moreover, the introduction of plant microbiota has been proposed as an alternative approach to minimize chemical inputs and enhance plant nutrient utilization (Bhardwaj *et al.*, 2014).

5.2 IMPORTANCE OF MICROBIOTA FOR PLANTS

The microbial constituents of the plant holobiont, referred to as the plant microbiota (encompassing all microorganisms) or the plant microbiome (encompassing all microbial genomes) within the rhizosphere, phyllosphere, and endosphere, play crucial roles in facilitating plant growth and maintaining plant health (Vorholt, 2012; Hardoim *et al.*, 2015; Brader *et al.*, 2017; Lemanceau *et al.*, 2017). Plants actively recruit microorganisms from various adjacent microbial reservoirs, including the soil/rhizosphere, phyllosphere (encompassing the leaf surface and its interaction with the external environment), anthosphere (surrounding flowers), spermosphere (germinated seed exterior), and carposphere (external fruit environment) (Hardoim *et al.*, 2015). The root microbiota, primarily acquired through horizontal transfer, originates from the soil environment, which harbours a diverse array of microorganisms, with a notable abundance of Acidobacteria, Verrucomicrobia, Bacteroidetes, Proteobacteria, Planctomycetes, and Actinobacteria (Fierer, 2017) (Figure 5.1). The rhizosphere, which refers to the thin layer of soil directly influenced by plant roots, is recognized as a highly active microbial hot spot and represents one of the most intricate ecosystems (Hiltner, 1904). The rhizosphere houses a significantly greater number of microbial genes compared to the number of plant genes (Mendes *et al.*, 2013). Extensive research has consistently demonstrated the profound impact of numerous plant-associated microorganisms on various aspects of plant biology, including seed germination, seedling vigour, growth and development, nutrient acquisition, disease resistance, and overall productivity (Mendes *et al.*, 2013).

Following the nomenclature employed for microorganisms inhabiting the human body (Qin *et al.*, 2010; Zhao, 2010; Gevers *et al.*, 2012), the combined assemblages of microorganisms associated with plants are commonly referred to as the plant microbiome or as an extension of the plant's genomic repertoire. From this perspective, plants can be perceived as superorganisms, wherein their microbiome plays a crucial role in contributing to specific functions and traits. In reciprocation, plants allocate their photosynthetically derived carbon into their immediate surroundings, encompassing the spermosphere, phyllosphere, rhizosphere, and mycorrhizosphere (Nelson, 2004;

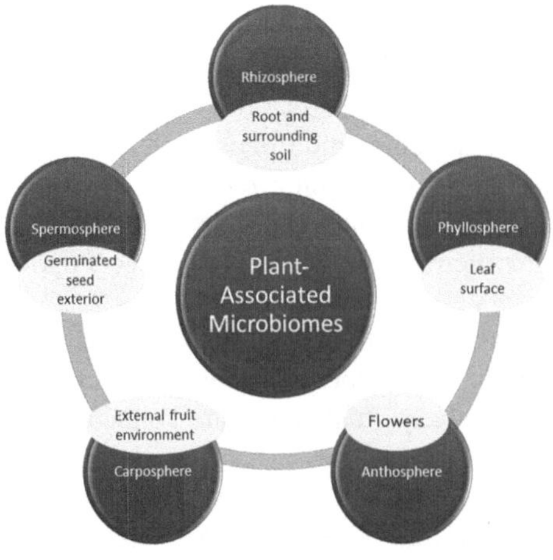

FIGURE 5.1 Different kinds of microbiomes associated with different plant parts.

Frey-Klett *et al.*, 2007; Raaijmakers *et al.*, 2009; Berendsen *et al.*, 2012; Vorholt, 2012). This carbon allocation nourishes the microbial community and exerts a profound influence on its composition and functional dynamics (Mendes *et al.*, 2013).

Traditionally, investigations into medicinal plants primarily emphasized their constituent components. However, in recent times, there has been a notable shift towards understanding the structure and functionality of various medicinal plant microbiomes. Surprisingly, it has come to light that not only do the plants themselves produce compounds possessing phytotherapeutic properties, but their affiliated microorganisms, especially endophytes, also exhibit this capability (Köberl *et al.*, 2013). Ongoing research consistently demonstrates that a significant proportion of natural products originate from microbial sources or arise through microbial interactions with the host organism from which they were derived (Gunatilaka, 2006).

Regarding numerous medicinal plants, it is postulated that the plant-associated microbiome, particularly the intricate community of the endomicrobiome, plays a direct or indirect role in the production of bioactive phytochemicals. However, at present, only a limited subset of microbial strains can be definitively linked to phytotherapeutic properties (Strobel and Daisy, 2003; Strobel *et al.*, 2004; Chandra, 2012; Miller *et al.*, 2012a, 2012b). The precise extent of their contribution to the acknowledged beneficial bioactivity of medicinal plants remains unclear at this stage (Köberl *et al.*, 2013).

5.3 CURRENT STATUS OF RESEARCH IN MICROBIOTA OF MEDICINAL PLANTS

Medicinal plants possess a distinct microbiome attributed to their unique and structurally diverse bioactive secondary metabolites, which are likely responsible for the notable specificity observed in the associated microorganisms (Qi *et al.*, 2012). Attempts have been made to study the associated microbiota of different plants of medicinal importance (Figure 5.2). Analysis of microbiomes of various Chinese medicinal plants, *Ainsliaea henryi Diels, Dioscorea opposita, Potentilla discolor Bge, Stellera chamaejasme* L., *Ophiopogon japonicus* (Thunb) Ker-Gawl., *Juncus effusus* L. var. decipiens Buchen., *Rhizoma arisaematis*, and others, has yielded intriguing findings (Li *et al.*, 2008;

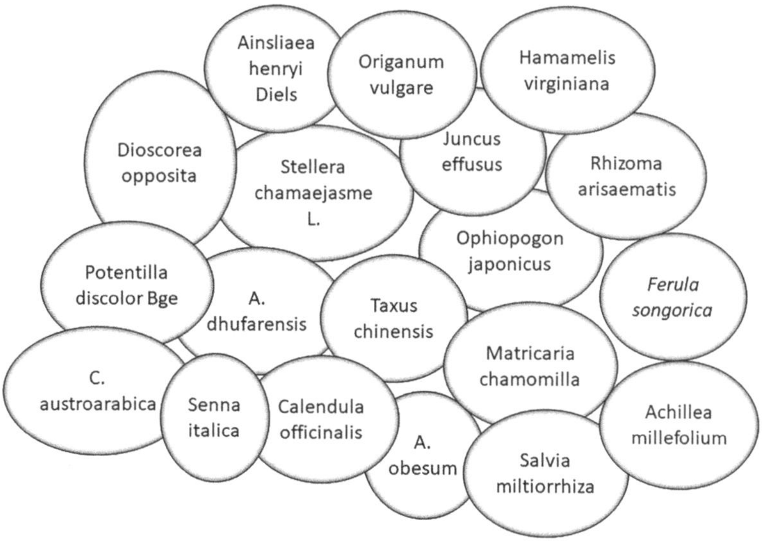

FIGURE 5.2 Different plants of medicinal importance for which attempts are taken to study their associated microbiota.

Zhao *et al.*, 2011), revealing that each plant hosts a distinct actinobacterial community. Furthermore, these plants exhibit a remarkably rich and diverse colonization of Actinobacteria in both their rhizospheric and endophytic environments, with the Actinobacteria displaying antimicrobial and antitumour properties (Zhao *et al.*, 2012). Gram-positive bacteria, often characterized by their ability to form spores, show promise as biological control agents (BCAs). Notably, the genus Streptomyces stands out as a recognized and distinctive reservoir of novel antibiotics (Goodfellow and Fiedler, 2010; Niraula *et al.*, 2010; Nachtigall *et al.*, 2011; Raaijmakers and Mazzola, 2012).

Study of root-associated microbiota has also been carried out in different medicinal plant species, including *Ferula songorica* (Liu *et al.*, 2016), *Stellera chamaejasme* (Jin *et al.*, 2014), and *Taxus chinensis* (Zhang *et al.*, 2015; Sun *et al.*, 2017). In a study, the bacterial communities inhabiting *Astragali radix*, which is derived from the roots of *Astragalus membranaceus* and *A. mongholicus*, were analysed (Sun *et al.*, 2017). In another study, microbiomes of three plant species, *A. dhufarensis, C. austroarabica*, and *A. obesum* were investigated. These plants were collected from previously unexplored areas of arid land and exhibit distinct ecological and medicinal significance (Khan *et al.*, 2020). The microbiome of different medicinal plants, such as *Origanum vulgare* (Pontonio *et al.*, 2018; Castronovo *et al.*, 2021), *Matricaria chamomilla*, and *Calendula officinalis* (Köberl *et al.*, 2019), as well as *Salvia miltiorrhiza* (Chen *et al.*, 2018; Huang *et al.*, 2018), was also studied (Sauer *et al.*, 2021). Further, the microbiome of *Achillea millefolium* and *Hamamelis virginiana* was also analysed by Sauer et al (2021). Metagenomic analysis was carried out to study the diversity of endophytic bacterial communities associated with the leaves and roots of the desert medicinal plant *Senna italica* (Alsaedi *et al.*, 2022). In another study conducted by Senn *et al.* (2022), the focus was on the elucidation of the role played by different microbes inhabiting the root zone of the medicinal plant *Datura inoxia*.

5.3.1 Role of Plant-Associated Microbiota in Growth and Development

Numerous rhizospheric microbes engage in beneficial interactions with their host plants through various mechanisms. These interactions can lead to the direct promotion of plant growth by enhancing nutrient uptake and stimulating hormonal activities. Additionally, these microbes can indirectly contribute to plant health by suppressing the growth of phytopathogens. Prominent examples of such beneficial microbes include biofertilizers, which are microorganisms that provide nutrients to the plant. Among these microorganisms, symbiotic root-nodulating rhizobia has gained considerable recognition as the primary nitrogen-fixing agent. The beneficial effects of these interactions have been extensively studied and documented (Berg, 2009; Lugtenberg and Kamilova, 2009). In addition to symbiotic root-nodulating rhizobia, other microbial biofertilizers, such as mycorrhizal fungi and various rhizobacteria, possess the capability to solubilize plant-accessible phosphate derived from either organic or inorganic phosphate sources (Lugtenberg *et al.*, 2002). These microorganisms can enhance plant growth through hormonal stimulation, earning them the name 'phytostimulators'. Diverse rhizobacteria, such as *Burkholderia cepacia, Staphylococcus epidermidis*, and strains belonging to the *Bacillus subtilis* group, exert a growth-promoting effect on plants through the emission of volatile organic compounds (VOCs; Ryu *et al.*, 2003; Kai *et al.*, 2007; Effmert *et al.*, 2012; Bitas *et al.*, 2013). VOCs are small molecules with low molecular weight (<300 Da) and high vapor pressures, enabling them to diffuse extensively through the porous soil structure and water-filled pores (Kai *et al.*, 2007; Insam and Seewald, 2010). This indirect promotion of plant growth can occur through the biological control of phytopathogens.

Activation of plant defence mechanisms, known as induced systemic resistance (ISR), represents an alternative strategy to mitigate the activity of pathogenic microorganisms, and this response can be induced by specific non-pathogenic rhizobacteria. Various bacterial components, including flagella, lipopolysaccharides, siderophores, and volatile organic compounds, are believed to play a role in triggering the signalling pathway for non-pathogenic rhizobacteria-mediated ISR (van Loon *et al.*, 1998; Lugtenberg and Kamilova, 2009).

The significance of the microbiome in enabling medicinal plants to thrive under environmentally challenging conditions was evident from a study where three medicinal plant species, *Matricaria chamomilla* L., *Calendula officinalis* L., and *Salicornia distichum*, exhibited successful growth on desert soil, which was abundant in fungal phytopathogens (Köberl *et al.*, 2013; Castronovo *et al.*, 2021). This successful adaptation was attributed to the presence of endophytic bacterial communities, particularly strains belonging to the Bacillus genus, that exhibited antagonistic activity against soil-borne phytopathogens. Several studies have investigated the efficacy of endophytic plant growth-promoting rhizobacteria (PGPR) in various important medicinal plants. For instance, Ahmed *et al.* (2014) conducted studies on several prominent medicinal plants, including *Cymbopogon citratus, Majorana hortensis, Marrubium vulgare, Matricaria chamomilla, Melilotus officinalis, Melissa officinalis, Ocimum basilicum, Ocimum syriacum, Quisqualis indica, Solidago virgaurea*, and *Thymus vulgaris*. They reported that PGPR strains belonging to Bacillus, Azotobacter, *Pseudomonas fluorescens*, and Actinomycetes exhibited activity against fungal, bacterial, viral, and nematode pathogens associated with various diseases in these medicinal plants. Singh and Pathak (2015), in their study on medicinal plants such as *Atropa belladonna, Cassia angustifolia, Catharanthus roseus, Dioscorea* spp., *Glycyrrhiza glabra, Hyoscyamus niger, Panax ginseng, Papaver somniferum, Plantago major, Plectranthus barbatus, Podophyllum peltatum, Rauwolfia serpentina*, and *Withania somnifera*, also found that PGPR strains exhibited activity against phytopathogens causing significant yield losses, including *Fusarium oxysporum, Rhizoctonia solani*, and the nematode *Meloidogyne incognita* (Castronovo *et al.*, 2021).

The rhizosphere microbiome plays a crucial role in modulating the nutrient availability and acquisition of plants. Prominent examples of this phenomenon include the nitrogen-fixing symbiosis established between plants and rhizobia, as well as the mycorrhizal associations that enhance phosphorus uptake efficiency (Hawkins *et al.*, 2000; Richardson *et al.*, 2009; Miransari, 2011). Rhizosphere microorganisms play a significant role in facilitating the uptake of essential trace elements, such as iron, by plants. Iron is abundant in soil, but its availability for microbial growth is limited under neutral to alkaline conditions, as it predominantly exists in the insoluble ferric oxide form. Microbes face iron scarcity in their habitats and must regulate intracellular iron concentrations to avoid toxicity and ensure proper cellular function. To overcome this challenge, bacteria employ various strategies, including the secretion of siderophores, which are high-affinity iron-chelating compounds (Lindsay and Schwab, 1982; Andrews *et al.*, 2003; Buckling *et al.*, 2007; Hider and Kong, 2010). Siderophores enable microorganisms to scavenge and solubilize iron from the surrounding environment, making it accessible for microbial growth and potentially enhancing its bioavailability to plants.

Bacterial communities in soil ecosystems play a crucial role in the release of essential nutritive cations from soil minerals, which are not only important for bacterial nutrition but also plant nutrition. Among these bacteria, mineral weathering bacteria have been isolated from diverse environments, with a notable presence in the rhizosphere and ectomycorrhizosphere. These bacteria have been found to contribute significantly to plant growth, particularly in nutrient-poor soils. The process of mineral weathering involves the breakdown and dissolution of mineral components, resulting in the release of nutrients such as cations. Bacterial-mediated mineral weathering is facilitated through various mechanisms, including the production of organic acids, siderophores, and chelating agents that aid in mineral solubilization (Mendes *et al.*, 2013).

5.3.2 Role of Plant-Associated Microbiota in Survival and Adaptation in Adverse Environmental Conditions

The contribution of the rhizosphere microbiome to the survival of different plant species in extreme environments is evident from many different studies (Jorquera *et al.*, 2012). In various agricultural systems, soil salinity can significantly impact plant productivity due to osmotic and drought stresses. However, halotolerant bacteria have demonstrated their ability to thrive under high-salt conditions and, when associated with host plants, exhibit traits that promote plant growth. These bacteria are adapted to salt-stress environments and can establish beneficial interactions with plants by contributing to their physiological adaptation to saline conditions. For instance, *Achromobacter piechaudii* ARV8, a bacterium isolated from arid and saline soil, has been found to enhance the biomass of tomato and pepper seedlings when subjected to transient drought stress (Mayak *et al.*, 2004a, 2004b). Similarly, under conditions of flooding, rhizobacteria have been shown to support plant growth (Grichko and Glick, 2001). Native legumes in high arctic regions have demonstrated the ability to form nodules and fix nitrogen at rates comparable to legumes in temperate climates, despite the challenging impact of low temperatures on these processes (Bordeleau and Prévost, 1994). The remarkable adaptation of arctic legumes to cold environments has garnered significant interest in the agricultural and horticultural sectors.

Abiotic factors, such as soil pH and high concentrations of toxic compounds, also significantly impact plant growth and pose considerable challenges in various agricultural systems worldwide. Soils characterized by low pH or contamination with pollutants are major issues. In the case of pH stress, research has demonstrated that corn plants grown in low-pH soils exhibited reduced foliar lesions when treated with a strain of *Pseudomonas fluorescens* that produces 2,4-diacetyl phloroglucinol (DAPG) (Raudales *et al.*, 2009). The presence of pollutants in soil has prompted the exploration of effective bioremediation methods as alternatives to traditional approaches like excavation and incineration. Among these methods, rhizoremediation, which combines phytoremediation with bioaugmentation, has emerged as a promising strategy for the remediation of polluted sites (Kuiper *et al.*, 2004). During rhizoremediation, plants release root exudates that stimulate the survival and activity of rhizobacteria capable of degrading pollutants.

In a nutshell, the rhizosphere microbiome comprises microorganisms that have the potential to mitigate both biotic and abiotic stresses on plants, offering a viable and environmentally friendly alternative to genetic engineering and conventional plant breeding approaches.

5.4 SIGNIFICANT ACTIVITIES EXHIBITED BY MICROBIOTA OF DIFFERENT MEDICINAL PLANTS

The microbial resource exerts significant influence on the various life activities of medicinal plants from multiple perspectives. Endophytes, microorganisms inhabiting the rhizosphere, and other environmental microorganisms all contribute to vital aspects of medicinal plant growth, development, plant yield, and clinical efficacy (Table 5.1). The microbiota plays a pivotal role in influencing the biosynthesis of active compounds within medicinal plants by activating specific metabolic pathways. Through this interaction, they stimulate host plants to enhance their resistance against environmental stresses by promoting the accumulation of secondary metabolites. Moreover, microorganisms establish intricate interactions with their host plants, leading to long-term, targeted selection outcomes that enhance the plants' capacity to adapt and thrive in their respective environments (Ouyang *et al.*, 2023). A multitude of studies have consistently demonstrated the capacity of plant-associated microbes to exert significant influence on crucial traits encompassing growth, disease resistance, tolerance to abiotic stress, water retention, and the synthesis of plant growth-promoting hormones. These findings underscore the integral role played by plant-associated microbes in shaping the overall health, resilience, and development of plants (Turner *et al.*, 2013; Köberl *et al.*, 2013).

The microbiome of medicinal plants possesses the potential to directly influence the metabolome of the host, thereby potentially impacting the effectiveness and therapeutic properties of herbal medicine (Huang *et al.*, 2018). Research findings have elucidated that the composition of secondary metabolites in medicinal plants of the same species can exhibit variations based on their cultivation location. This variability may be attributed, at least in part, to the distinct microbial communities associated with these plants across different sites. The presence and composition of these site-specific microbial populations have been identified as potential factors contributing to the divergent secondary metabolite profiles observed in medicinal plants grown in different locations (Köberl *et al.*, 2013).

TABLE 5.1
Significant Activities Shown by Different Medicinal Plant–Associated Microorganism(s)

Microorganism(s)	Host Plant	Medicinal Activity	Reference
Endophytic strains of Actinobacteria, primarily Streptomyces	*Lychnophora ericoides*	Significant activity against bacteria and yeast and potential cytotoxicity against human cancer cell lines	Conti *et al.*, 2016
Endophytic Bacillus strain	*Andrographis paniculata* Nees	Antimicrobial activity	Roy *et al.*, 2016
Bacterial isolates	*Echinacea purpurea*	Inhibitory effects against human opportunistic pathogens	Chiellini *et al.*, 2017
Bacillus strain	*Bacopa monnieri* L.	Inhibitory properties against phytopathogenic fungi (*Rhizoctonia* sp., *Sclerotium* sp., and *Phytophthora* sp.)	Jasim *et al.*, 2016
Endophytic bacteria, *Paenibacillus polymyxa* EJS-3	*Stemona japonica*	Antioxidant activity	Castronovo *et al.*, 2021
Bacterial endophytes	Holy basil (*O. tenuiflorum*)	Prompt ISR (induced systemic resistance) against *R. solani* in rice	Sahu *et al.*, 2020
Streptomyces sp. NRRL 30562	*Kennedia nigriscans*	Produces a novel class of bioactive compounds (munumbicins)	Castillo *et al.*, 2002
Endophytic strain *Streptomyces* sp. LJK 109	Alpinia galanga (L.) Willd	Shows suppression of macrophage production of various inflammatory mediators	Taechowisan *et al.*, 2012; Castronovo *et al.*, 2021

Endophytic strains of Actinobacteria, primarily Streptomyces, isolated from the Brazilian medicinal plant *Lychnophora ericoides*, exhibited significant activity against not only bacteria and yeast but also against human cancer cell lines, indicating their potential cytotoxicity (Conti *et al.*, 2016). Similarly, a broad-spectrum antimicrobial activity was demonstrated by another endophytic Bacillus strain, isolated from the ancient oriental medicinal plant *Andrographis paniculata* Nees (Roy *et al.*, 2016). This particular strain exhibited the ability to inhibit the growth of bacterial pathogens, including *B. subtilis*, *B. cereus*, *Vibrio parahaemolyticus*, *Aeromonas caviae*, *Proteus vulgaris*, and *P. aeruginosa* (Castronovo *et al.*, 2021). Notably, bacterial isolates obtained from the medicinal plant *Echinacea purpurea* displayed inhibitory effects against human opportunistic pathogens belonging to the *Burkholderia cepacia* complex (Bcc), which are known to cause severe infections in immunocompromised individuals (Chiellini *et al.*, 2017). A Bacillus strain derived from the stems of the Indian traditional medicinal plant *Bacopa monnieri* L. exhibited inhibitory properties against the growth of phytopathogenic fungi, including *Rhizoctonia* sp., *Sclerotium* sp., and *Phytophthora* sp. (Jasim *et al.*, 2016). Furthermore, this particular bacterium exhibited inhibitory effects against pathogenic organisms, including *Escherichia coli, Salmonella enterica* Tiphy, *B. subtilis, S. aureus*, and *Klebsiella pneumoniae.* Notably, the bioactive fractions of the bacterial strain extracts contained significant antibiotics such as surfactin, iturin, and fengycin. Strain *Rheinheimera* sp. EpRS3 displayed complete inhibition against a range of analysed strains of *B. multivorans* and *B. cenocepacia*, as well as other clinically significant human pathogens, including *Acinetobacter baumannii* N50 and *A. baumannii* YMCR 363 (Presta *et al.*, 2017). The antioxidant potential of endophytic bacteria was demonstrated by *Paenibacillus polymyxa* EJS-3, isolated from the Chinese medicinal plant *Stemona japonica* (Castronovo *et al.*, 2021). Sahu *et al.*, 2020 showed that bacterial endophytes extracted from the holy basil (*O. tenuiflorum*) promote plant growth and induce ISR against *R. solani* in rice.

Comprehensive investigations into the bioactivity of medicinal plant microbiota have unveiled a multitude of bioactive molecules with diverse functionalities, revealing the presence of significant antibiotic compounds that have been previously characterized, as well as novel entities (Castronovo *et al.*, 2021). Carbazole derivatives derived from the endophytic strain *Streptomyces* sp. LJK 109 (Taechowisan *et al.*, 2012) exhibited a dose-dependent suppression of macrophage production of various inflammatory mediators, including NO, PGE2, TNF-α, IL-1β, IL-6, and IL-10 (Castronovo *et al.*, 2021). A novel class of bioactive compounds named munumbicins was isolated from *Streptomyces* sp. NRRL 30562 , which was extracted from the stem tissue of the medicinal plant *Kennedia nigriscans* (Castillo *et al.*, 2002). The aforementioned examples underscore the significant potential of bacteria associated with medicinal plants in producing bioactive molecules with diverse applications.

5.5 APPROACHES FOR SHAPING THE MICROBIOME OF PLANT RHIZOSPHERE

Nowadays, there is considerable interest in devising strategies that can reshape the rhizosphere microbiome to favour microorganisms that enhance plant productivity and prevent the proliferation of harmful pathogens (Figure 5.3). Different research studies have suggested that the rhizosphere microbiome is influenced by two primary factors: plant genotype and soil type. The genotype of the plant plays a crucial role in recruiting specific members of the soil microbiome, as exemplified by the malic-acid-mediated stimulation of beneficial *B. subtilis* demonstrated by Rudrappa *et al.* (2008). Moreover, studies conducted by Berg and Smalla (2009) and Bakker *et al.* (2012) have highlighted the impact of soil type on shaping the rhizosphere microbiome. Soils serve as intricate and diverse ecosystems, exerting profound effects on plant physiology and root exudation patterns and subsequently shaping the composition of the rhizosphere microbiome.

Several strategies have been proposed to manipulate the microbial composition of the rhizosphere and modulate microbial activities. One such approach involves altering the quantity and/or quality of root exudates through plant breeding or genetic modification to influence the rhizosphere microbiome. This strategy, commonly known as 'rhizosphere engineering', has garnered significant attention in the field (Oger *et al.*, 2004; Ryan *et al.*, 2009; Bakker *et al.*, 2012). However, successful implementation of this approach necessitates a comprehensive understanding of the chemical complexity of rhizodeposits and the intricate molecular interactions occurring within the rhizosphere. Despite advances in the understanding of communication and interactions within the rhizosphere, there remains a scarcity of breeding programs that explicitly consider rhizosphere-related traits

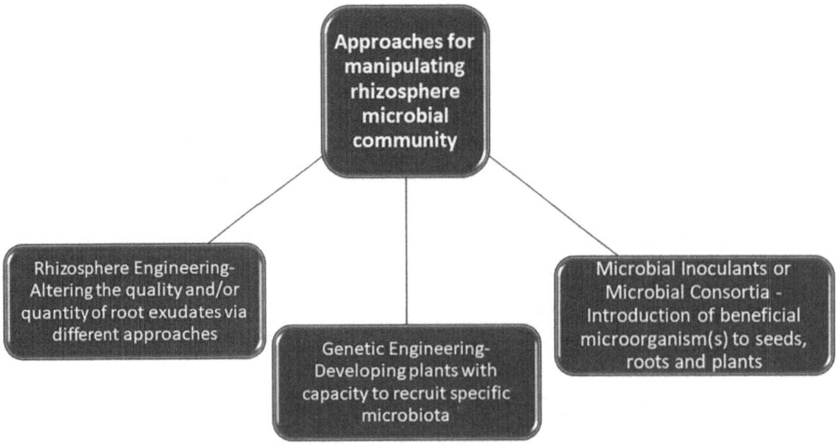

FIGURE 5.3 Different approaches to manipulate or alter plant associated microbiota.

and root exudation characteristics (Smith and Goodman, 1999; Smith *et al.*, 1999; Rengel, 2002; Wissuwa *et al.*, 2009). To date, only a limited number of studies have incorporated the evaluation of plant lines based on their interactions with the soil microbiome (Bakker *et al.*, 2012).

Genetic engineering holds promise as an alternative to conventional plant breeding, although it is a time-intensive process. Various genetic engineering approaches have been employed to modify the rhizosphere pH (Gevaudant *et al.*, 2007; Yang *et al.*, 2007), regulate organic anion efflux (Tesfaye *et al.*, 2001; Anoop *et al.*, 2003; Delhaize *et al.*, 2004; Sasaki *et al.*, 2004; Li *et al.*, 2005), and facilitate citrate efflux (Furukawa *et al.*, 2007; Wang *et al.*, 2007). Transgenic plants engineered to secrete specific signalling molecules have demonstrated the capacity for communication with rhizosphere microorganisms (Oger *et al.*, 1997, 2000). Notably, plants release chemical compounds that disrupt bacterial quorum sensing (Gao *et al.*, 2003; Bauer and Mathesius, 2004). However, the current understanding of root exudation *in situ* remains limited, and the composition of root exudates is primarily determined under hydroponic culture conditions. Consequently, the identification of specific targets for plant breeding is still a challenge. To fully leverage the potential of genetic engineering for rhizosphere manipulation, it is crucial to enhance understanding of root exudation under natural soil conditions and to identify stable and reliable targets for plant breeding. Additionally, the intricate interplay between the plant genotype, environmental factors, and root exudation dynamics necessitates a comprehensive approach to unravelling the complexities of rhizosphere communication and optimizing genetic engineering strategies.

Alternative strategies for manipulating the rhizosphere microbiome include the introduction of beneficial microorganisms into the soil, seeds, and planting materials, as well as the stimulation of resident beneficial rhizosphere microorganisms through soil or crop management practices. Numerous beneficial strains of rhizobacteria and fungi with diverse characteristics have been employed to enhance plant productivity by introducing them into the soil or applying them to seeds or planting materials (Bhattacharyya and Jha, 2012). This approach of reshaping the rhizosphere microbiome by incorporating beneficial microorganisms that confer protection against pathogen infections in the host plant bears similarities to the use of probiotics in human health applications. Despite the potential benefits of introducing microbial strains into new environments, many of these strains encounter challenges in terms of survival and establishing sufficient population densities in the rhizosphere to effectively control soilborne pathogens (Raaijmakers *et al.*, 2009). To address these limitations, the development of genetically engineered microbial strains, known as designer probiotics, has been proposed (Paton *et al.*, 2006; Picard and Bosco, 2008; Picard *et al.*, 2008; Berlec, 2012; Mendes *et al.*, 2013).

Instead of employing a single microorganism, the utilization of assemblages comprising multiple microorganisms with complementary or synergistic traits is expected to offer a more effective and consistent outcome. Bakker *et al.* (2012) have highlighted the potential of such consortia to expedite niche saturation within the rhizosphere microbiome and competitively exclude pathogens. The concept of concerted rhizosphere communities, which involves the coordinated action of multiple microorganisms, is gaining increasing recognition. However, the identification or selection of suitable consortium members poses challenges due to issues related to compatibility among the constituents (Mendes *et al.*, 2013).

5.6 CONCLUSION AND IMPLICATIONS FOR FUTURE RESEARCH

Given the growing concern over the rising prevalence of antibiotic resistance in medically significant bacteria, medicinal plants harbouring antibacterial properties are of significant value as bioresources for discovering novel active metabolites (Palombo and Semple, 2001). From the previous discussion, it is evident that the microbiota of medicinal plants serves as a valuable source for the isolation of biocontrol agents effective against both human and plant pathogens. The vast range of bioactive compounds produced by medicinal plant–associated bacteria highlights their promising role in addressing various challenges related to human and plant health.

To fully harness the potential of beneficial microorganisms, a deeper understanding is required regarding the mechanisms underlying the communication between microbiota and the host plant. Fundamental knowledge is needed to elucidate the molecular and metabolic changes induced in plants in response to beneficial microorganisms, as well as their impacts on the population dynamics and virulence of plant pathogenic microorganisms. In conjunction with conducting an extensive phylogenetic examination of the microbiome, it is imperative to move beyond mere identification of microbial communities and delve into elucidating the specific microbial entities that actively participate in different stages of plant and root growth. Moreover, it is crucial to ascertain the functions and pathways exhibited by these microorganisms in a temporally and spatially resolved manner.

The ability of plant-associated microbiota to promote plant growth and improve stress tolerance represents a promising avenue for enhancing agricultural productivity in different climatic regions. Further exploration of the specific interactions between microbiota and host plants in different environments can contribute to a better understanding of the intricate mechanisms involved and will aid in the development of sustainable agricultural practices.

Acknowledgments: Ratna Prabha and Kumar M are thankful to the Project 'Mining Agricultural Microbiome Datasets for Antibiotic Resistance Genes (ARG) Diversity and Prediction of Microbial Resistome'.

REFERENCES

Ahmed, E.A., Hassan, E.A., Tobgy, K.M.K.E., Ramadan, E.M. 2014. Evaluation of rhizobacteria of some medicinal plants for plant growth promotion and biological control. *Ann Agric Sci* 59: 273–280.

Alsaedi, Z.S., Ashy, R.A., Shami, A.Y., Majeed, M.A., Alswat, A.M., Baz, L., Baeshen, M.N., Jalal, R.S. 2022. Metagenomic study of the communities of bacterial endophytes in the desert plant Senna Italica and their role in abiotic stress resistance in the plant. *Braz J Biol* 82: e267584. doi: 10.1590/1519-6984.267584.

Andrews, S.C., Robinson, A.K., Rodríguez-Quiñones, F. 2003. Bacterial iron homeostasis. *FEMS Microbiol Rev* 27: 215–237.

Anoop, V.M., Basu, U., McCammon, M.T., McAlister-Henn, L., Taylor, G.J. 2003. Modulation of citrate metabolism alters aluminum tolerance in yeast and transgenic canola overexpressing a mitochondrial citrate synthase. *Plant Physiol* 132: 2205–2217.

Bais, H.P., Weir, T.L., Perry, L.G., Gilroy, S., Vivanco, J.M. 2006. The role of root exudates in rhizosphere interactions with plants and other organisms. *Annu Rev Plant Biol* 57: 233–266. doi: 10.1146/annurev.arplant.57.032905.105159.

Bakker, M.G., Manter, D.K., Sheflin, A.M., Weir, T.L., Vivanco, J.M. 2012. Harnessing the rhizosphere microbiome through plant breeding and agricultural management. *Plant Soil* 360: 1–13.

Bauer, W.D., Mathesius, U. 2004. Plant responses to bacterial quorum sensing signals. *Curr Opin Plant Biol* 7: 429–433.

Berendsen, R.L., Pieterse, C.M.J., Bakker, P. 2012. The rhizosphere microbiome and plant health. *Trends Plant Sci* 17: 478–486.

Berg, G. 2009. Plant-microbe interactions promoting plant growth and health: Perspectives for controlled use of microorganisms in agriculture. *Appl Microbiol Biotechnol* 84: 11–18. doi: 10.1007/s00253-009-2092-7.

Berg, G., Smalla, K. 2009. Plant species and soil type cooperatively shape the structure and function of microbial communities in the rhizosphere. *FEMS Microbiol Ecol* 68: 1–13.

Berlec, A. 2012. Novel techniques and findings in the study of plant microbiota: Search for plant probiotics. *Plant Sci* 193: 96–102.

Bhardwaj, D., Ansari, M., Sahoo, R., Tuteja, N. 2014. Biofertilizers function as key players in sustainable agriculture by improving soil fertility, plant tolerance, and crop productivity. *Microb Cell Factories* 13: 66.

Bhattacharyya, P.N., Jha, D.K. 2012. Plant growth-promoting rhizobacteria (PGPR): Emergence in agriculture. *World J Microbiol Biotechnol* 28: 1327–1350.

Bitas, V., Kim, H.S., Bennett, J.W., Kang, S. 2013. Sniffing on microbes: Diverse roles of microbial volatile organic compounds in plant health. *Mol Plant Microbe Interact* 26: 835–843. doi: 10.1094/MPMI-10-12-0249-CR.

Bordeleau, L., Prévost, D. 1994. Nodulation and nitrogen fixation in extreme environments. *Plant Soil* 161: 115–125.

Brader, G., Company, S., Vescio, K., Mitter, B., Trognitz, F., Ma, L.-J. 2017. Ecology and genomic insights into plant-pathogenic and plant-nonpathogenic endophytes. *Annu Rev Phytopathol* 55: 61–83.

Buckling, A., Harrison, F., Vos, M., Brockhurst, M.A., Gardner, A., West, S.A., Griffin, A. 2007. Siderophore-mediated cooperation and virulence in Pseudomonas aeruginosa. *FEMS Microbiol Ecol* 62: 135–141.

Bulgarelli, D., Schlaeppi, K., Spaepen, S., Ver Loren van Themaat, E., Schulze-Lefert, P. 2013. Structure and functions of the bacterial microbiota of plants. *Annu Rev Plant Biol* 64: 807–838

Castillo, U., Strobel, G.A., Ford, E.J., Hess, W.M., Porter, H., Jensen, J.B., *et al.* 2002. Munumbicins, wide-spectrum antibiotics produced by Streptomyces NRRL 30562, endophytic on Kennedia nigriscans. *Microbiology* 148: 2675–2685.

Castronovo, L.M., Vassallo, A., Mengoni, A., Miceli, E., Bogani, P., Firenzuoli, F., Fani, R., Maggini, V. 2021. Medicinal plants and their bacterial microbiota: A review on antimicrobial compounds production for plant and human health. *Pathogens* 10(2): 106. doi: 10.3390/pathogens10020106.

Chandra, S. 2012. Endophytic fungi: Novel sources of anticancer lead molecules. *Appl Microbiol Biotechnol* 95: 47–59. doi: 10.1007/s00253-012-4128-7.

Chen, H., Wu, H., Yan, B., Zhao, H., Liu, F., Zhang, H., *et al.* 2018. Core microbiome of medicinal plant Salvia miltiorrhiza seed: A rich reservoir of beneficial microbes for secondary metabolism? *Int J Mol Sci* 19: 672. doi: 10.3390/ijms19030672.

Chiellini, C., Maida, I., Maggini, V., Bosi, E., Mocali, S., Emiliani, G., Perrin, E., Firenzuoli, F., Mengoni, A., Fani, R. 2017. Preliminary data on antibacterial activity of *Echinacea purpurea*-associated bacterial communities against Burkholderia cepacia complex strains, opportunistic pathogens of cystic fibrosis patients. *Microbiol Res* 196: 34–43.

Compant, S., Clément, C., Sessitsch, A. 2010. Plant growth-promoting bacteria in the rhizo- and endosphere of plants: Their role, colonization, mechanisms involved and prospects for utilization. *Soil Biol Biochem* 42: 669–678. doi: 10.1016/j.soilbio.2009.11.024.

Conti, R., Chagas, F.O., Caraballo-Rodriguez, A.M., Melo, W.G., do Nascimento, A.M., Cavalcanti, B.C., de Moraes, M.O., Pessoa, C., Costa-Lotufo, L.V., Krogh, R., Andricopulo, A.D., Lopes, N.P., Pupo, M.T. 2016. Endophytic Actinobacteria from the Brazilian medicinal plant *Lychnophora ericoides* Mart. and the biological potential of their secondary metabolites. *Chem Biodivers* 13: 727–736.

Delhaize, E., Ryan, P.R., Hebb, D.M., Yamamoto, Y., Sasaki, T., Matsumoto, H. 2004. Engineering high-level aluminum tolerance in barley with the ALMT1 gene. *P Natl Acad Sci USA* 101: 15249–15254.

Doornbos, R.F., van Loon, L.C., Bakker, P.A. 2012. Impact of root exudates and plant defense signaling on bacterial communities in the rhizosphere. *Agron Sustain Dev* 32: 227–243. doi: 10.1007/s13593-011-0028-y.

Effmert, U., Kalderas, J., Warnke, R., Piechulla, B. 2012. Volatile mediated interactions between bacteria and fungi in the soil. *J Chem Ecol* 38: 665–703.

Fierer, N. 2017. Embracing the unknown: Disentangling the complexities of the soil microbiome. *Nature Rev Microbiol* 15: 579–590.

Frey-Klett, P., Garbaye, J., Tarkka, M. 2007. The mycorrhiza helper bacteria revisited. *New Phytol* 176: 22–36.

Furukawa, J., Yamaji, N., Wang, H., *et al.* 2007. An aluminum-activated citrate transporter in barley. *Plant Cell Physiol* 48: 1081–1091.

Gao, M.S., Teplitski, M., Robinson, J.B., Bauer, W.D. 2003. Production of substances by Medicago truncatula that affect bacterial quorum sensing. *Mol Plant Microbe Interact* 16: 827–834.

Gevaudant, F., Duby, G., von Stedingk, E., Zhao, R., Morsomme, P., Boutry, M. 2007. Expression of a constitutively activated plasma membrane H + -ATPase alters plant development and increases salt tolerance. *Plant Physiol* 144: 1763–1776.

Gevers, D., Knight, R., Petrosino, J.F., *et al.* 2012. The human microbiome project: A community resource for the healthy human microbiome. *PLoS Biol* 10: e1001377.

Goodfellow, M., Fiedler, H.P. 2010. A guide to successful bioprospecting: Informed by actinobacterial systematics. *Antonie Van Leeuwenhoek* 98: 119–142. doi: 10.1007/s10482-010-9460-2.

Grichko, V.P., Glick, B.R. 2001. Amelioration of flooding stress by ACC deaminase-containing plant growth-promoting bacteria. *Plant Physiol Biochem* 39: 11–17.

Gunatilaka, A.A. 2006. Natural products from plant-associated microorganisms: Distribution, structural diversity, bioactivity, and implications of their occurrence. *J Nat Prod* 69: 509–526. doi: 10.1021/np058128n.

Hardoim, P.R., van Overbeek, L.S., Berg, G., Pirttilä, A.M., Compant, S., Campisano, A., *et al.* 2015. The hidden world within plants: Ecological and evolutionary considerations for defining functioning of microbial endophytes. *MMBR*, 79: 293–320.

Hartmann, A., Rothballer, M., Schmid, M. 2008. Lorenz Hiltner, a pioneer in rhizosphere microbial ecology and soil bacteriology research. *Plant Soil* 312: 7–14. doi: 10.1007/s11104-007-9514-z.

Hawkins, H.J., Johansen, A., George, E. 2000. Uptake and transport of organic and inorganic nitrogen by arbuscular mycorrhizal fungi. *Plant Soil* 226: 275–285.

Hider, R.C., Kong, X. 2010. Chemistry and biology of siderophores. *Nat Prod Rep* 27: 637–657.

Hiltner, L. 1904. Über neuere Erfahrungen und Probleme auf dem Gebiete der Bodenbakteriologie unter besonderer Berücksichtigung der Gründüngung und Brache. *Arb DLG* 98: 59–78.

Huang, W., Long, C., Lam, E. 2018. Roles of plant-associated microbiota in traditional herbal medicine. *Trends Plant Sci* 23: 559–562. doi: 10.1016/j.tplants.2018.05.003.

Insam, H., Seewald, M.S.A. 2010. Volatile organic compounds (VOCs) in soils. *Biol Fertil Soils* 46: 199–213.

Jasim, B., Sreelakshmi, S., Mathew, J., Radhakrishnan, E.K. 2016. Identification of endophytic *Bacillus mojavensis* with highly specialized broad-spectrum antibacterial activity. *3 Biotech* 6.

Jin, H., Yang, X.Y., Yan, Z.Q., Liu, Q., Li, X.Z., Chen, J.X., Zhang, D.H., Zeng, L.M., Qin, B. 2014. Characterization of rhizosphere and endophytic bacterial communities from leaves, stems and roots of medicinal Stellera chamaejasme L. *Syst Appl Microbiol* 37: 376–385.

Jorquera, M.A., Shaharoona, B., Nadeem, S.M., de la Luz Mora, M., Crowley, D.E. 2012. Plant growth-promoting rhizobacteria associated with ancient clones of creosote bush (Larrea tridentata). *Microb Ecol* 64: 1008–1017.

Kai, M., Effmert, U., Berg, G., Piechulla, B. 2007. Volatiles of bacterial antagonists inhibit mycelial growth of the plant pathogen *Rhizoctonia solani*. *Arch Microbiol* 187: 351–360.

Khan, A.L., Asaf, S., Abed, R.M.M., Chai, Y.N., Al-Rawahi, A.N., Mohanta, T.K., Al-Rawahi, A., Schachtman, D.P., Al-Harrasi, A. 2020. Rhizosphere microbiome of arid land medicinal plants and extra cellular enzymes contribute to their abundance. *Agronomy & Horticulture – Faculty Publications* 1328.

Köberl, M., Erschen, S., Etemadi, M., White, R.A., El-Arabi, T.F., Berg, G. 2019. Deciphering the microbiome shift during fermentation of medicinal plants. *Sci Rep* 9: 13461. doi: 10.1038/s41598-019-49799-2.

Köberl, M., Schmidt, R., Ramadan, E.M., Bauer, R., Berg, G. 2013. The microbiome of medicinal plants: Diversity and importance for plant growth, quality, and health. *Front Microbiol* 4: 400. doi: 10.3389/fmicb.2013.00400.

Kuiper, I., Lagendijk, E.L., Bloemberg, G.V., Lugtenberg, B.J.J. 2004. Rhizoremediation: A beneficial plant-microbe interaction. *Mol Plant Microbe Interact* 17: 6–15.

Lemanceau, P., Blouin, M., Muller, D., Moënne-Loccoz, Y. 2017. Let the core microbiota be functional. *Trends Plant Sci* 22: 583–595.

Li, J., Zhao, G.Z., Chen, H.H., Wang, H.B., Qin, S., Zhu, W.Y., *et al.* 2008. Antitumour and antimicrobial activities of endophytic streptomycetes from pharmaceutical plants in the rainforest. *Lett Appl Microbiol* 47: 574–580. doi: 10.1111/j.1472-765X.2008.02470.x.

Li, J.Y., Xu, R.K., Xiao, S.C., Ji, G.L. 2005. Effect of low-molecular-weight organic anions on exchangeable aluminum capacity of variable charge soils. *J Colloid Interface Sci* 284: 393–399.

Lindsay, W.L., Schwab, A.P. 1982. The chemistry of iron in soils and its availability to plants. *J Plant Nutr* 5: 821–840.

Liu, Y.H., Guo, J.W., Salam, N., Li, L., Zhang, Y.G., Han, J., Mohamad, O.A., Li, W.J. 2016. Culturable endophytic bacteria associated with medicinal plant Ferula songorica: Molecular phylogeny, distribution and screening for industrially important traits. *3 Biotech* 6: 209.

Lugtenberg, B., Chin-A-Woeng, T.F., Bloemberg, G.V. 2002. Microbe-plant interactions: Principles and mechanisms. *Antonie Van Leeuwenhoek* 81: 373–383. doi: 10.1023/A:1020596903142.

Lugtenberg, B., Kamilova, F. 2009. Plant-growth-promoting rhizobacteria. *Annu Rev Microbiol* 63: 541–556.

Mayak, S., Tirosh, T., Glick, B.R. 2004a. Plant growth-promoting bacteria that confer resistance to water stress in tomatoes and peppers. *Plant Sci* 166: 525–530.

Mayak, S., Tirosh, T., Glick, B.R. 2004b. Plant growth-promoting bacteria confer resistance in tomato plants to salt stress. *Plant Physiol Biochem* 42: 565–572.

Mendes, R., Garbeva, P., Raaijmakers, J.M. 2013. The rhizosphere microbiome: Significance of plant beneficial, plant pathogenic, and human pathogenic microorganisms. *FEMS Microbiol Rev* 37: 634–663. doi: 10.1111/1574-6976.12028.

Mendes, R., Kruijt, M., de Bruijn, I., *et al.* 2011. Deciphering the rhizosphere microbiome for disease-suppressive bacteria. *Science* 332: 1097–1100.

Miller, K.I., Qing, C., Sze, D.M., Neilan, B.A. 2012a. Investigation of the biosynthetic potential of endophytes in traditional Chinese anticancer herbs. *PLoS One* 7: e35953. doi: 10.1371/journal.pone.0035953.

Miller, K.I., Qing, C., Sze, D.M., Roufogalis, B.D., Neilan, B.A. 2012b. Culturable endophytes of medicinal plants and the genetic basis for their bioactivity. *Microb Ecol* 64: 431–449. doi: 10.1007/s00248-012-0044-8.

Miransari, M. 2011. Arbuscular mycorrhizal fungi and nitrogen uptake. *Arch Microbiol* 193: 77–81.

Nachtigall, J., Kulik, A., Helaly, S., Bull, A.T., Goodfellow, M., Asenjo, J.A., *et al.* 2011. Atacamycins A-C, 22-membered antitumor macrolactones produced by Streptomyces sp. C38. *J Antibiot (Tokyo)* 264: 775–780. doi: 10.1038/ja.2011.96.

Nelson, E.B. 2004. Microbial dynamics and interactions in the spermosphere. *Annu Rev Phytopathol* 42: 271–309.

Niraula, N.P., Kim, S.H., Sohng, J.K., Kim, E.S. 2010. Biotechnological doxorubicin production: Pathway and regulation engineering of strains for enhanced production. *Appl Microbiol Biotechnol* 87: 1187–1194. doi: 10.1007/s00253-010-2675-3.

Oger, P.M., Mansouri, H., Dessaux, Y. 2000. Effect of crop rotation and soil cover on alteration of the soil microflora generated by the culture of transgenic plants producing opines. *Mol Ecol* 9: 881–890.

Oger, P.M., Mansouri, H., Nesme, X., Dessaux, Y. 2004. Engineering root exudation of lotus toward the production of two novel carbon compounds leads to the selection of distinct microbial populations in the rhizosphere. *Microb Ecol* 47: 96–103.

Oger, P.M., Petit, A., Dessaux, Y. 1997. Genetically engineered plants producing opines alter their biological environment. *Nat Biotechnol* 15: 369–372.

Ouyang, Y., Cheng, Q., Cheng, C., Tang, Z., Huang, Y., Tan, E., Ma, S., Lin, X., Xie, Y., Zhou, H. 2023. Effects of plants-associated microbiota on cultivation and quality of Chinese herbal medicines. *Chinese Herbal Medicines* 2023. doi: 10.1016/j.chmed.2022.12.004.

Palombo, E.A., Semple, S.J. 2001. Antibacterial activity of traditional Australian medicinal plants. *J Ethnopharmacol* 77: 151–157. doi: 10.1016/S0378-8741(01)00290-2.

Paton, A.W., Morona, R., Paton, J.C. 2006. Designer probiotics for prevention of enteric infections. *Nat Rev Microbiol* 4: 193–200.

Picard, C., Baruffa, E., Bosco, M. 2008. Enrichment and diversity of plant-probiotic microorganisms in the rhizosphere of hybrid maize during four growth cycles. *Soil Biol Biochem* 40: 106–115.

Picard, C., Bosco, M. 2008. Genotypic and phenotypic diversity in populations of plant-probiotic Pseudomonas spp. colonizing roots. *Naturwissenschaften* 95: 1–16.

Pontonio, E., Di Cagno, R., Tarraf, W., Filannino, P., De Mastro, G., Gobbetti, M. 2018. Dynamic and assembly of epiphyte and endophyte lactic acid bacteria during the life cycle of Origanum vulgare L. *Front Microbiol* 9: 1372. doi: 10.3389/fmicb.2018.01372.

Presta, L., Bosi, E., Fondi, M., Maida, I., Perrin, E., Miceli, E., Maggini, V., Bogani, P., Firenzuoli, F., Di Pilato, V., *et al.* 2017. Phenotypic and genomic characterization of the antimicrobial producer Rheinheimera sp. EpRS3 isolated from the medicinal plant Echinacea purpurea: Insights into its biotechnological relevance. *Res Microbiol* 168: 293–305.

Qi, X., Wang, E., Xing, M., Zhao, W., Chen, X. 2012. Rhizosphere and non-rhizosphere bacterial community composition of the wild medicinal plant *Rumex patientia*. *World J Microbiol Biotechnol* 28: 2257–2265. doi: 10.1007/s11274-012-1033-2.

Qin, J., Li, R., Raes, J., *et al.* 2010. A human gut microbial gene catalog established by metagenomic sequencing. *Nature* 464: 59–65.

Raaijmakers, J., Mazzola, M. 2012. Diversity and natural functions of antibiotics produced by beneficial and pathogenic soil bacteria. *Annu Rev Phytopathol* 50: 403–424.

Raaijmakers, J.M., Paulitz, T.C., Steinberg, C., Alabouvette, C., Moënne-Loccoz, Y. 2009. The rhizosphere: A playground and battlefield for soilborne pathogens and beneficial microorganisms. *Plant Soil* 321: 341–361.

Raudales, R.E., Stone, E., McSpadden Gardener, B.B. 2009. Seed treatment with 2,4-diacetyl phloroglucinol-producing pseudomonads improves crop health in low-pH soils by altering patterns of nutrient uptake. *Phytopathology* 99: 506–511.

Rengel, Z. 2002. Breeding for better symbiosis. *Plant Soil* 245: 147–162.

Richardson, A.E., Barea, J.M., McNeill, A.M., Prigent-Combaret, C. 2009. Acquisition of phosphorus and nitrogen in the rhizosphere and plant growth promotion by microorganisms. *Plant Soil* 321: 305–339.

Roy, S., Yasmin, S., Ghosh, S., Bhattacharya, S., Banerjee, D. 2016. Anti-infective metabolites of a newly isolated *Bacillus thuringiensis* KL1 associated with Kalmegh (*Andrographis paniculata* Nees.), a traditional medicinal herb. *Microbiol Insights* 9: MBI.S33394.

Rudrappa, T., Czymmek, K.J., Pare, P.W., Bais, H.P. 2008. Root-secreted malic acid recruits beneficial soil bacteria. *Plant Physiol* 148: 1547–1556.

Ryan, P.R., Dessaux, Y., Thomashow, L.S., Weller, D.M. 2009. Rhizosphere engineering and management for sustainable agriculture. *Plant Soil* 321: 363–383.

Ryu, C.M., Farag, M.A., Hu, C.H., Reddy, M.S., Wei, H.X., Paré, P.W., *et al.* 2003. Bacterial volatiles promote growth in Arabidopsis. *Proc Natl Acad Sci USA* 100: 4927–4932. doi: 10.1073/pnas.0730845100.

Sahu, P.K., Singh, S., Gupta, A.R., Gupta, A., Singh, U.B., Manzar, N., Bhowmik, A., Singh, H.V., Saxena, A.K. 2020. Endophytic bacilli from medicinal-aromatic perennial Holy basil (*Ocimum tenuiflorum* L.) modulate plant growth promotion and induced systemic resistance against Rhizoctonia solani in rice (*Oryza sativa* L.) *Biol Control* 150: 104353. doi: 10.1016/j.biocontrol.2020.104353.

Sasaki, T., Yamamoto, Y., Ezaki, B., *et al.* 2004. A wheat gene encoding an aluminum-activated malate transporter. *Plant J* 37: 645–653.

Sauer, S., Dlugosch, L., Kammerer, D.R., Stintzing, F.C., Simon, M. 2021. The microbiome of the medicinal plants *Achillea millefolium* L. and *Hamamelis virginiana* L. *Front Microbiol* 12: 696398. doi: 10.3389/fmicb.2021.696398.

Senn, S., Pangell, K., Bowerman, A.L. 2022. Metagenomic insights into the composition and function of microbes associated with the rootzone of *Datura inoxia*. *BioTech* 11(1): 1. doi: 10.3390/biotech11010001.

Singh, S.K., Pathak, R. 2015. *Ecological manipulations of rhizobacteria for curbing medicinal plant diseases*, Springer: Cham, Switzerland, 217–230.

Smalla, K., Wieland, G., Buchner, A., Zock, A., Parzy, J., Kaiser, S., *et al.* 2001. Bulk and rhizosphere soil bacterial communities studied by denaturing gradient gel electrophoresis: Plant-dependent enrichment and seasonal shifts revealed. *Appl Envrion Microbiol* 67: 4742–4751. doi: 10.1128/AEM.67.10.4742-4751.2001.

Smith, K.P., Goodman, R.M. 1999. Host variation for interactions with beneficial plant-associated microbes. *Annu Rev Phytopathol* 37: 473–491.

Smith, K.P., Handelsman, J., Goodman, R.M. 1999. Genetic basis in plants for interactions with disease-suppressive bacteria. *P Natl Acad Sci USA* 96: 4786–4790.

Strobel, G., Daisy, B. 2003. Bioprospecting for microbial endophytes and their natural products. *Microbiol Mol Biol Rev* 67: 491–502. doi: 10.1128/MMBR.67.4.491-502.2003.

Strobel, G., Daisy, B., Castillo, U., Harper, J. 2004. Natural products from endophytic microorganisms. *J Nat Prod* 67: 257–268. doi: 10.1021/np030397v.

Sun, H., Kang, B., Chai, Z., *et al.* 2017. Characterization of root-associated microbiota in medicinal plants *Astragalus membranaceus* and *Astragalus mongholicus*. *Ann Microbiol* 67: 587–599. doi: 10.1007/s13213-017-1285-z.

Taechowisan, T., Chanaphat, S., Ruensamran, W., Phutdhawong, W.S. 2012. Antifungal activity of 3-methylcarbazoles from *Streptomyces* sp. LJK109, an endophyte in *Alpinia galanga*. *J Appl Pharm Sci* 2: 255820.

Tesfaye, M., Temple, S.J., Allan, D.L., Vance, C.P., Samac, D.A. 2001. Overexpression of malate dehydrogenase in transgenic alfalfa enhances organic acid synthesis and confers tolerance to aluminum. *Plant Physiol* 127: 1836–1844.

Turner, T.R., James, E.K., Poole, P.S. 2013. The plant microbiome. *Genome Biol* 14: 209. doi: 10.1186/gb-2013-14-6-209.

van Loon, L.C., Bakker, P.A., Pieterse, C.M. 1998. Systemic resistance induced by rhizosphere bacteria. *Annu Rev Phytopathol* 36: 453–483. doi: 10.1146/annurev.phyto.36.1.453.

Vorholt, J.A. 2012. Microbial life in the phyllosphere. *Nat Rev Microbiol* 10: 828–840.

Wang, J., Raman, H., Zhou, M., *et al.* 2007. High-resolution mapping of the Alp locus and identification of a candidate gene HvMATE controlling aluminum tolerance in barley (*Hordeum vulgare* L.). *Theor Appl Genet* 115: 265–276.

Wissuwa, M., Mazzola, M., Picard, C. 2009. Novel approaches in plant breeding for rhizosphere-related traits. *Plant Soil* 321: 409–430.

Yang, H., Knapp, J., Koirala, P., *et al.* 2007. Enhanced phosphorus nutrition in monocots and dicots over-expressing a phosphorus-responsive type IH+-pyrophosphatase. *Plant Biotechnol J* 5: 735–745.

Zhang, Q., Liu, H., Sun, G., Wilson, I.W., Wu, J., Hoffman, A., Cheng, J., Qiu, D. 2015. Baseline survey of root-associated microbes of Taxus chinensis (Pilger) Rehd. *PLoS One* 10: e0123026.

Zhao, K., Penttinen, P., Chen, Q., Guan, T., Lindström, K., Ao, X., *et al.* 2012. The rhizospheres of traditional medicinal plants in Panxi, China, host a diverse selection of actinobacteria with antimicrobial properties. *Appl Microbiol Biotechnol* 94: 1321–1335. doi: 10.1007/s00253-011-3862-6.

Zhao, K., Penttinen, P., Guan, T., Xiao, J., Chen, Q., Xu, J., *et al.* 2011. The diversity and anti-microbial activity of endophytic actinomycetes isolated from medicinal plants in Panxi plateau, *China Curr Microbiol* 62: 182–190. doi: 10.1007/s00284-010-9685-3.

Zhao, L. 2010. The tale of our other genome. *Nature* 465: 879–880.

6 Microbiomes
A Road Map of Medicinal Plant Growth, Disease Control, and Abiotic Stress Management

Nadia Riaz, Zubaida Yousaf, Afifa Younas, Sajjad Hyder, Arusa Aftab, Zainab Maqbool, Zainab Shahzadi, and Samina Hanif

6.1 INTRODUCTION: PLANT-ASSOCIATED MICROORGANISMS

From the very beginning, medicinal plants have been used to treat diseases. Food and medicinal plants and their parts such as vegetables, cereals, mushrooms, beverages, flowers, spices, and traditional medicinal herbs are the main sources of exogenous antioxidants (Li *et al.*, 2016), while natural antioxidants are derived from agricultural by-products. Polyphenols and vitamins are plant-derived natural antioxidants (Manach *et al.*, 2004). These natural antioxidants show numerous biological effects, such as antibacterial, antiviral, and anticancer (Arathi *et al.*, 2015; Zhou *et al.*, 2016). Since ancient times, traditional herbal medicines and their derivatives have been the main source of medicine (Ekor, 2014); however, plants and their parts are vital sources of medicinal compounds that help in curing various diseases in one way or another. Pharmaceutical companies worldwide are finding ways to develop novel medicinal resources with the use of plants (Pan *et al.*, 2010). Today, herbal drugs are preferred over allopathic dugs due to their reliability and lower side effects (Humber, 2001). However, some problems with medicinal plants are their diverse nature and climate change. The soil microbiome plays an important role in influencing important traits such as growth, yield, resistance to pathogens, abiotic tolerance, water-holding capacity, and production of regulatory phytohormones (Turner *et al.*, 2013; Köberl *et al.*, 2013).

In nature, most of the microbial community associated with plants is composed of prokaryotes, particularly bacterial domain members. The plants' associated bacteria are responsible for the regulation of plant growth, nutrition, and defence-related activities, as illustrated in Figure 6.1 (Gnanamanickam, 2006). Research has uncovered soil composition of a large bacterial count, ranging between 10^8 and 10^9. Mutualistic relationships in the soil between plants and their related bacterial communities are relatively widespread. These associations include rhizobacteria (bacteria that reside in the soil surrounding the root region and thus are greatly influenced by root exudate production), bacterial epiphytes (living on the root surface), and bacterial endophytes (bacteria that live inside the root) (Vessey, 2003; Berg *et al.*, 2014; Poudel *et al.*, 2021). It is also possible for plant-associated bacteria (PAB) to be found aboveground in the phyllosphere on plant organs, including leaves and floral parts. They can also be found on structures that survived previous pathogen attacks, where they are referred to as hyperparasites, regarding sclerotia (Poudel *et al.*, 2021).

DOI: 10.1201/b23155-6

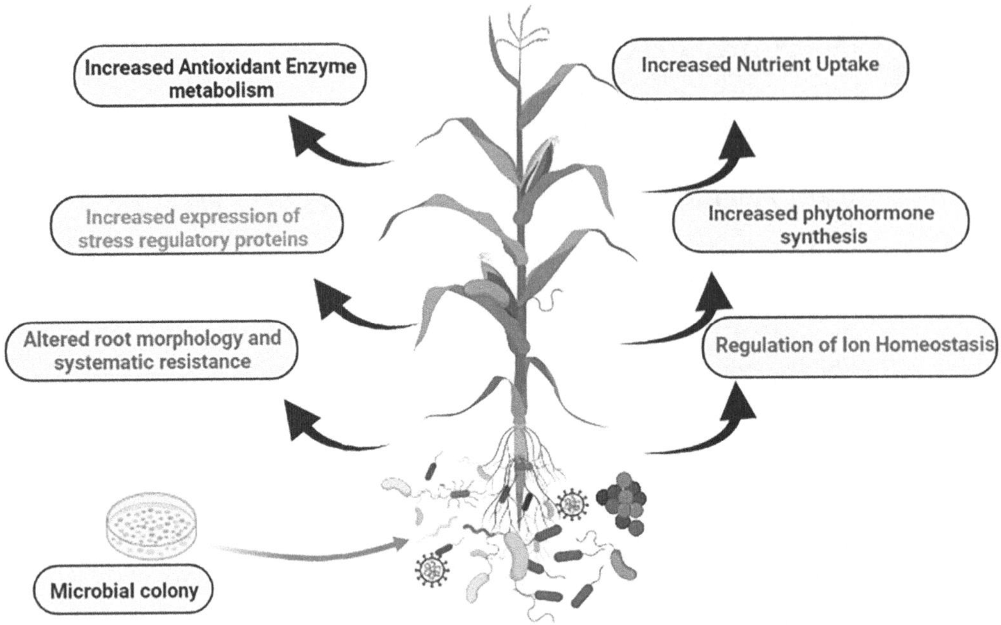

FIGURE 6.1 Illustration of microbes associated with plant protection and defence metabolism.

6.2 BIO-STIMULANTS AND THE GREEN REVOLUTION

Plant-associated bacteria could be utilized as biological control agents (antibiotics, antifungal agents) for targeting plant infections and can also stimulate the synthesis of antimicrobial substances like plant-based hormones and siderophores, altering the plant's immune response to promote normal growth and development of plants. When it comes to preventing yield losses induced by plant stress and climate change, bio-stimulants (BSs) may be one of the most promising options today. A wide range of plant-beneficial substances is included in bio-stimulants, which exclude chemical fertilizers and pesticides (García-García *et al.*, 2020). Du Jardin (2015) described bio-stimulators as any substance or microbe that is intended to promote plant growth, induce a stress response, and increase crop production, irrespective of their nutrient content (Figure 6.1). BSs are key regulators of the plant life process to promote growth and resource utilization efficiency, whether plants are subjected to stress or non-stress circumstances (Du Jardin, 2012).

Bio-stimulants are frequently used in systems to enhance plant physiological processes to protect plants against biotic and abiotic stress (Yakhin *et al.*, 2017). Applying an appropriate amount to crops could stimulate growth, development, and/or responses to stress factors such as water stress, salinity, and toxic elements like Al, Cd, and As (Do Santos *et al.*, 2012). Globally, to date, no attempt has been made to classify these biostimulants. However, a few main categories are generally accepted by scientists, and these categories cover both compounds and microbes (Du Jardin, 2012; Calvo *et al.*, 2014). Microorganisms include useful bacteria, such as PGPR, and beneficial fungi, among other types of organisms. They may be free-living, rhizospheric, or endophytic PGPB, depending on their environment.

The green revolution has adopted initiatives to increase the yield and quality of crops. These initiatives are (1) expansion of cultivated land, (2) double-cropping annual systems, (3) selection of high-yielding germplasm, (4) systematic use of fertilizers and pesticides, (5) availability of expanded irrigation amenities, and (6) advanced crop protection strategies (Brainerd and Menon, 2014; John and Babu, 2021). Hence, a 'fresh' green revolution, also known as the bio-revolution, is needed to generate greater output than input to maintain a positive impact on the environment.

Therefore, biological inputs such as phytomicrobiome could be an alternative beneficial source for the improvement of crop yield rather than the use of chemicals causing environmental pollution (Timmusk *et al.*, 2017).

6.3 MICROBIAL ANTAGONISM

Biological control of plant diseases depends on the categories of parasitism (which may be accomplished by the synthesis of volatile or nonvolatile antibiotics), competition (for nutrients and space), and hyper-parasitism (Singh and Faull, 2020). These are a few of the forms of antagonism that can be identified using a dual culture system on agar plates of the target and the possible antagonist. Biological control of soil-borne phytopathogens is provided by a large variety of saprophytic bacteria that compete for nutrients and space in the rhizosphere. Microbial competition has frequently been considered to occur outside of the plants, in the phylloplane, rhizosphere, or bulk soil. Pathogens enter and persist in the phyllosphere in opposition to one another due to competition for resources and available space, antibiotic production, and hyperparasitism, among other factors.

Rapid nutrient uptake by epiphytic microorganisms causes a quick decline in the amount of nutrients available to the pathogens. Basu *et al.* (2021) examined a significant group of helpful, root-colonizing bacteria called plant growth-promoting rhizobacteria (PGPRs) that is active in both the bulk soil and rhizosphere of the plants. By facilitating tolerance to biotic and abiotic stress, they encourage plant growth and assist in the nourishment of host plants. PGPRs are viewed as an environmentally beneficial substitute for harmful chemical fertilizers because of their active growth-promoting actions in a variety of plant species, including garlic (Zhuang *et al.*, 2021), ginseng (Sun *et al.*, 2018), and milk thistle (Zahra *et al.*, 2021).

The application of PGPR is an environmentally sound technique for increasing agricultural yields. These processes include suppressing phytopathogens, solubilizing minerals for easy uptake by plants, and modifying hormonal and nutritional homeostasis (Nazir *et al.*, 2018). PGPRs promote plant growth directly by hormone production or indirectly, such as by controlling well-known phytopathogens (Figure 6.2). Because of their enormous potential to suppress plant diseases, PGPRs, particularly strains of genera like *Bacillus*, *Pseudomonas*, and *Burkholderia*, have drawn a lot of interest from scientific communities working in agriculture. Moreover, several PGPRs are sold as bioinoculants to enhance plant growth and suppress the pathogens of important crops (Gouda *et al.*, 2018).

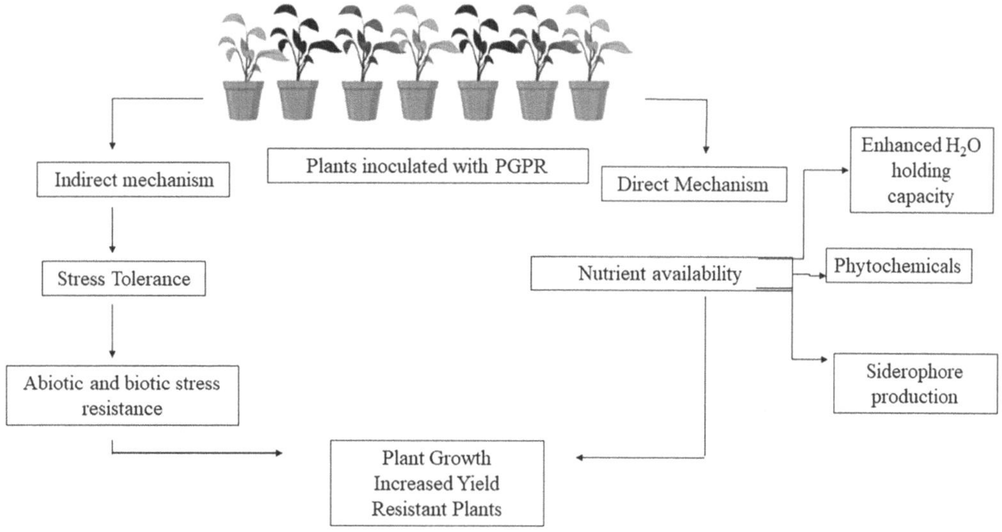

FIGURE 6.2 Direct and indirect mechanism of action of plant growth promoting rhizobacteria (PGPRs).

6.3.1 Direct Antagonism

In the direct mode of action, PGPRs improve plant growth by fixing atmospheric nitrogen; solubilizing essential minerals from the soil; and regulating important plant hormones such as auxins, cytokinins, abscisic acid, gibberellins, and ethylene (Kalam *et al.*, 2020). Strategies like seed bacterization and soil inoculation either alone or in consortium form ensure nutrient availability and their cycling. They facilitate seedling growth, seed germination, and crop yield improvement. Several *Pseudomonas* spp. have displayed remarkable improvement in seed germination and sapling growth in medicinal plants (Deshwal *et al.*, 2011; Santoro *et al.*, 2015).

A successful PGPR-based formulation should possess attributes like growth-enhancing ability, safety toward the ecosystem, compatibility with other partnering organisms, rhizosphere competence, extensive competitive saprophytic ability, broad-spectrum action, and ease of mass production. *Arthrobacter, Burkholderia, Azospirillum, Azotobacter, Pseudomonas, Klebsiella, Alcaligenes, Enterobacter, Bacillus*, and *Serratia* are a few of the bacteria species that improve plant growth (Mohanty *et al.*, 2021).

6.3.2 Indirect Antagonism

PGPRs indirectly improve plant health by controlling deleterious pathogens by employing various mechanisms such as parasitism, competing for nutrients and space within the rhizosphere, production of antagonistic substances (hydrogen cyanide, siderophores, antibiotics, and certain metabolites) and lytic enzymes (chitinases, glucanases, and proteases), and inducing systemic resistance in plants against invading microbes (Meena *et al.*, 2020; Sayyed *et al.*, 2019). The production of lytic enzymes, antibiotics, and bacteriocins aids in destroying the pathogenic cells, while the production of siderophores binds the iron, making it unavailable to phytopathogens, which ultimately excludes the phytopathogen from the niche. Bacterial genera such as *Bacillus*, *Burkholderia*, *Enterobacter, Pseudomonas*, *Staphylococcus*, and *Stenotrophomonas* are well explored for their antagonistic potential. In a previous study, a combination of three bacterial agents (*B. cereus*, *B. subtilis*, and *Serratia* spp.) was reported to control the phytophthora blight disease in sweet peppers and improve the quality of the fruits (Zhang *et al.*, 2019). Similarly, *B. subtilis*, *P. chlororaphis*, and *P. fluorescens* suppressed the mycelial growth of *Corynespora casiicola*, associated with stem blight disease in *Phyllanthus amarus* (Mathiyazhagan *et al.*, 2004). These bacteria have been well described for a broad range of antagonistic activities to control phytopathogens (Tariq *et al.*, 2017).

6.4 ROLE OF MICROBES IN PLANT HEALTH

Microorganisms are ubiquitous and present in every living entity on the globe (Kirchman, 2018). Microbes are found in all areas of a plant, even the phylloplane areas, phyllosphere, or rhizosphere. The rhizosphere is a source of different biological and chemical activity. Plant exudates are released in this region, and microbes use them for energy (Prashar *et al.*, 2014). The microbial species residing in the plant rhizosphere are the most important since they directly affect plant development, health, and nutrition. In the last ten years, the scientific community has paid great attention to research on rhizosphere-based microbial communities due to their significant contribution to eco-friendly sustainable agriculture.

Of all the rhizospheric bacterial communities, PGPRs are efficient root colonizing agents that improve plant growth and development by increasing nutrient acquisition directly or indirectly (Prasad *et al.*, 2019). PGPRs indirectly improve plant growth by producing secondary metabolites, antibiotics, cyanides, hormone-like compounds, IAA, and exopolysaccharides; performing siderophore synthesis; and controlling the soil-borne pathogens (Chandra *et al.*, 2020). Various PGPR strains have been reported to improve plant health by improving root and shoot length, leaf width, fresh and dry root and shoot weight, chlorophyll contents, and plant biomass and increase significant gross productivity by solubilizing soluble nutrients from the rhizosphere.

Bacteria belonging to *Arthrobacter, Azoarcus, Arthrobacter, Acetobacter, Azospirillum, Alcaligenes, Acinetobacter, Azotobacter, Beijerinckia, Burkholderia, Derxia, Gluconacetobacter, Klebsiella, Ochrobactrum, Bacillus, Enterobacter, Pantoae, Pseudomonas, Rhodococcus, Serratia, Herbaspirillum, Stenotrophomonas, Lactobacillus, Paenobacillus*, and *Zoogloea* genera are well known to fix atmospheric nitrogen, carbon, and potassium and provide phosphate solubilization (Vega-Celedón *et al.*, 2021).

Potential PGPRs are extensively used in agriculture, notably in the biogeochemical mineral cycle and biofertilizers (Shah *et al.*, 2021). In addition to this, plant-associated microbes, specifically PGPRs, improve soil health and fertility (Kumar *et al.*, 2019). Plants help PGPRs expand by producing a variety of storage materials and root exudates, which PGPRs use for their sustainable survival (Kausar *et al.*, 2018). PGPRs develop endophytic or exophytic symbiotic relationships with plants, such as mycorrhizal nitrogen-fixing root nodulation (He *et al.*, 2018), and also improve plant growth and health (Neshat *et al.*, 2022).

PGPRs efficiently contribute to enhancing plant growth in various crops when grown under various stress conditions. For instance, a study reported the application of *Paenibacillus lentimorbus* to improve nutrient deficiency-induced stress in chickpea plants. The results showed that bacterial strain B-30488 with many PGP traits helped in improving the growth and development of chickpea plants by enhancing nutrient uptake ability in nutrient-deficit soil (Bisht *et al.*, 2019). Similarly, Kuzmicheva *et al.* (2017) concluded that the utilization of *Pseudomonas oryzihabitans* on the three soybean cultivars, Nice-Mecha, Bara, and Svapa, showed a significant improvement in plant growth and increased root biomass. Recently, Thanuja *et al.* (2020) conducted an interactive study on carrom seeds and examined the effects of organic manure along with PGPRs (*Azospirillum lipoferium, Bacillus megathurium*, and *Frateuria aurantia*) and micronutrients on yield and quality of carrom seeds. The results showed that growth, cereal production, quality elements of carrom, and high yield were significantly improved. It is evident from previous studies that PGPRs not only trigger plant growth by producing PGP traits but also improve plant health.

6.5 ROLE OF MICROBES IN PLANT DISEASE CONTROL

Global agriculture productivity and associated food systems are significantly declining due to the invasion of various pests and pathogens (Savary *et al.*, 2019). Phytopathogenic diseases account for at least 30–40% of yield loss, costing $200 billion per year globally (Sawicka and Egbuna, 2020). To combat plant diseases, synthetic pesticides and antibiotics have been utilized in a non-discriminative manner. Unfortunately, continuous applications of these substances have resulted in several issues, including environmental contamination, soil salinization, and resistance induction in phytopathogens (Rodríguez *et al.*, 2020). Also, ever-increasing pesticide residues in the soil and water are harmful to humans and the environment (Lamichhane *et al.*, 2017). Therefore, a proximate, feasible long-term technique to combat phytopathogens is to switch from chemical control to biological control. The application of microbial antagonists (biocontrol agents) to control weeds, insects, and plant diseases is known as biocontrol (Ali *et al.*, 2020).

Due to their accessibility, affordability, and easy mode of application, PGPRs have received major consideration. These bacteria have been reported to have a tremendous beneficial impact on agricultural production whether employed alone or in consortia form (Dinesh *et al.*, 2015). The possibility of using PGPRs as biocontrol agents has increased because of their antagonistic action against a vast variety of phytopathogens (Lamsal *et al.*, 2013). These biocontrol agents (BCAs) can function in a variety of ways: directly or indirectly inducing plant defence mechanisms (Gupta *et al.*, 2015). Many BCAs can produce plant secondary metabolites such as antibiotics, cell wall-degrading enzymes, volatile organic compounds (VOC), and secondary metabolites on or around the surface of the host plant (Ali *et al.*, 2020). Many of these substances are just as efficient as chemical insecticides. Particularly, well-reported species of *Pseudomonas*, *Trichoderma*, and *Bacillus* produce secondary metabolites that are important in controlling plant diseases (Jayaprakashvel and Mathivanan, 2011).

Siderophore production plays a vital role in suppressing the phytopathogens. Siderophore production by PGPRs restricts the growth of phytopathogens by depriving them of the available iron. As a result of this, iron deficiency prevents pathogen growth by obstructing vital processes including sporulation and nucleic acid formation (Sulochana *et al.*, 2014). Many researchers isolated siderophore-producing bacteria from the rhizosphere that belong to the various genera, including *Pseudomonas, Bradyrhizobium, Rhizobium, Streptomyces,* and *Serratia* (Kuffner *et al.*, 2008). *Pseudomonas* strains defend plants from pathogens by inducing systemic resistance in plants. After being exposed to biotic stimuli, plants can develop increased levels of resistance to diseases, which is conferred by numerous distinct PGPRs (Singh e*t al.*, 2016). Systemic resistance, also known as induced systematic resistance and systemic acquired resistance, is involved in this process. Similarly, elicitors are the chemicals released by microorganisms that activate these types of systemic resistance. Elicitors typically include N-acyl homoserine, a signal molecule, cell wall polysaccharides, cyclic lipopeptides, salicylic acid, siderophores, and antibiotics (Pérez-Montaño *et al.*, 2014).

In a previous study, three antagonistic bacterial strains (*P. fluorescens*, *B. subtilis*, and *B. pumilus*) were effectively used to suppress the white rot disease in onion and garlic (Elshahawy *et al.*, 2018). Similarly, in another study, PGPRs belonging to *Pseudomonas* spp. with siderophore and phosphate solubilization ability were found effective in controlling the root rot disease of *N. sativa* caused by *Fusarium solani* (Al-Sman *et al.*, 2019). Many other studies have explored the potential use of PGPRs in suppressing various pathogenic diseases in medicinally important plants such as chamomile (Schmidt *et al.*, 2014), ginseng (Sun *et al.*, 2019), cannabis (Balthazar *et al.*, 2022), valerian (Ghods-Alavi *et al.*, 2012), and *Mentha arvensis* (Pandey *et al.*, 2011). Based on the multifaceted impacts, PGPRs could be employed in developing effective integrated disease management strategies to manage plant diseases, crop health, and yield improvement.

6.6 ROLE OF MICROBES IN ABIOTIC STRESS MANAGEMENT

Stress is an adverse circumstance or substance that affects a plant's metabolism and subsequently its process of growth and development. The two types of environmental stress that can exist are abiotic stress and biotic stress. Abiotic stress on plants reduces crop yield and quality, which causes agriculture to suffer substantial financial losses (Bhandari *et al.*, 2023).

In many agricultural regions across the world, natural conditions, combinations of two or more stresses, such as salinity and heat or drought and salt, and combinations of drought with extremely high temperatures or light levels, are widespread and could affect crop productivity (Suzuki *et al.*, 2014). By 2050, the world's population is predicted to reach 9.7 billion, resulting in a 70% increase in the amount of food needed to feed everyone while simultaneously depleting natural resources (Masood *et al.*, 2004; Cole *et al.*, 2018). Added external constraints brought on by climate change that have a direct impact on agricultural productivity have seriously affected food security. It affects the majority of the world's poor people as well as those whose livelihoods depend on agriculture on a global scale. Crop plants are currently finding it more difficult to adapt to changing climatic conditions via their basic biological mechanisms, necessitating the search for alternative methods. Consequently, crop enhancement research has benefited greatly from the development of microbial technology (Lipper *et al.*, 2014; Meena *et al.*, 2017).

Abiotic stress is the type of stress that specific environmental variables cause in plants. It consists of the nonliving components of the ecosystem that have an impact on the living components. Temperature, water, salts, vital nutrients, and pH are the primary abiotic elements that affect plant growth in an agro-ecosystem. All of these elements are interconnected, and many of them are affected by climate change. For instance, more rainfall causes flooding, but less or no rainfall causes drought, which has a detrimental effect on crops (Enebe and Babalola, 2018).

Abiotic stresses can be handled by naturally occurring microorganisms from a variety of habitats recognizing their immense metabolic capabilities. Microbial interactions with plants in the natural ecosystem can control local and systemic responses in their host to constantly changing environmental conditions (Meena *et al.*, 2017). Agriculturally significant crops have successfully used stress-tolerant wild plant isolates of plant growth-promoting bacteria as inoculants (Coleman-Derr and Tringe, 2014).

Abiotic stressors alter the growth- and stress-related phytohormones that plants produce, impairing their ability to function normally. Growth-promoting microbes employ numerous means and methods for plant development, growth enhancement, and tolerance of abiotic stresses (Table 6.1). In stressful situations, plant growth-promoting bacteria directly affect plant growth and development by producing phytohormones such as auxins (indole-3-acetic acid [IAA]), gibberellins, ethylene, and cytokinins (Fahad *et al.*, 2015).

TABLE 6.1
Tolerance to Abiotic Stress in Medicinal Plants Mediated by Microbes

Plant Species	Microbes	Type of Abiotic Stress	References
Panicum turgidum	*Rhizophagus intraradices, Funneliformis mosseae, Claroideoglomus etunicatum*	Salinity	Hashem *et al.* (2015)
Pisum Sativum L.	*Funneliformis mosseae, Rhizoglomus intraradices, Gigaspora* sp. and *Rhizoglomus fasciculatum*		Parihar *et al.* (2020)
Cucumis sativus cv. Dasher II	Arbuscular mycorrhizal fungi (AMF)		Hashem *et al.* (2018)
Zea mays L.	Kocuria rhizophila		Li *et al.* (2020)
Helianthus annuus L.	*Rhizophagus intraradices, Funneliformis mosseae, Claroideoglomus etunicatum*	Heavy metal: cadmium (Cd)	Abd_allah *et al.* (2015)
Robinia pseudoacacia L.	*Rhizophagus intraradices, Funneliformis mosseae*	Heavy metal: lead (Pb)	Yang *et al.* (2015)
Poncirus trifoliata	*Glomus versiforme*	Iron (Fe)	Li *et al.* (2015)
Glycine max L.	*Paecilomyces formosus*	Heavy metals (Ni, Cd, and Al), drought, high-temperature	Bilal *et al.* (2020)
Nicotiana tabacum L. cv Petit Havanna	*Rhizophagus intraradices*	Phosphorus (P) deficiency	Del-Saz *et al.* (2017)
Sorghum bicolor L.	*Glomus intraradices, Glomus mosseae, Glomus aggregatum, Glomus etunicatum*	Iron deficiency	Prity *et al.* (2020)
Solanum melongena L	*Funneliformis mosseae, Rhizophagus irregularis, Claroideoglomus etunicatum,* and *Diversispora versiformis*	Low temperature	Pasbani *et al.* (2020)
Solanum lycopersicum L. cv. Zongza 9	*Funneliformis mosseae*		Liu *et al.* (2015)
Cucumis sativus L. cv. Zhongnong No. 26	*Rhizophagus irregularis*		Ma *et al.* (2016)
Colobanthus quitensis	Fungal endophytes	UV-B radiation	Barrera *et al.* (2020)
Solanum lycopersicum cv. ACE 55	*Fusarium solani*	Water	Kavroulakis *et al.* (2018)
Zea mays L.	*Glomus versiforme*	Drought	Begum *et al.* (2019)
Triticum aestivum and *Zea mays*	*Bacillus* sp. *and Enterobacter* sp.		Jochum *et al.* (2020)
Cicer arietinum L.	*Bacillus subtilis, Bacillus thuringiensis,* and *Bacillus megaterium*		Khan *et al.* (2019)
Triticum aestivum	*Rhizophagus intraradices, Funneliformis mosseae, Funneliformis geosporum*		Mathur *et al.* (2018)

The two climatic conditions that most likely have the greatest global impact on crop growth and production are probably high temperatures and droughts. These stresses result in a variety of physiological changes that have an impact on crop output and quality. By creating phytohormones and facilitating plant growth in contaminated soils, microbial endophytes can act as growth-promoting agents for plants. Endophytes use a variety of methods to control plant development under challenging circumstances such as salt, drought, heat, heavy metal stress, and nutritional stress (Eid *et al.*, 2019).

6.6.1 Salinity Stress

More than 20% of agricultural soil is threatened by the salinity problem One of the most significant environmental factors limiting agricultural productivity globally is salinity stress Zhu, 2000). By a variety of processes, salt stress can cause morphological, physiological, biochemical, and metabolic changes that ultimately affect plant growth, development, and productivity (Ma *et al.*, 2020). Salinity significantly lowers crop productivity and contributes to the abandonment of agricultural fields and aquifers (Dodd and Perez-Alfocea, 2012).

Interactions between plants and microbes are regarded as a crucial factor in determining ecological dynamics. Introducing salt-tolerant microorganisms that promote crop growth is an alternative method of increasing crop salt tolerance. Many crops grown in a variety of root-zone salinities have benefited from the growth-promoting effects of soil microbes (Rodriguez and Redman, 2008). It has been discovered that PGPBs, which include both endophytes and rhizobacteria, are efficient at reducing salinity stress (Shinwari *et al.*, 2019). Endophytic microorganisms can modify metabolic and phytohormone signalling, as well as improving growth characteristics. Because they are comparatively insulated from the harsh soil environment under high salt, drought, or other stress conditions, endophytes constitute a special issue for increased crop adaptation to stress (Sturz *et al.*, 2000).

Growth, seed production, and quality improvement of *Camelina sativa* can be facilitated on marginal land that is unsuitable for cultivation due to high salinity by using plant growth–promoting bacteria that produce ACC deaminase enzyme or transgenic plants that express the corresponding acdS gene (Heydarian *et al.*, 2016). A predictive model of the stress tolerance mechanism in carrot plants is shown in Figure 6.3.

6.6.2 Drought Stress

The production and viability of plants are profoundly affected by drought. Plant metabolism and physiology processes are lowered. It inhibits nutrient availability and transport under conditions of water scarcity and slows down plant growth, development, nodulation, yield, and chlorophyll content (Enebe and Babalola, 2018; Vurukonda *et al.*, 2016). Also, it causes an imbalance between the rate of electron transport and the activity of lowering power for metabolic consumption, which increases reactive oxygen species (ROS) and oxidative stress (Beck *et al.*, 2007; Kasim *et al.*, 2013).

Moreover, ROS cause modifications to membrane function and structure, lipid peroxidation, and enzyme stability (Tiwari *et al.*, 2016). By controlling the concentration of antioxidant enzymes, PGPBs contribute to a plant's ability to cope with abiotic stresses by moderating its antioxidant machinery (Ghosh *et al.*, 2018). Rhizobacteria that promote plant development reduce the effects of drought by altering the biochemistry and physiology of plants in a process known as rhizobacterial-induced drought endurance and resilience (RIDER).

Exopolysaccharides (EPS), matching organic solutes, such as proteins, polysaccharides, and polyamines, are secreted during this process, and volatile organic compounds, heat-shock proteins, and dehydrins may also be produced (Kaushal and Wani, 2016). By sustaining plant development, membrane integrity, and enzyme stability as well as successfully controlling the water potential and nutrient uptake through increased root surface area, these systems assist plants in surviving drought stress (Kumar and Verma, 2018; Vacheron *et al.*, 2013).

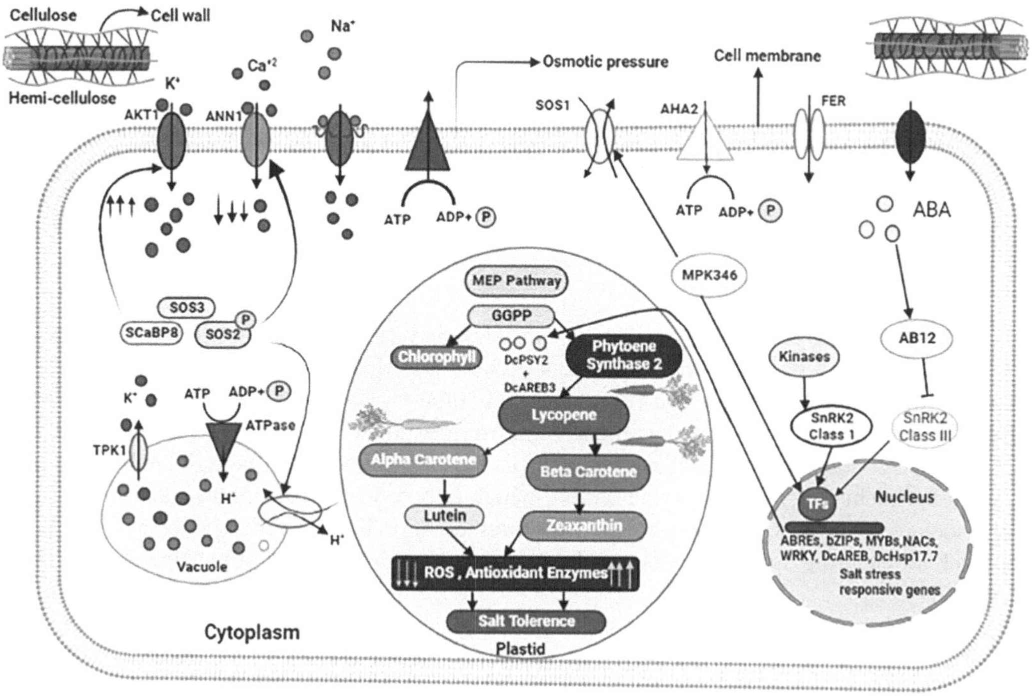

FIGURE 6.3 Predictive model of salt stress tolerance mechanism in *Daucus carota* L. The SOS signalling pathway composed of SOS3 and SOS2 Ca^{2+} binding proteins (SCaBP8), upon sensing salt stress via Ca^{2+} signalling channels in cytoplasm, starts to regulate ion homeostasis by extruding Na^+ ions in vacuole by using ATPases. SCaBP8 arbitrates the influx of K^+ ions into the cytosol under salt stress. Salt stress induces ABA accumulation and activates SnRk2 Class III protein kinases. ABA boosts the expression of the *DcPSY2* gene by binding to the DcAREB3 transcription factor to ABA-responsive elements (ABREs). *DcPSY2* is the gene responsible for the biosynthesis of phytoene synthase 2 required for carotenoid production. Under higher levels of salinity (150 mM NaCl), phytoene enhances the production of antioxidant pigments lycopene, β carotene, α carotene, and ζ carotene, which all help in elevating ROS and increase salt tolerance.

6.6.3 Extreme Temperature Stress

The growth of plants is negatively impacted by extreme temperatures, and high temperatures cause severe damage to cellular proteins that are widely denaturized and aggregated, which results in cell death. In contrast, low temperatures affect metabolism by inhibiting enzyme activities, interacting macromolecules, altering protein structure, and modifying the characteristics of membranes (Janmohammadi *et al.*, 2015).

The severity of heat and cold stress has increased due to the ever-changing environmental circumstances. The continuously evolving environmental conditions have made heat and cold stress more severe. Microbes that have evolved to live in a cold or hot environment are better able to counteract the negative effects of temperature stress. Microbes have sophisticated enzymatic machinery that controls their metabolism to adjust to changing temperatures and enables them to preserve the integrity of their membranes and the stability of their enzymes. Under these conditions, heat- and cold-shock proteins are overexpressed. Défense against heat stress is provided by molecular chaperons (Alam *et al.*, 2017; Kumar and Verma, 2018).

Trehalose, a molecule found in bacteria, can prevent protein denaturation under high temperatures by creating a gel-like web that shields plants from the dehydration brought on by heat stress. Moreover, it contributes to salinity and drought stress (Shameer and Prasad, 2018).

Cold-adapted microorganisms in high-altitude agroecosystems have enormous potential to assist plants in surviving difficult climatic conditions. Psychrophilic and psychrotolerant bacteria like *Arthrobacter, Pseudomonas, Brevundimonas, Citricoccus, Jeotgalicoccus, Methylobacterium, Bacillus, Exiguobacterium, Sanguibacter, Sporosarcina, Staphylococcus, Providencia, Bosea, Psychrobacter, Burkholderia, Janthinobacterium, Aeromicrobium, Pantoea, Aeromonas, Plantibacter, Sphingobacterium, Variovorax, Rhodococcus, Janibacter*, and *Kocuria* have characteristics that induced plant growth (Yadav *et al.*, 2015).

6.6.4 Heavy Metal Pollution Stress

Heavy metal deposition in soil and water is caused by industrialization, excessive pesticide and fertilizer usage, and poor waste management techniques. Heavy metals can build in soils for hundreds of years due to their non-recyclable and stubborn nature. They penetrate the bodies of plants and animals, causing detrimental environmental health effects (Seneviratne *et al.*, 2017). These contaminants create serious risks to the environment and human health. A viable remediation technology is phytoremediation, which is based on the combined action of plants and associated microbes. Microbe-assisted phytoremediation has been recognized as an efficient approach for cleaning up heavy metal–polluted soils and organic pollutant biodegradation (Feng *et al.*, 2017).

Although bacteria can bind heavy metals to their negatively charged functional groups all over their cell wall, which allows for plenty of interaction with positive ions, particularly metal ions, and binds them, this phenomenon is known as metal biosorption. Heavy metals are toxic to plants, but bacteria can neutralize the metal toxicity (Syed and Chinthala, 2015). When applied to contaminated soils, microbial species with high metal resistance have demonstrated the ability to immobilize hazardous metals, lowering their toxicity to plants or crops. Bacillus and Pseudomonas species of microbes are excellent biosorbents for a variety of heavy metals (Wani *et al.*, 2008).

Enzymatic detoxification, metal mobilization, bioaccumulation, immobilization, volatilization, and EPS complexation are all methods used by PGPBs to combat heavy metal stress, in addition to phytohormone synthesis, phosphate solubilization, siderophore, ACC deaminase, and nitrogen fixation (Glick, 2010; Ma *et al.*, 2016). Metal solubility and availability in soil can be influenced by microbes. Metal contaminants cannot be decomposed and must thus be removed or stabilized in the soil. In stress conditions, the mechanism of nutrient mobilization and uptake via metal-chelating siderophores and enzymes involved in phosphate solubilization also facilitates heavy metal uptake (Ullah *et al.*, 2015).

Chelating chemicals produced by growth-promoting microorganisms, such as siderophores, may lower soil pH and improve metal solubility through complex formation. Organic acids generated by these bacteria, such as citric acid, gluconic acid, and oxalic acid, can boost metal mobility to ensure its uptake and accumulation in plant shoots, a process known as phytoextraction (Abou-Shanab *et al.*, 2019). Plant development and growth-promoting bacteria may minimize metal availability in a process known as phytostabilization, which is especially essential in severely metal-contaminated soils. Changes in metal speciation, metals adhering to their cell walls, or exclusion by precipitation are all involved. A method known as phytomanagement that combines several phytotechnologies with a cost-effective and sustainable application might greatly aid the rehabilitation of metal-contaminated fields (Burges *et al.*, 2018; Kong and Glick, 2017).

In addition to sodium chloride stress, there are salts such as sodium carbonate and sodium hydrogen carbonate that are harmful to crops at high concentrations. Nutrient deficiencies and osmotic imbalance are brought on by high pH in alkaline soils because it decreases the bioavailability of vital macro- and micronutrients including phosphorus, manganese, zinc, copper, and iron (Chen *et al.*, 2011).

6.7 POTENTIAL APPLICATION OF MICROORGANISM

In medicinal plants, microbes are a major biota in soil and play a significant role in the renewal of ecosystems. The rhizosphere is a highly active zone mainly characterized by the interaction of plant root systems and soil microbes (Figure 6.4). These microbes are bacteria, fungi, and actinomycetes, contained around the plant root system. Microbes play a significant role in the soil environment at different levels, such as physical, chemical, and biological levels (Shao *et al.*, 2021). The rhizosphere microbiology is directly and indirectly involved in plant survival, growth, development, and metabolism. There are many applications of microbes: they act as bio-fertilizers and decomposers, in plant–microbe interactions, and in regulation of soil minerals by microbes (Wang, 2022).

Plant root and microbe interactions are ubiquitous and completely symbiotic relationships without any harm, such as mycorrhizal interactions. Arbuscular mycorrhizal fungi (AMF) were reported for P solubilization and mobilization. AMF inhabit terrestrial plant roots and increase the surface area of roots for absorption of water and minerals. Active solubilization and mobilization have been demonstrated in the field and greenhouse with various plants, such as cereals, legumes, and medicinal plants (Turrini *et al.*, 2018). *Funneliformis mosseae, Glomus etunicatum,* and *Rhizophagus irregularis* are examples of AMF inoculants (Basiru *et al.*, 2020).

Organic farming is a basic need, increasing pollutant-free medicinal and cash crop production. Microorganisms act as biofertilizers which increase soil fertility, nutrients, and quality of soil. Inorganic fertilizers form cation complexes in different ecosystems and become inaccessible for plant use. Different regions of the globe, such as east and west Africa, expressed NPK deficiency (Bationo *et al.*, 2012). Therefore, the existence of microbiomes is significant and can help to mobilize and solubility of organic P and K, leading to the K and P geochemical cycle. They can mineralize micronutrients through different hydrolytic chain reactions and complex enzymatic reactions. The inorganic compounds hydrolysed into organic acids, such as gluconic, lactic, and oxalic acids, maintain the cationic and anionic balance in soil (Sharma *et al.*, 2013). Phosphorus- and

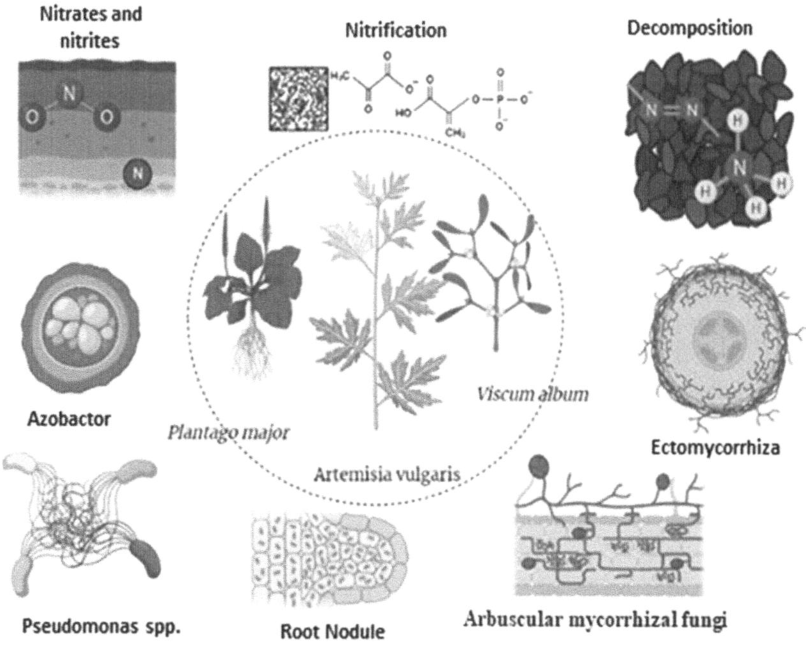

FIGURE 6.4 Microbe-mediated growth of medicinal plants.

potassium-solubilizing microbes include the bacteria *Burkholderia* spp., *Pseudomonas* spp., and *Enterobacter* spp. and fungi *Trichoderma* spp., *Aspergillus* spp., and *Penicillium* spp. (Selvi *et al.*, 2017).

The bacterial endophytes are abundant in nature and symbiotically inhabit any internal part of the medicinal plants such as root, stem, leaves, and seed. Bacterial endophytes play a major role in increasing plant growth and development, improving accessibility of major mineral nutrients that can produce phytohormones, such as indole-3-acetic acid. These interactions also protect plants from pathogenic attack by producing enzymes like cellulases and chitinases, which destroy fungal cell walls or hyphae (Borah *et al.*, 2020). Bacterial endophytes improve plant biomass and root and shoot lengths (Vijayabharathi *et al.*, 2018).

Beneficial microbes can also work in bioremediation in contaminated soil. The main function of soil biota is bioremediation of polluted or contaminated soil due to microbial metabolic activities degrading the poisonous chemicals. Moreover, microbes have a great ability to clean up pollutants, stimulate plant growth, and increase the bioavailability of minerals. The microbiome improves the soil fertility or conditions before the cultivation of medicinal plants (Smułek *et al.*, 2020). Soil microorganisms act as decomposers and regulate soil fertility in different ways. Microbes can trigger and promote the availability of soil nutrients; nitrogen-fixing bacteria fix N and transform the N compounds, and extracellular secretions stabilize the soil particles and increase organic matter by decomposing plant and animal remains (Liang *et al.*, 2019).

6.8 MICROORGANISMS AS BIOFERTILIZERS

The microbiome is an astonishing environment to increase soil fertility and ultimately improve food production without affecting the environment. Plant growth-promoting microbes (PGPMs) and soil fertility–improving and nitrogen-fixing (e.g., nitrifying and denitrifying) bacteria exist in the rhizosphere, penetrate plant roots, and improve their development. Their positive effect is attained through the mobilization of micro-nutrients, nitrogen fixation, and the production of endogenous plant hormones. Microorganisms as biofertilizers are a recent approach in microbial functional diversity, colonizing capabilities, and mode of action application. Biofertilizers can improve phytohormones' protection from pathogens and recovery from environmental stress. Plant growth–promoting microbe formulations are called biofertilizers. These fertilizers can be used instead of chemical fertilizers, increasing plant growth and development. They are ecofriendly and more reliable for sustainable plant growth.

A current more critical issue is the poor quality of biofertilizer products, which results in a lack of commercialization. For improved production, the desired quality and sustainability should be achieved. Several experiments have been done on biofertilizers. For the development of biofertilizers, it is necessary to improve the knowledge about the plant and microbe physiology of soil texture. The most beneficial indication for the production of biofertilizers is mass multiplication (Babalola and Glick, 2012). Soil microbes improve plant nutrition as well. They can change insoluble nutrients through biological processes. Other than chemical fertilizers, microbial fertilizers, solubilization, and mineralization are the best ways to increase plant availability of micronutrients. In a normal environment, various microbes in the rhizosphere and soil are operative for phosphorus release (Bhattacharyya and Jha, 2012). The main focus of consumers for medicinal plants and cash crops is the health, quality, quantity, and nutritional value of products. So, biofertilizers are a good choice that can improve soil fertility and plant production without affecting health and the environment. Scientists must transfer knowledge to the local farmers and continue to learn about biofertilizers. The maximum benefit can be achieved by biofertilizers by using an appropriate inoculum (which contains living microbes or colonies of microorganisms). The highest efficacy of inoculant includes selection of suitable microbes for micronutrients, culturing and isolation of effective or active microbes, characterization of selected microbes, optimum growth conditions, field study, production, and quality control check. Each step needs attention and precautionary measures so

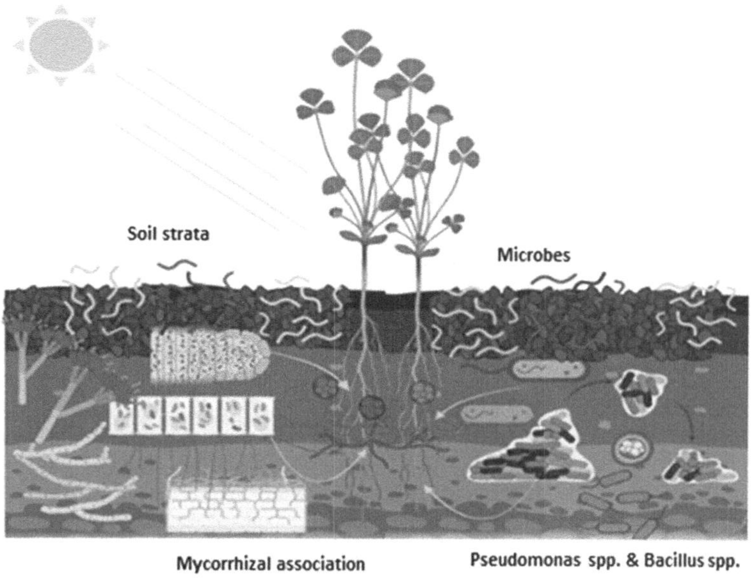

FIGURE 6.5 Microbes as biofertilizers in soil substratum.

that the process results of biofertilizer achieve the desired quality (Vassilev *et al.*, 2015). For further research, they pay attention to the molecular mechanism of microbes and their impact on plants, plant nutrition, and soil fertility management (Figure 6.5).

6.9 CONCLUSIONS AND FUTURE PERSPECTIVES

Medicinal plants are valuable resources that are widely used because of their strong antiviral, anti-inflammatory, antioxidant, and other effects for the treatment and prevention of diseases. The quality and production of medicinal plants are severely hampered by various factors, including soil deterioration; heavy metal pollution; and excessive use of pesticides, herbicides, fungicides, insecticides, and fertilizers. The relationship between plants and microbes offers an effective, environmentally friendly way for plants to handle stressful environmental situations. The use of beneficial microbes has been considered a possible replacement for traditional pesticides and fertilizers, as microbes are economically efficient, environment friendly, and non-toxic. The application of microbes helps plants mitigate the adverse effects of various biotic and abiotic stresses by improving soil fertility, suppressing pathogens, generating systemic resistance, and removing soil pollutants. The effect of the microbiota on the development and efficacy of many medicinally important species requires further study.

REFERENCES

Abd_allah, E. F., Abeer, H., Alqarawi, A. A. and Alwathnani, H. A. (2015). Alleviation of adverse impact of cadmium stress in sunflower (*Helianthus annuus* L.) by arbuscular mycorrhizal fungi. *Pakistan Journal of Botany*, *47*, 785–795.

Abou-Shanab, R. A., El-Sheekh, M. M. and Sadowsky, M. J. (2019). Role of rhizobacteria in phytoremediation of metal-impacted sites. *Emerging and Eco-Friendly Approaches for Waste Management*, 299–328.

Alam, M. A., Seetharam, K., Zaidi, P. H., Dinesh, A., Vinayan, M. T. and Nath, U. K. (2017). Dissecting heat stress tolerance in tropical maize (Zea mays L.). *Field Crops Research*, *204*, 110–119.

Ali, S., Hameed, S., Shahid, M., Iqbal, M., Lazarovits, G. and Imran, A. (2020). Functional characterization of potential PGPR exhibiting broad-spectrum antifungal activity. *Microbiological Research*, *232*, 126389.

Al-Sman, M. K., Abo-Elyousr, K. A., Eraky, A. and El-Zawahry, A. (2019). Efficiency of *Pseudomonas* spp.-based formulation for controlling root rot disease of black cumin under greenhouse and field conditions. *Archives of Phytopathology and Plant Protection*, *52*(19–20), 1313–1325.

Arathi, B. P., Raghavendra, R. S. P., Vijay, K., Baskaran, V. and Lakshminarayana, R. (2015). Metabolomics of carotenoids: The challenges and prospects – a review. *Trends in Food Science and Technology*, *45*, 105–117. doi: 10.1016/j.tifs.2015.06.003.

Babalola, O. O. and Glick, B. R. (2012). The use of microbial inoculants in African agriculture: Current practice and prospects. *Journal of Food, Agriculture and Environment*, *10*(3–4), 540–549.

Balthazar, C., Novinscak, A., Cantin, G., Joly, D. L. and Filion, M. (2022). Biocontrol activity of *Bacillus* spp. and *Pseudomonas* spp. against Botrytis cinerea and other cannabis fungal pathogens. *Phytopathology*®, 112(3), 549–560.

Barrera, A., Hereme, R., Ruiz-Lara, S., Larrondo, L. F., Gunde, P. E. and Pollmann, S. (2020). Fungal endophytes enhance the photoprotective mechanisms and photochemical efficiency in the antarctic *Colobanthus quitensis* (Kunth) Bartl. exposed to UV-B radiation. *Frontiers in Ecology and Evolution*, *8*, 122. doi: 10.3389/fevo.2020.00122.

Basiru, S., Mwanza, H. P. and Hijri, M. (2020). Analysis of arbuscular mycorrhizal fungal inoculant benchmarks. *Microorganisms*, *9*, 81.

Basu, A., Prasad, P., Das, S. N., Kalam, S., Sayyed, R. Z., Reddy, M. S. and El Enshasy, H. (2021). Plant growth promoting rhizobacteria (PGPR) as green bioinoculants: Recent developments, constraints, and prospects. *Sustainability*, *13*(3), 1140.

Bationo, A., Hartemink, A., Lungu, O., Naimi, M., Okoth, P., Smaling, E., Thiombiano, L. and Waswa, B. (2012). Knowing the African soils to improve fertilizer recommendations. In *Improving Soil Fertility Recommendations in Africa Using the Decision Support System for Agrotechnology Transfer (DSSAT)* (pp. 19–42). Springer, Dordrecht.

Beck, E. H., Fettig, S., Knake, C., Hartig, K. and Bhattarai, T. (2007). Specific and unspecific responses of plants to cold and drought stress. *Journal of Biosciences*, *32*, 501–510.

Begum, N., Ahanger, M. A., Su, Y., Lei, Y., Mustafa, N. S. A. and Ahmad, P. (2019). Improved drought tolerance by AMF inoculation in maize (Zea mays) involves physiological and biochemical implications. *Plants*, *8*, 579. doi: 10.3390/plants8120579.

Berg, G., Grube, M., Schloter, M. and Smalla, K. (2014). Unraveling the plant microbiome: Looking back and future perspectives. *Frontiers in Microbiology*, *5*, 148. doi: 10.3389/fmicb.2014.00148.

Bhandari, S., Kukreja, S., Singh, B., Kumar, V., Gautam, S., Sharma, V. and Goutam, U. 2023. Role of beneficial microbes in biotic and abiotic stress. In *Relationship Between Microbes and the Environment for Sustainable Ecosystem Services* (Vol. 3, pp. 243–259), Elsevier.

Bhattacharyya, P. N. and Jha, D. K. (2012). Plant growth-promoting rhizobacteria (PGPR): Emergence in agriculture. *World Journal of Microbiology and Biotechnology*, *28*, 1327–1350.

Bilal, S., Shahzad, R., Imran, M., Jan, R., Kim, K. M. and Lee, I. J. (2020). Synergistic association of endophytic fungi enhances Glycine max L. resilience to combined abiotic stresses: Heavy metals, high temperature, and drought stress. *Industrial Crops and Products*, *2019*, 111931. doi: 10.1016/j.indcrop.2019.111931.

Bisht, N., Tiwari, S., Singh, P. C., Niranjan, A. and Chauhan, P. S. (2019). A multifaceted rhizobacterium *Paenibacillus lentimorbus* alleviates nutrient deficiency-induced stress in Cicer arietinum L. *Microbiological Research*, *223*, 110–119.

Borah, A. and Thakur, D. (2020). Phylogenetic and functional characterization of culturable endophytic actinobacteria associated with Camellia spp. for growth promotion in commercial tea cultivars. *Frontiers in Microbiology*, *11*, 318.

Brainerd, E. and Menon, N. (2014). Seasonal effects of water quality: The hidden costs of the Green Revolution to infant and child health in India. *Journal of Development and Economics*, *107*, 49–64. doi: 10.1016/j.jdeveco.2013.11.004.

Burges, A., Alkorta, I., Epelde, L. and Garbisu, C. (2018). From phytoremediation of soil contaminants to phytomanagement of ecosystem services in metal-contaminated sites. *International Journal of Phytoremediation*, *20*(4), 384–397.

Calvo, P., Nelson, L. and Kloepper, J. W. (2014). Agricultural uses of plant biostimulants. *Plant and Soil*, *383*(1), 3–41.

Chandra, H., Kumari, P., Bisht, R., Prasad, R. and Yadav, S. (2020). Plant growth-promoting Pseudomonas aeruginosa from *Valeriana wallichii* displays antagonistic potential 91 against three phytopathogenic fungi. *Molecular Biology Reports*, *47*(8), 6015–6026.

Chen, L., Yin, H., Xu, J. and Liu, X. (2011). Enhanced antioxidative responses of a salt-resistant wheat cultivar facilitate its adaptation to salt stress. *African Journal of Biotechnology*, *10*(74), 16884–16886.

Cole, M. B., Augustin, M. A., Robertson, M. J. and Manners, J. M. (2018). The science of food security. *NPI Science of Food*, *2*(1), 14.

Coleman-Derr, D. and Tringe, S. G. (2014). Building the crops of tomorrow: Advantages of symbiont-based approaches to improving abiotic stress tolerance. *Frontiers in Microbiology*, *5*, 283.

Del-Saz, N. F., Romero-Munar, A., Cawthray, G. R., Aroca, R., Baraza, E. and Flexas, J. (2017). Arbuscular mycorrhizal fungus colonization in Nicotiana tabacum decreases the rate of both carboxylate exudation and root respiration and increases plant growth under phosphorus limitation. *Plant Soil*, *416*, 97–106. doi: 10.1007/s11104-017-3188-y.

Deshwal, V. K., Devi, M. S., Bhajanka, N., Mistri, J., Bose, A. and Saini, N. (2011). *Pseudomonas aeruginosa* strains and their role in plant growth promotion in medicinal plants. *Global Journal of Applied Agricultural Research*, *1*(49–55), 5.

Dinesh, R., Anandaraj, M., Kumar, A., Bini, Y. K., Subila, K. P. and Aravind, R. (2015). Isolation, characterization, and evaluation of multi-trait plant growth-promoting rhizobacteria for their growth-promoting and disease-suppressing effects on ginger. *Microbiological Research*, *173*, 34–43.

Do Santos, C. A. C., Peixoto, C. P., Vieira, E. L., Carvalho, E. V. and Peixoto, V. A. B. (2012). Ação da cinetina, ácido indolbutírico e ácido giberélico na emergência do girassol sob estresse por alumínio. *Comunicata Scientiae*, *3*(3), 206–209.

Dodd, I. C. and Perez-Alfocea, F. (2012). Microbial amelioration of crop salinity stress. *Journal of Experimental Botany*, *63*(9), 3415–3428.

Du Jardin, P. (2012). *The Science of Plant Biostimulants – A Bibliographic Analysis, Ad Hoc Study Report.* European Commission.

Du Jardin, P. (2015). Plant biostimulants: Definition, concept, main categories and regulation. *Scientia Horticulturae*, *196*, 3–14.

Eid, A. M., Salim, S. S., Hassan, S. E. D., Ismail, M. A. and Fouda, A. (2019). Role of endophytes in plant health and abiotic stress management. In *Microbiome in Plant Health and Disease: Challenges and Opportunities* (pp. 119–144), Springer.

Ekor, M. (2014). The growing use of herbal medicines: Issues relating to adverse reactions and challenges in monitoring safety. *Frontiers in Pharmacology*, *4*, 177.

Elshahawy, I. E., Saied, N. M., Abd-El-Kareem, F. and Morsy, A. A. (2018). Field application of selected bacterial strains and their combinations for controlling onion and garlic white rot disease caused by *Stromatinia cepivora*. *Journal of Plant Pathology*, *100*, 493–503.

Enebe, M. C. and Babalola, O. O. (2018). The influence of plant growth-promoting rhizobacteria in plant tolerance to abiotic stress: A survival strategy. *Applied Microbiology and Biotechnology*, *102*, 7821–7835.

Fahad, S., Hussain, S., Bano, A., Saud, S., Hassan, S., Shan, D. and Huang, J. (2015). Potential role of phytohormones and plant growth-promoting rhizobacteria in abiotic stresses: Consequences for changing environment. *Environmental Science and Pollution Research*, *22*, 4907–4921.

Feng, N. X., Yu, J., Zhao, H. M., Cheng, Y. T., Mo, C. H., Cai, Q. Y. and Wong, M. H. (2017). Efficient phytoremediation of organic contaminants in soils using plant – endophyte partnerships. *Science of the Total Environment*, *583*, 352–368.

García-García, A. L., García-Machado, F. J., Borges, A. A., Morales-Sierra, S., Boto, A. and Jiménez-Arias, D.(2020). Pure organic active compounds against abiotic stress: A biostimulant overview. *Frontiers in Plant Sciences*, *11*, 575829. doi: 10.3389/fpls.2020.575829.

Ghods-Alavi, B. S., Ahmadzadeh, M., Behboudi, K. and Jamali, S. (2012). Biocontrol of rhizome soft rot (*Pectobacterium carotovorum*) on valerian by *Pseudomonas* spp. under in vitro and greenhouse conditions. *Journal of Agricultural Technology*, *8*(6), 1913–1923.

Ghosh, D., Sen, S. and Mohapatra, S. (2018). Drought-mitigating Pseudomonas putida GAP-P45 modulates proline turnover and oxidative status in Arabidopsis thaliana under water stress. *Annals of Microbiology*, *68*, 579–594.

Glick, B. R. (2010). Using soil bacteria to facilitate phytoremediation. *Biotechnology Advances*, *28*(3), 367–374.

Gnanamanickam, S. S. (2006). *Plant-Associated Bacteria | SpringerLink.* Springer. doi: 10.1007/978-1-4020-4538-7.

Gouda, S., Kerry, R. G., Das, G., Paramithiotis, S., Shin, H. S. and Patra, J. K. (2018). Revitalization of plant growth promoting rhizobacteria for sustainable development in agriculture. *Microbiological Research*, *206*, 131–140.

Gupta, G., Parihar, S. S., Ahirwar, N. K., Snehi, S. K. and Singh, V. (2015). Plant growth promoting rhizobacteria (PGPR): Current and prospects for development of sustainable agriculture. *Journal of Microbial & Biochemical Technology*, *7*(2), 096–102.

Hashem, A., Abd_Allah, E. F., Alqarawi, A. A., Aldubise, A. and Egamberdieva, D. (2015). Arbuscular mycorrhizal fungi enhance salinity tolerance of *Panicum turgidum* Forssk by altering photosynthetic and antioxidant pathways. *Journal of Plant Interactions*, *10*, 230–242. doi: 10.1080/17429145.2015.1052025.

Hashem, A., Alqarawi, A. A., Radhakrishnan, R., Al-Arjani, A. B. F. and Aldehaish, H. A. (2018). Arbuscular mycorrhizal fungi regulate the oxidative system, hormones, and ionic equilibrium to trigger salt stress tolerance in Cucumis sativus L. *Saudi Journal of Biological Sciences*, *25*, 1102–1114. doi: 10.1016/j.sjbs.2018.03.009.

He, L., Liu, Y., Guo, Y., Shen, K., Hui, H. and Tan, Z. (2018). Diversity of intestinal bacterial lactase gene in antibiotics-induced diarrhea mice treated with Chinese herbs compound Qi Wei Bai Zhu San. *3 Biotechnology*, *8*, 1–8.

Heydarian, Z., Yu, M., Gruber, M., Glick, B. R., Zhou, R. and Hegedus, D. D. (2016). Inoculation of soil with plant growth-promoting bacteria producing 1-aminocyclopropane-1-carboxylate deaminase or expression of the corresponding acdS gene in transgenic plants increases salinity tolerance in Camelina sativa. *Frontiers in Microbiology*, *7*, 1966.

Humber, J. M. (2001). The role of complementary and alternative medicine: Accommodating pluralism. *Journal of American Medical Association*, *288*(13), 1655–1656.

Janmohammadi, M., Zolla, L. and Rinalducci, S. (2015). Low temperature tolerance in plants: Changes at the protein level. *Phytochemistry*, 117, 76–89.

Jayaprakashvel, M. and Mathivanan, N. (2011). Management of plant diseases by microbial metabolites. *Bacteria in Agrobiology: Plant Nutrient Management,* 237–265.

Jochum, M. D., McWilliams, K. L., Borrego, E. J., Kolomiets, M. V., Niu, G. and Pierson, E. A. (2020). Bioprospecting plant growth-promoting rhizobacteria that mitigate drought stress in grasses. *Frontiers in Microbiology*, *10*, 2106. doi: 10.3389/fmicb.2019.02106.

John, D. A. and Babu, G. R. (2021). *Lessons From the Aftermaths of Green Revolution on Food System and Health* (Vol. 5, pp. 644–559). doi: 10.3389/fsufs.2021.644559.

Kalam, S., Basu, A. and Podile, A. R. (2020). Functional and molecular characterization of plant growth promoting Bacillus isolates from tomato rhizosphere. *Heliyon*, *6*(8), e04734.

Kasim, W. A., Osman, M. E., Omar, M. N., Abd El-Daim, I. A., Bejai, S. and Meijer, J. (2013). Control of drought stress in wheat using plant-growth-promoting bacteria. *Journal of Plant Growth Regulation*, *32*, 122–130.

Kausar, R., Choudhary, M. I., Akram, M. I. and Rashid, M. (2018). Response of groundnut (Arachis hypogaea L.) to plant growth promoting Rhizobacteria in degraded soils. *African Journal of Agricultural Research*, *13*(17), 904–910.

Kaushal, M. and Wani, S. P. (2016). Plant-growth-promoting rhizobacteria: Drought stress alleviators to ameliorate crop production in drylands. *Annals of Microbiology*, *66*, 35–42.

Kavroulakis, N., Doupis, G., Papadakis, I. E., Ehaliotis, C. and Papadopoulou, K. K. (2018). Tolerance of tomato plants to water stress is improved by the root endophyte *Fusarium solani* FsK. *Rhizosphere*, *6*, 77–85.

Khan, N., Bano, A., Rahman, M. A., Guo, J., Kang, Z. and Babar, M. A. (2019). Comparative physiological and metabolic analysis reveals a complex mechanism involved in drought tolerance in Chickpeas (Cicer arietinum L.) Induced by PGPR and PGRs. *Scientific Reports*, *9*, 2097.

Kirchman, D. L. (2018). The physical-chemical environment of microbes. *Processes in Microbial Ecology*, 35–52.

Köberl, M., Schmidt, R., Ramadan, E., Bauer, R. and Berg, G. (2013). The microbiome of medicinal plants: Diversity and importance for plant growth, quality and health. *Frontiers in Microbiology*, *4*.

Kong, Z. and Glick, B. R. (2017). The role of plant growth-promoting bacteria in metal phytoremediation. *Advances in Microbial Physiology*, *71*, 97–132.

Kuffner, M., Puschenreiter, M., Wieshammer, G., Gorfer, M. and Sessitsch, A. (2008). Rhizosphere bacteria affect the growth and metal uptake of heavy metal accumulating willows. *Plant and Soil*, *304*, 35–44.

Kumar, A., Kumari, M. and Swarupa, P. (2019). Characterization of pH-dependent growth response of agriculturally important microbes for development of plant growth promoting bacterial consortium. *Journal of Pure and Applied Microbiology*, (2), 1053–1062.

Kumar, A. and Verma, J. P. (2018). Does plant-microbe interaction confer stress tolerance in plants: A review? *Microbiological Research*, *207*, 41–52.

Kuzmicheva, Y. V., Shaposhnikov, A. I., Petrova, S. N., Makarova, N. M., Tychinskaya, I. L., Puhalsky, J. V. and Belimov, A. A. (2017). Variety-specific relationships between effects of rhizobacteria on root exudation, growth, and nutrient uptake of soybean. *Plant and Soil*, *419*(1), 83–96.

Lamichhane, J. R., Dürr, C., Schwanck, A. A., Robin, M.-H., Sarthou, J.-P. and Cellier, V. (2017). Integrated management of damping-off diseases. A review. *Agronomy for Sustainable Development*, *37*, 10.

Lamsal, K., Kim, S. W., Kim, Y. S. and Lee, Y. S. (2013). Biocontrol of late blight and plant growth promotion in tomato using rhizobacterial isolates. *Journal of Microbiology and Biotechnology*, *23*(7), 897–904.

Li, J. F., He, X. H., Li, H., Zheng, W. J., Liu, J. F. and Wang, M. Y. (2015). Arbuscular mycorrhizal fungi increase growth and phenolic synthesis in Poncirus trifoliata under iron deficiency. *Scientia Horticultuare*, *183*, 87–92. doi: 10.1016/j.scienta.2014.12.015.

Li, X., Sun, P., Zhang, Y., Jin, C. and Guan, C. (2020). A novel PGPR strain *Kocuria rhizophila* Y1 enhances salt stress tolerance in maize by regulating phytohormone levels, nutrient acquisition, redox potential, ion homeostasis, photosynthetic capacity, and stress-responsiveness. *Environmental and Experimental Botany*, *174*, 104023. doi: 10.1016/j.envexpbot.2020.104023.

Li, Y., Zhang, J. J., Xu, D. P., Zhou, T., Zhou, Y., Li, S., Li, H. B. (2016). Bioactivities and health benefits of wild fruits. *International Journal of Molecular Sciences*, *17*, 1258.

Liang, C., Amelung, W., Lehmann, J. and Kastner, M. (2019). Quantitative assessment of microbial necromass contribution to soil organic matter. *Global Change Biology*, *25*, 3578–3590.

Lipper, L., Thornton, P., Campbell, B. M., Baedeker, T., Braimoh, A., Bwalya, M. and Torquebiau, E. F. (2014). Climate-smart agriculture for food security. *Nature Climate Change*, *4*(12), 1068–1072.

Liu, H., Yu, C., Li, H., Ouyang, B., Wang, T., Zhang, J., Wang, X. and Ye, Z. (2015). Overexpression of ShDHN, a dehydrin gene from Solanum habrochaites enhances tolerance to multiple abiotic stresses in tomato. *Plant Science*, 231, 198–211.

Ma, J., Janoušková, M., Li, Y., Yu, X., Yan, Y. and Zou, Z. (2015). Impact of arbuscular mycorrhizal fungi (AMF) on cucumber growth and phosphorus uptake under cold stress. *Functional Plant Biology*, *42*, 1158–1167. doi: 10.1071/FP15106.

Ma, Y., Dias, M. C. and Freitas, H. (2020). Drought and salinity stress responses and microbe-induced tolerance in plants. *Frontiers in Plant Science*, *11*, 591911.

Ma, Y., Rajkumar, M., Zhang, C. and Freitas, H. (2016). Beneficial role of bacterial endophytes in heavy metal phytoremediation. *Journal of Environmental Management*, *174*, 14–25.

Manach, C., Scalbert, A., Morand, C., Remesy, C. and Jimenez, L. (2004). Polyphenols: Food sources and bioavailability. *American Journal of Clinical Nutrition*, *79*, 727–747.

Masood, S., Seiji, Y. A., Shinwari, Z. K. and Anwar, R. (2004). Mapping quantitative trait loci (QTLs) for salt tolerance in rice (Oryza sativa) using RFLPs. *Pakistan Journal of Botany*, *36*(4), 825–834.

Mathiyazhagan, S., Kavitha, K., Nakkeeran, S., Chandrasekar, G., Manian, K., Renukadevi, P. and Fernando, W. G. D. (2004). PGPR mediated management of stem blight of *Phyllanthus amarus* (Schum and Thonn) caused by *Corynespora cassiicola* (Berk and Curt) Wei. *Archives of Phytopathology and Plant Protection*, *37*(3), 183–199.

Mathur, S., Tomar, R. S. and Jajoo, A. (2018). Arbuscular Mycorrhizal fungi (AMF) protect the photosynthetic apparatus of wheat under drought stress. *Photosynthesis Research*, *39*, 227–238.

Meena, K. K., Sorty, A. M., Bitla, U. M., Choudhary, K., Gupta, P., Pareek, A. and Minhas, P. S. (2017). Abiotic stress responses and microbe-mediated mitigation in plants: The omics strategies. *Frontiers in Plant Science*, *8*, 172.

Meena, M., Swapnil, P., Divyanshu, K., Kumar, S., Tripathi, Y. N., Zehra, A. and Upadhyay, R. S. (2020). PGPR-mediated induction of systemic resistance and physiochemical alterations in plants against the pathogens: Current perspectives. *Journal of Basic Microbiology*, *60*(10), 828–861.

Mohanty, P., Singh, P. K., Chakraborty, D., Mishra, S. and Pattnaik, R. (2021). Insight into the role of PGPR in sustainable agriculture and environment. *Frontiers in Sustainable Food Systems*, *5*, 667150.

Nazir, N., Kamili, A. N. and Shah, D. (2018). Mechanism of plant growth promoting rhizobacteria (PGPR) in enhancing plant growth-a review. *International Journal of Management Technology and Engineering*, *8*, 709–721.

Neshat, M., Abbasi, A., Hosseinzadeh, A., Sarikhani, M. R., Dadashi Chavan, D. and Rasoulnia, A. (2022). Plant growth-promoting bacteria (PGPR) induce antioxidant tolerance against salinity stress through biochemical and physiological mechanisms. *Physiology and Molecular Biology of Plants*, *28*(2), 347–361.

Pan, S. Y., Pan, S., Yu, Z. L., Ma, D. L., Chen, S. B., Fong, W. F., Han, Y. F. and Ko, K. M. (2010). New perspectives on innovative drug discovery: An overview. *International Journal of Pharmacy and Pharmaceutical Sciences*, *13*(3), 450–471.

Pandey, R., Mishra, A. K., Tiwari, S., Singh, H. N. and Kalra, A. (2011). Enhanced tolerance of Mentha arvensis against *Meloidogyne incognita* (Kofoid and White) Chitwood through mutualistic endophytes and PGPRs. *Journal of Plant Interactions*, *6*(4), 247–253.

Parihar, M., Rakshit, A., Rana, K., Tiwari, G. and Jatav, S. S. (2020). The effect of arbuscular mycorrhizal fungi inoculation in mitigating salt stress of Pea (Pisum Sativum L.). *Communications in Soil Science and Plant Analysis*, 1–15. doi: 10.1080/00103624.2020.1784917.

Pasbani, B., Salimi, A., Aliasgharzad, N. and Hajiboland, R. (2020). Colonization with arbuscular mycorrhizal fungi mitigates cold stress through the improvement of antioxidant defense and accumulation of protecting molecules in eggplants. *Scientia Horticulturare*, *272*, 109575.

Pérez-Montaño, F., Alías-Villegas, C., Bellogín, R. A., Del Cerro, P., Espuny, M. R., Jiménez-Guerrero, I. and Cubo, T. (2014). Plant growth promotion in cereal and leguminous agricultural important plants: From microorganism capacities to crop production. *Microbiological Research*, *169*(5–6), 325–336.

Poudel, M., Mendes, R., Costa, L. A. S., Bueno, C. G., Meng, Y., Folimonova, S. Y., Garrett, K. A. and Martins, S. J. (2021). The role of plant-associated bacteria, fungi, and viruses in drought stress mitigation. *Frontiers in Microbiology*, 12, 743512.

Prasad, M., Srinivasan, R., Chaudhary, M., Choudhary, M. and Jat, L. K. (2019). Plant growth promoting rhizobacteria (PGPR) for sustainable agriculture: Perspectives and challenges. In *PGPR Amelioration in Sustainable Agriculture* (pp. 129–157). Woodhead Publishing.

Prashar, P., Kapoor, N. and Sachdeva, S. (2014). Rhizosphere: Its structure, bacterial diversity, and significance. *Reviews in Environmental Science and BioTechnology*, *13*, 63–77.

Prity, S. A., Sajib, S. A., Das, U., Rahman, M. M., Haider, S. A. and Kabir, A. H. (2020). Arbuscular mycorrhizal fungi mitigate Fe deficiency symptoms in sorghum through phytosiderophore-mediated Fe mobilization and restoration of redox status. *Protoplasma*, *257*, 1373–1385.

Rodríguez, M., Torres, M., Blanco, L., Béjar, V., Sampedro, I. and Llamas, I. (2020). Plant growth-promoting activity and quorum quenching-mediated biocontrol of bacterial phytopathogens by *Pseudomonas segetis* strain P6. *Scientific Reports*, *10*(1), 1–12.

Rodriguez, R. and Redman, R. (2008). More than 400 million years of evolution and some plants still can't make it on their own: Plant stress tolerance via fungal symbiosis. *Journal of Experimental Botany*, *59*(5), 1109–1114.

Santoro, M. V., Cappellari, L. D. R., Giordano, W. and Banchio, E. (2015). Plant growth-promoting effects of native *Pseudomonas* strains on Mentha piperita (peppermint): An in vitro study. *Plant Biology*, *17*(6), 1218–1226.

Savary, S., Willocquet, L., Pethybridge, S. J., Esker, P., McRoberts, N. and Nelson, A. (2019). The global burden of pathogens and pests on major food crops. *Nature Ecology & Evolution*, *3*(3), 430–439.

Sawicka, B. and Egbuna, C. (2020). Pests of crops and control measures. In *Natural Remedies for Pest, Disease and Weed Control* (pp. 1–16). Academic Press.

Sayyed, R. Z., Seifi, S., Patel, P. R., Shaikh, S. S., Jadhav, H. P. and Enshasy, H. E. (2019). Siderophore production in groundnut rhizosphere isolates, Achromobacter sp. RZS2 is influenced by physicochemical factors and metal ions. *Environmental Sustainability*, *2*(2), 117–124.

Schmidt, R., Köberl, M., Mostafa, A., Ramadan, E. M., Monschein, M., Jensen, K. B. and Berg, G. (2014). Effects of bacterial inoculants on the indigenous microbiome and secondary metabolites of chamomile plants. Frontiers in Microbiology, *5*, 64.

Selvi, K. B., Paul, J. J. A., Vijaya, V. and Saraswathi, K. (2017). Analyzing the efficacy of phosphate solubilizing microorganisms by enrichment culture techniques. *Biochemistry and Molecular Biology Journal*, *3*(1), 1–7.

Seneviratne, M., Seneviratne, G., Madawala, H. M. S. P. and Vithanage, M. (2017). Role of rhizospheric microbes in heavy metal uptake by plants. In *Agro-Environmental Sustainability: Volume 2: Managing Environmental Pollution* (pp. 147–163), Springer.

Shah, A., Nazari, M., Antar, M., Msimbira, L. A., Naamala, J., Lyu, D. and Smith, D. L. (2021). PGPR in agriculture: A sustainable approach to increasing climate change resilience. *Frontiers in Sustainable Food Systems*, *5*, 667546.

Shameer, S. and Prasad, T. N. V. K. V. (2018). Plant growth promoting rhizobacteria for sustainable agricultural practices with special reference to biotic and abiotic stresses. *Plant Growth Regulation*, *84*, 603–615.

Shao, Q. Y., Dong, C. B., Han, Y. F. and Liang, Z. Q. (2021). Advances in the study of plant rhizosphere microorganisms groups. *Journal of Plant Nutrition and Fertilizer*, *27*(01), 144–152.

Sharma, S. B., Sayyed, R. Z., Trivedi, M. H. and Gobi, T. A. (2013). Phosphate solubilizing microbes: Sustainable approach for managing phosphorus deficiency in agricultural soils. *SpringerPlus*, *2*, 1–14.

Shinwari, Z. K., Tanveer, F. and Iqrar, I. (2019). Role of microbes in plant health, disease management, and abiotic stress management. In *Microbiome in Plant Health and Disease: Challenges and Opportunities* (pp. 231–250), Springer.

Singh, J. and Faull, J. L. (2020). Antagonism and biological control. In *Biocontrol of Plant Diseases* (pp. 167–177). CRC Press.

Singh, M., Dotaniya, M. L., Mishra, A., Dotaniya, C. K., Regar, K. L. and Lata, M. (2016). Role of biofertilizers in conservation agriculture. In *Conservation Agriculture: An Approach to Combat Climate Change in Indian Himalaya* (pp. 113–134), Springer.

Smułek, W., Sydow, M., Zabielska-Matejuk, J. and Kaczorek, E. (2020). Bacteria involved in biodegradation of creosote PAH-A case study of long-term contaminated industrial area. *Ecotoxicology and Environmental Safety*, *187*, 109843.

Sturz, A. V., Christie, B. R. and Nowak, J. (2000). Bacterial endophytes: Potential role in developing sustainable systems of crop production. *Critical Reviews in Plant Sciences*, *19*(1), 1–30.

Sulochana, M. B., Jayachandra, S. Y., Kumar, S. K. A. and Dayanand, A. (2014). Antifungal attributes of siderophore produced by the *Pseudomonas aeruginosa* JAS-25. *Journal of Basic Microbiology*, *54*(5), 418–424.

Sun, Z., Yang, L. M., Han, M., Han, Z. M., Yang, L., Cheng, L. and Lv, Z. L. (2019). Biological control ginseng grey mold and plant colonization by antagonistic bacteria isolated from rhizospheric soil of Panax ginseng Meyer. *Biological Control*, *138*, 104048.

Sun, Z., Yang, L. M., Zhang, L. and Han, M. (2018). An investigation of Panax ginseng Meyer growth promotion and the biocontrol potential of antagonistic bacteria against ginseng black spot. *Journal of Ginseng Research*, *42*(3), 304–311.

Suzuki, N., Rivero, R. M., Shulaev, V., Blumwald, E. and Mittler, R. (2014). Abiotic and biotic stress combinations. *New Phytologist*, *203*(1), 32–43.

Syed, S. and Chinthala, P. (2015). Heavy metal detoxification by different Bacillus species isolated from solar salterns. *Scientifica*, *2015*.

Tariq, M., Noman, M., Ahmed, T., Hameed, A., Manzoor, N. and Zafar, M. (2017). Antagonistic features displayed by plant growth promoting rhizobacteria (PGPR): A review. *Journal of Plant Science Phytopathol*ogy, *1*(1), 038–43.

Thanuja, G. S., Sreenivasulu, B., Yuvaraj, K. M., Giridhar, K., Umakrishna, U. and Latha, P. (2020). Effect of organic manures, plant growth promoting rhizobacteria and micronutrients on growth, grain yield, and quality of Ajwain (*Trachyspermum ammi* L.). *Journal of Pharmacognosy and Phytochemistry*, *9*(5), 1864–1870.

Timmusk, S., Paalme, V., Pavlicek, T., Bergquist, J., Vangala, A., Danilas, T. and Nevo, E. (2017). Bacterial distribution in the rhizosphere of wild barley under contrasting microclimates. *PLoS One*. doi: 10.1371/journal.pone.0017968.

Tiwari, S., Lata, C., Chauhan, P. S. and Nautiyal, C. S. (2016). Pseudomonas putida attunes morphophysiological, biochemical, and molecular responses in *Cicer arietinum* L. during drought stress and recovery. *Plant Physiology and Biochemistry*, *99*, 108–117.

Turner, T., James, E. and Poole, P. (2013). The plant microbiome. *Genome Biology*, *14*(6).

Turrini, A., Avio, L., Giovannetti, M. and Agnolucci, M. (2018). Functional complementarity of arbuscular mycorrhizal fungi and associated microbiota: The challenge of translational research. *Frontiers in Plant Science*, *9*, 1407.

Ullah, A., Heng, S., Munis, M. F. H., Fahad, S. and Yang, X. (2015). Phytoremediation of heavy metals assisted by plant growth promoting (PGP) bacteria: A review. *Environmental and Experimental Botany*, *117*, 28–40.

Vacheron, J., Desbrosses, G., Bouffaud, M. L., Touraine, B., Moënne-Loccoz, Y., Muller, D. and Prigent-Combaret, C. (2013). Plant growth-promoting rhizobacteria and root system functioning. *Frontiers in plant science*, *4*, 356.

Vassilev, N., Vassileva, M., Lopez, A., Martos, V., Reyes, A., Maksimovic, I., Eichler-Löbermann, B. and Malusà, E. (2015). Unexploited potential of some biotechnological techniques for biofertilizer production and formulation. *Applied Microbiology and Biotechnology*, *99*(12), 4983–4996.

Vega-Celedón, P., Bravo, G., Velásquez, A., Cid, F. P., Valenzuela, M., Ramírez, I. and Seeger, M. (2021). Microbial diversity of psychrotolerant bacteria isolated from wild flora of Andes mountains and Patagonia of Chile towards the selection of plant growth-promoting bacterial consortia to alleviate cold stress in plants. *Microorganisms*, *9*(3), 538.

Vessey, J. K. (2003). Plant growth promotes rhizobacteria as biofertilizers. *Plant and Soil*, *255*, 571–586.

Vijayabharathi, R., Gopalakrishnan, S., Sathya, A., Srinivas, V. and Sharma, M. (2018). Deciphering the tri-dimensional effect of endophytic *Streptomyces* sp. on chickpea for plant growth promotion, helper effect with *Mesorhizobium ciceri*, and host-plant resistance induction against *Botrytis cinerea*. *Microbial Pathogenesis*, *122*, 98–107.

Vurukonda, S. S. K. P., Vardharajula, S., Shrivastava, M. and SkZ, A. (2016). Enhancement of drought stress tolerance in crops by plant growth promoting rhizobacteria. *Microbiological Research*, *184*, 13–24.

Wang, G., Ren, Y., Bai, X., Su, Y. and Han, J. (2022). Contributions of beneficial microorganisms in soil remediation and quality improvement of medicinal plants. *Plants*, *11*(23), 3200.

Wani, P. A., Khan, M. S. and Zaidi, A. (2008). Effect of metal-tolerant plant growth-promoting Rhizobium on the performance of pea grown in metal-amended soil. *Archives of Environmental Contamination and Toxicology*, *55*, 33–42.

Yadav, A. N., Sachan, S. G., Verma, P. and Saxena, A. K. (2015). Prospecting cold deserts of north-western Himalayas for microbial diversity and plant growth promoting attributes. *Journal of Bioscience and Bioengineering*, *119*(6), 683–693.

Yakhin, O. I., Lubyanov, A. A., Yakhin, I. A. and Brown, P. H. (2017). Biostimulants in plant science: A global perspective. *Frontiers in Plant Science*, *7*, 2049.

Yang, Y., Han, X., Liang, Y., Ghosh, A., Chen, J. and Tang, M. (2015). The combined effects of Arbuscular Mycorrhizal Fungi (AMF) and lead (Pb) stress on Pb accumulation, plant growth parameters, photosynthesis, and antioxidant enzymes in Robinia pseudoacacia L. *PLoS One*, *10*, e0145726.

Zahra, N., Wahid, A., Hafeez, M. B., Shaukat, K., Shahzad, S., Shah, T. and Alyemeni, M. N. (2021). Plant growth promoters alleviate oxidative damage and improve the growth of milk thistle (*Silybum marianum* L.) under salinity stress. *Journal of Plant Growth Regulation*, 1–26.

Zhang, L. N., Wang, D. C., Hu, Q., Dai, X. Q., Xie, Y. S., Li, Q. and Guo, J. H. (2019). Consortium of plant growth-promoting rhizobacteria strains suppresses sweet pepper disease by altering the rhizosphere microbiota. *Frontiers in Microbiology*, *10*, 1668.

Zhou, Y., Zheng, J., Li, Y., Xu, D. P., Li, S., Chen, Y. M. and Li, H. B. (2016). Natural polyphenols for prevention and treatment of cancer. *Nutrients*, *8*, 515. doi: 10.3390/nu8080515.

Zhu, J. K. (2000). Overexpression of a delta-pyrroline-5-carboxylate synthetase gene and analysis of tolerance to water and salt stress in transgenic rice. *Trends in Plant Sciences*, *6*, 66–72.

Zhuang, L., Li, Y., Wang, Z., Yu, Y., Zhang, N., Yang, C. and Wang, Q. (2021). Synthetic community with six *Pseudomonas* strains screened from garlic rhizosphere microbiome promotes plant growth. *Microbial Biotechnology*, *14*(2), 488–502.

7 Heavy Metal Accumulation in Medicinal Plants and Its Health Impacts

A Case Study of African Countries

Hubert Olivier Dossou-Yovo, Muhammad Nauman Khan, Alevcan Kaplan, and Brice Sinsin

7.1 INTRODUCTION

Plant species harvested from various ecosystems provide medicinal, perfume, and food products (Demir *et al.*, 2022; Dossou-Yovo *et al.*, 2021, 2022a, 2022b; Vodouhe *et al.*, 2016). The medicinal uses of plant species have been reported for many decades and remain important for more than 80% of the world population (Cunningham, 1993; Dossou-Yovo *et al.*, 2014; Khan and Badshah, 2019). Both herbaceous and ligneous plant species are exploited by people to treat a wide range of diseases and disorders (Dossou-Yovo *et al.*, 2021, 2022b; Dougnon *et al.*, 2021). Apart from the ethnobotanical uses widely reported for medicinal plants, many pharmacological investigations have been undertaken to validate indigenous knowledge (Fagbohun *et al.*, 2020, 2021; Mahomoodally *et al.*, 2021; Ozturk *et al.*, 2022). The importance of medicinal plants has resulted in high pressure on their populations within natural ecosystems, and some species that previously were common have become threatened or locally extinct (Leaman, 2001). As a result, in addition to naturally grown plants, more medicinal plant species are cultivated in commercial gardens and purchased by herbal medicine traders and other consumers (Dossou-Yovo *et al.*, 2021, 2022b). This cultivation contributes to their conservation in Africa and many other parts of the world.

The impacts of heavy metals on human and animal health have been widely reported. For example, Street (2012) mentioned poisoning from heavy metal contamination in plants causing health complications. Most of the toxic metals in Africa and other developing nations are caused by industrialization coupled with poor effluent disposal and management, with major elements being zinc (Zn), lead (Pb), aluminium (Al), cadmium (Cd), nickel (Ni), iron (Fe), manganese (Mn), and arsenic (As), described by Okereafor *et al.* (2020). Heavy metals in foods can accumulate in humans, with harmful effects on health (Hejna *et al.*, 2018; Naz *et al.*, 2022). Elsewhere, Okem *et al.* (2014) reported levels of heavy metals in some medicinal plants higher than the permissible limits of the World Health Organization (WHO) and mentioned that low antimicrobial activity was induced by high levels of heavy metals. Hlihor *et al.* (2022) highlighted the accumulation of heavy metals in human tissues. There have also been many investigations into heavy metal toxicity from the consumption of animal products. For instance, Sharma and Street (1980) concluded that consumption of liver and kidney from animals exposed to cadmium should be avoided. More recently, Hejna *et al.* (2018) advised the control of the intake of heavy metals by animals to reduce their quantity in animal tissues. Such studies reveal heavy metals in both plants and animals with potential adverse impacts on humans. As a result, there is an increasing interest in studies based on heavy metal contamination in plants worldwide (Fagbohun *et al.*, 2020, 2021). This chapter surveys articles, book chapters, and technical reports to collate information on the content of heavy metals in cultivated

DOI: 10.1201/b23155-7

medicinal plants in Africa. Furthermore, the impacts of such metals on human health as well as techniques for their elimination in cultivated medicinal plants are evaluated. Medicinal plant consumers in Africa do not have access to Western knowledge on heavy metal contamination in such plants, although it is a public health issue. This chapter will thus serve not only to increase awareness of the risks associated with the consumption of medicinal plants containing heavy metals but also to motivate further studies on heavy metal accumulation in medicinal plants and their effects on humans.

7.1.1 Diversity of and Heavy Metal Occurrence in Cultivated Medicinal Plants in Africa

Heavy metals have been reported in a wide range of medicinal plants, both naturally grown and cultivated. Street (2012) mentioned two African medicinal plants, *Datura metal* L. (Solanaceae) as a cobalt (Co) and nickel accumulator and use as a phytomonitor for these elements in soil and *Datura innoxia* Miller, Gard. as tolerant of heavy metals. Similarly, the same author mentioned *Helichrysum candolleanum* H. Buek (Asteraceae) and *Blepharis diversispina* C. B. Clarke (Acanthaceae) as containing high heavy metal concentrations. While surveying heavy metals in 27 medicinal plants in Ghana, Annan *et al.* (2010) found cadmium in all samples, with *Ocimum canum* (Lamiaceae), *Clausena amisata* (Rutaceae), and *Rauvolfia vomitoria* (Apocynaceae) also showing toxic levels of iron. Contrary to the expectation that the cultivation of medicinal plants may help control the chemical content of plants, many agrochemicals contain heavy metals that may enter the soil and move into cultivated plants (Street, 2012). In Benin, farmers cultivate or conserve naturally grown plants of *Phyllanthus amarus* Schum. & Thonn. (Phyllanthaceae) (Dossou-Yovo *et al.*, 2021), and it was reported that a certain level of cadmium stress in the species induces higher activity of phyllanthin and hypophyllanthin (Street, 2012). Vaikosen and Alade (2017) considered nine species of wild and cultivated medicinal plants from a garden in Nigeria, including *Zea mays* L. (Poaceae), *Ocimum gratissimum* L. (Lamiaceae), and *Talinum triangulare* (Jacq.) Wild. (Taliceae). They found Cd and Ni in all the species studied, whereas lead was reported in only 11% of them. They concluded that the levels of heavy metals in medicinal plants depend on the soil characteristics of the collection sites. All these studies show that heavy metal contamination in medicinal plants is a severe threat to humans (Luo *et al.*, 2021). Using atomic absorption spectrometry on both cultivated and wild medicinal plants sold in a Ghanaian market, Nkansah *et al.* (2016) reported permissible limits of Fe, Pb, Zn, and Cd in 15 herbs, while Cd was found at toxic levels in four samples. There is thus evidence of medicinal plants from diverse botanical groups containing heavy metals at concentrations influenced by the growing environment and the biochemical characteristics of the herbs. Species belonging to botanical families Acanthaceae, Apocynaceae, Asteraceae, Lamiaceae, Phyllanthaceae, Poaceae, Rutaceae, Solanaceae, and Talinaceae are all reported as accumulating a wide range of heavy metals. Pb, Fe, Cd, Ni, and Zn are the main elements found and are present in permissible and sometimes non-permissible levels in cultivated medicinal plants in Africa. Similar to the results reported previously on Lamiaceae species in Africa, Angelova *et al.* (2006) found high levels of Pb, Zn, and Cd in the leaves of three cultivated Lamiaceae species in Bulgaria (*Mentha piperita* L., *Salvia officinalis* L., and *Salvia sclarea* L.). The study was undertaken in a polluted region, and concentrations of heavy metals were above the permissible limits. Elsewhere, research on Apocynaceae plants revealed heavy metal accumulation in a polluted city (Venkateshwar *et al.*, 2004). Knowledge related to the impact of heavy metals on human health is expected to alert people to the threat that they represent for communities in Africa and elsewhere when present in medicinal plants. There is a need for more research on heavy metals in a greater number of medicinal species concerning pollution gradients in Africa. Medicinal herbs commercialized in African markets may be cultivated locally or collected from the wild but may also be international imports (Nkansah *et al.*, 2016). Figures 7.1 and 7.2 show the mechanisms of heavy metal transport and their accumulation in plants. This schematic model for the translocation of heavy metals reveals that the process

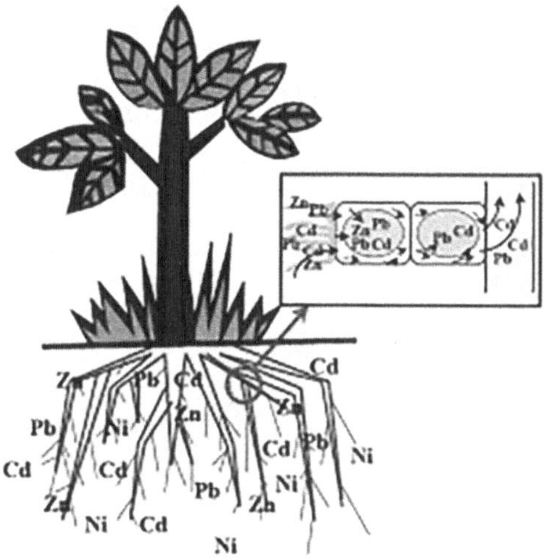

FIGURE 7.1 A schematic model for translocation of heavy metal ions in plants (Chen *et al.*, 2021).

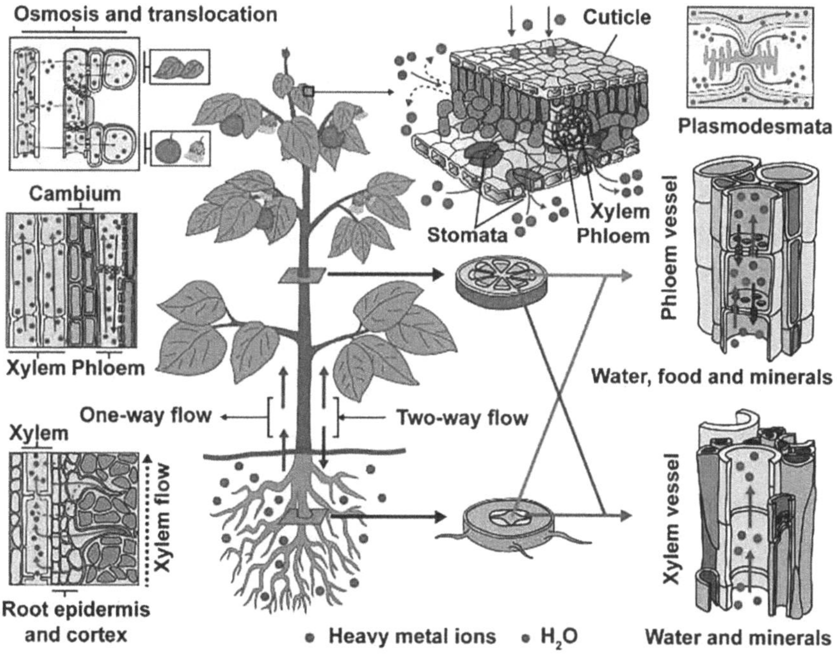

FIGURE 7.2 Heavy metal absorption and accumulation in the root and shoot of the plant (Adiloğlu, 2018).

starts from the root system, with osmosis as a major phenomenon. Heavy metals move along the plant tissues, with water playing a great role. The one-way and two-way flows reported in the figures highlight that heavy metals coming from the soil to the plant organs also return to the same soil in remarkable proportion.

Therefore, research should also focus on market surveys assessing the levels of heavy metals in medicinal plants to limit their negative impacts. Similarly, international trading of medicinal herbs from Africa has become a priority activity for herbalists and other traders. As a result, heavy metals contained in medicinal herbs in Africa can affect consumers of such medicines worldwide.

7.1.2 Impacts of Heavy Metals Recorded in Cultivated Medicinal Plants in Africa on Human Health and Techniques to Limit Contamination

Excessive exposure to a range of heavy metals is harmful to human health. There is widespread use of a plethora of herbal remedies across Africa to treat a wide range of diseases and disorders (Okaiyeto and Oguntibeju, 2021). This, together with limited testing of products in unregulated, informal markets for medicinal plants, makes many African populations particularly susceptible to the adverse health effects of heavy metals. Nickel is a hard silvery-white metal that causes skin irritation, and human exposure to this element as well as its accumulation cause cancer and harms the lungs, stomach, and kidneys (Genchi *et al.*, 2020). High doses and accumulation of iron lead to upset stomach, constipation, nausea, vomiting, and diarrhoea (Abbaspour *et al.*, 2014). Pregnant women, children, and older persons with such disorders are at high risk, as permanent constipation can lead to liver disease. However, it is important to highlight that when iron is present in herbal medicines within permissible limits, it helps avoid Fe deficiency in diets and reduces the risk of anaemia. According to the World Health Organization, the median dietary intake of iron is approximately 16 to 18 mg/day for men and 12 mg/day for women, with a tolerable upper intake level (UL) for adults being 45 mg/day. Pb has very harmful effects on human health. Consumption or absorption through the skin can lead to brain and nervous system damage, delaying growth and development in children (Collin *et al.*, 2022). When women consume medicinal herbs containing traces of Pb, it enters the blood and can lead to an increased risk of miscarriage during pregnancy (Jelliffe-Pawlowski *et al.*, 2006). Zn is an essential element for human health (Plum *et al.*, 2010). Zn deficiency induces dermatitis and inhibits reproductive function (Gibson, 2012; Prasad, 2008). Consumption of herbal medicines with permissible levels of Zn in Africa helps to avoid its deficiency. The World Health Organization gives a recommended dietary allowance (RDA) for Zn of 11 mg/day for men and 8 mg/day for women, and owing to their lower average body weights, lower Zn intake is recommended for infants (2–3 mg/day) and children (5–9 mg/day). According to the Office of Dietary Supplements (ODS), taking too much zinc can be harmful and may cause nausea, diarrhoea, and headaches. Similarly, Plum *et al.* (2010) reported the same diseases in addition to abdominal pain concerning Zn toxicity. It is important to highlight that even at low concentrations,

TABLE 7.1
Permissible Levels of Major Heavy Metals Found in Cultivated Medicinal Plants in Africa

Heavy Metals	Chemical Symbol	Permissible Levels
Copper	Cu	The tolerable upper intake level is the maximum daily intake unlikely to cause harmful effects on health. The UL for copper for adults 19+ years and those pregnant and lactating is 10,000 micrograms daily (WHO/FAO)
Lead	Pb	93 μg day−1 for adults (for an average body weight of 78.1 kg) and 30 μg day−1 for children (for a body weight of 25 kg) (Sharma and Dubey, 2005)
Zinc	Zn	The recommended dietary allowance for zinc is 11 mg/day for men and 8 mg/day for women, while lower zinc intake is recommended for infants (2–3 mg/day) and children (5–9 mg/day) because of their lower average body weights (Plum *et al.*, 2010)
Cadmium	Cd	Cadmium levels can be measured in the blood, urine, hair, nail, and saliva samples; based on FAO/WHO rules, its permitted level in plants is 0.2 mg/kg
Nickel	Ni	A value of 0.6 μg Ni/m^3 (annual basis) is recommended for nickel to protect the general public against carcinogenic and other potential hazardous effects associated with exposure to both soluble and less soluble nickel compounds (WHO/FAO)

greater than 5–10 μg Cd g–1 leaf dry weight, Cd is not only toxic to plants but also to human health (Gill and Tuteja, 2011). For instance, Cd toxicity in the plant *Phyllanthus amarus* will inhibit its therapeutic effects against diabetes, stomach aches, and intestinal infections (Dossou-Yovo *et al.*, 2021; Street, 2012). Exposure to Cd causes kidney disease and makes bones fragile. Moreover, Cd causes low birth weight in babies and high risks of abortion (Ikeh-Tawari *et al.*, 2013). Major heavy metals found in cultivated medicinal plants in Africa and the permissible levels for human health are reported in Table 7.1. Most herbal medicine consumers in Africa do not have information on the permitted levels of heavy metals. Moreover, they are less informed about the risks associated with cultivated medicinal plants. There are many approaches that farmers and other plant specialists may adopt to reduce the levels of heavy metals in cultivated medicinal herbs and therefore limit the risks to human health. These include the control and safe use of phytochemicals containing heavy metals, frequent watering of the aerial parts of cultivated medicinal herbs to limit their contact with heavy metal elements, and avoiding cultivation of medicinal plants in contaminated soils or soils with naturally high heavy metal contents. Figure 7.3 exhibits how heavy metals affect human health and shows that all human organs are damaged when such elements are consumed. Human exposure to heavy metals concerns not only the consumed medicinal plants but also many other

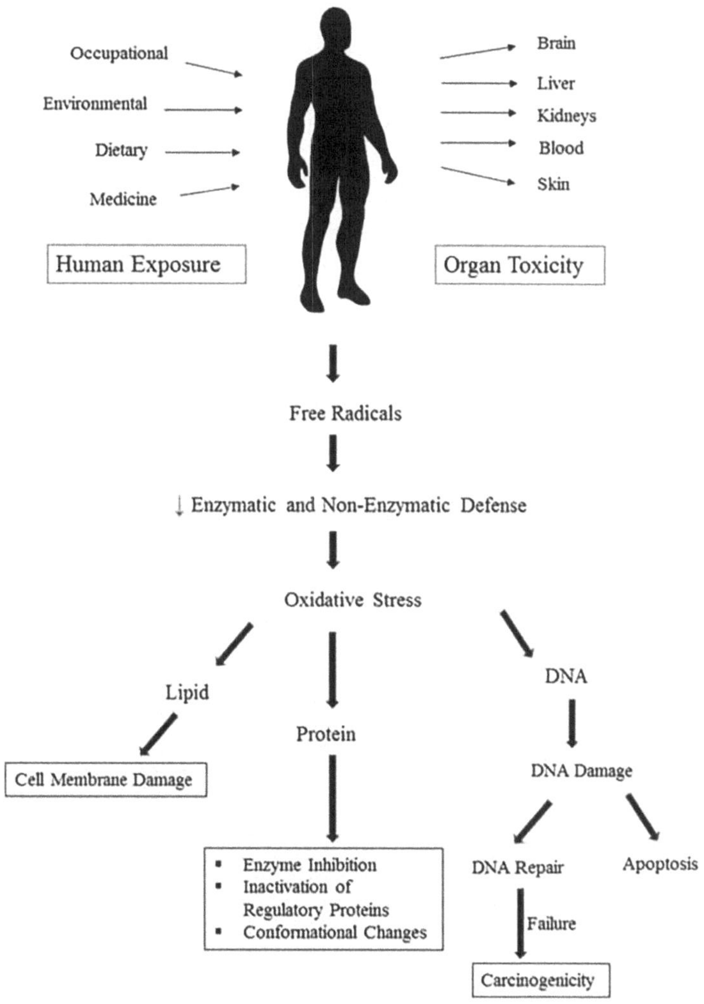

FIGURE 7.3 Toxic mechanisms of heavy metals found in cultivated medicinal plants in Africa with affected organs (Balali-Mood *et al.*, 2021).

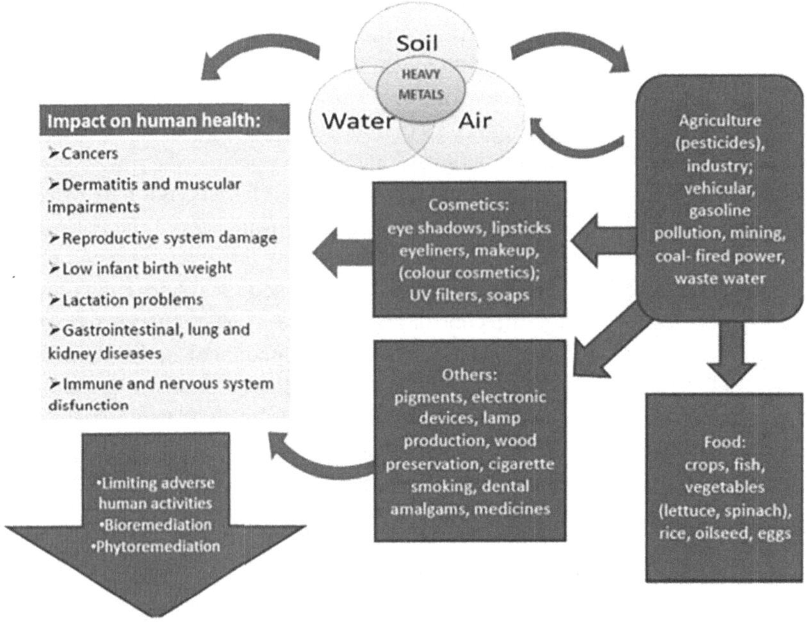

FIGURE 7.4 Schematic sources and impacts of heavy metals on human health (Briffa *et al.*, 2020).

factors, including occupation and diet. The same figure reports brain, liver, kidney, blood, and skin as organs affected by heavy metals. Their deterioration can induce death in humans. Similarly, Figure 7.4 shows skin diseases and many other cancers affecting various organs of human health. All of these show that heavy metals contained in cultivated medicinal plants represent severe threats to human survival.

7.2 CONCLUSIONS AND FUTURE PERSPECTIVES

Herbal medicines provide alternative health care in many countries, especially in Africa. Exposure to medicinal herbs containing unacceptable levels of heavy metals is likely to be widespread in Africa and to have harmful health effects, including vomiting, skin irritation, and lung and skin cancer. Pregnancy and fetal development are also particularly affected by heavy metals, leading to abortion or impaired development of children. However, some heavy metals such as Zn and Fe are also essential metals for human health when consumed in safe quantities. Fe, for example, found in cultivated medicinal herbs, is required to avoid anaemia, a common symptom of Fe deficiency. Medicinal plants cultivated in polluted areas and soils with high natural heavy metal contents in Africa are more likely to be contaminated by heavy metals. So, cultivation should be concentrated in less polluted areas and in soils with low heavy metal content to safeguard the health of consumers. Phytochemicals used in medicinal herb cultivation also contribute to metal contamination through the intake by plant roots and contact of these metals with the aerial parts of the plants. Therefore, phytochemical use should be controlled to limit contamination. Regarding the Sustainable Development Goals aiming to ensure good health and a safe environment for people all over the world, much research attention should be paid to the heavy metals contained in medicinal plants in many countries. In other words, investigations on the heavy metals contained in chemicals used to fertilize medicinal plants as well as levels of heavy metals in various cultivated medicinal plants should be better researched. Following a gradient of pollution, levels of heavy metals in cultivated medicinal plant leaves need to be studied in Africa and worldwide. Analyses should be undertaken on potential consumers of cultivated medicinal plants to state levels of contamination in heavy metals. Chemical studies should be done on soils where medicinal plants are cultivated in Africa, and techniques undertaken in Africa for heavy metal elimination need to be investigated.

Acknowledgements: The authors are thankful to Phil Harris (retired professor, University of Coventry, UK) for providing editing of the first draft of this chapter.

REFERENCES

Abbaspour, N., Hurrell, R., Kelishadi, R., 2014. Review on iron and its importance for human health. *Journal of Research in Medical Sciences: The Official Journal of Isfahan University of Medical Sciences* 19, 164.

Adiloğlu, S., 2018. Heavy metal removal with phytoremediation. *Advances in Bioremediation and Phytoremediation*, 115–126.

Angelova, V., Ivanov, K., Ivanova, R., 2006. Heavy metal content in plants from the family Lamiaceae cultivated in an industrially polluted region. *Journal of Herbs, Spices & Medicinal Plants* 11, 37–46.

Annan, K., Kojo, A.I., Cindy, A., Samuel, A.-N., Tunkumgnen, B.M., 2010. Profile of heavy metals in some medicinal plants from Ghana commonly used as components of herbal formulations. *Pharmacognosy Research* 2, 41.

Balali-Mood, M., Naseri, K., Tahergorabi, Z., Khazdair, M.R., Sadeghi, M., 2021. Toxic mechanisms of five heavy metals: Mercury, lead, chromium, cadmium, and arsenic. *Frontiers in Pharmacology*, 227.

Briffa, J., Sinagra, E., Blundell, R., 2020. Heavy metal pollution in the environment and their toxicological effects on humans. *Heliyon* 6, e04691.

Chen, Y.-G., Wang, Q., Wołowicz, A., Gładysz-Płaska, A., Wawrzkiewicz, M., Sofińska-Chmiel, W., Lv, G.-Y., Kołodyńska, D., Chen, S.-H., 2021. Medical plant extract purification from cadmium (II) using modified thermoplastic starch and ion exchangers. *Materials* 14, 4734.

Collin, M.S., Kumar Venkataraman, S., Vijayakumar, N., Kanimozhi, V., Arbaaz, S.M., Stacey, R.S., Anusha, J., Choudhary, R., Lvov, V., Tovar, G.I., 2022. Bioaccumulation of lead (Pb) and its effects on humans: A review. *Journal of Hazardous Materials Advances*, 100094.

Cunningham, A., 1993. *African medicinal plants*. United Nations Educational, Scientific and Cultural Organization: Paris, France.

Demir, N., Dalgiç, S., Kaplan, A., 2022. Investigation of some bioactivities and odor components of Jasminum officinale Linn. (Oleaceae): A valuable tool for cosmetic product design. *Commagene Journal of Biology* 6, 197–206.

Dossou-Yovo, H.O., Kindomihou, V., Vodouhè, F.G., Sinsin, B., 2021. Assessment of the diversity of medico-magic knowledge on four herbaceous species in Benin. *The Scientific World Journal* 2021, 1–11.

Dossou-Yovo, H.O., Vodouhe, F., Sinsin, B., 2014. Assessment of the medicinal uses of plant species found on termitaria in the Pendjari biosphere reserve in Benin. *Journal of Medicinal Plant Research* 8, 368–377.

Dossou-Yovo, H.O., Vodouhè, F.G., Kaplan, A., Sinsin, B., 2022a. Application of ethnobotanical indices in the utilization of five medicinal herbaceous plant species in Benin, West Africa. *Diversity* 14, 612.

Dossou-Yovo, H.O., Vodouhè, F.G., Kindomihou, V., Sinsin, B., 2022b. Investigating the use profile of Kigelia africana (Lam.) Benth. through market survey in Benin. *Conservation* 2, 275–285.

Dougnon, V., Agbodjento, E., Hounsa, E., Legba, B.B., Deguenon, E., Bohoungbe, N., Akotegnon, R., Klotoe, J.R., Dougnon, J., 2021. An ethnobotanical survey of seventeen plant species used against diarrhea and other diseases in southern Benin (West Africa). *Journal of Biological Research-Bollettino della Società Italiana di Biologia Sperimentale* 94.

Fagbohun, O.F., Joseph, J.S., Salami, O.A., Msagati, T.A., 2021. Exploration of modern chromatographic methods coupled to mass spectrometric techniques for trace element and chemical composition analyses in the leaf extracts of Kigelia africana. *Biological Trace Element Research* 199, 1633–1648.

Fagbohun, O.F., Olawoye, B., Ademakinwa, A.N., Jolayemi, K.A., Msagati, T.A., 2020. Metabolome modulatory effects of Kigelia africana (Lam.) Benth. fruit extracts on oxidative stress, hyperlipidaemic biomarkers in STZ-induced diabetic rats, and antidiabetic effects in 3T3 L1 adipocytes. *Journal of Pharmacy and Pharmacology* 72, 1798–1811.

Genchi, G., Carocci, A., Lauria, G., Sinicropi, M.S., Catalano, A., 2020. Nickel: Human health and environmental toxicology. *International Journal of Environmental Research and Public Health* 17, 679.

Gibson, R.S., 2012. Zinc deficiency and human health: Etiology, health consequences, and future solutions. *Plant and Soil* 361, 291–299.

Gill, S.S., Tuteja, N., 2011. Cadmium stress tolerance in crop plants: Probing the role of sulfur. *Plant Signaling & Behavior* 6, 215–222.
Hejna, M., Gottardo, D., Baldi, A., Dell'Orto, V., Cheli, F., Zaninelli, M., Rossi, L., 2018. Nutritional ecology of heavy metals. *Animal* 12, 2156–2170.
Hlihor, R.M., Roşca, M., Hagiu-Zaleschi, L., Simion, I.M., Daraban, G.M., Stoleru, V., 2022. Medicinal plant growth in heavy metals contaminated soils: Responses to metal stress and induced risks to human health. *Toxics* 10, 499.
Ikeh-Tawari, E.P., Anetor, J.I., Charles-Davies, M., 2013. Cadmium level in pregnancy, influence on neonatal birth weight, and possible amelioration by some essential trace elements. *Toxicology International* 20, 108.
Jelliffe-Pawlowski, L., Miles, S., Courtney, J., Materna, B., Charlton, V., 2006. Effect of magnitude and timing of maternal pregnancy blood lead (Pb) levels on birth outcomes. *Journal of Perinatology* 26, 154–162.
Khan, M.N., Badshah, L., 2019. Floristic diversity and utility of flora of district Charsadda, Khyber Pakhtunkhwa. *Acta Ecologica Sinica* 39, 306–320.
Leaman, D.J., 2001. Conservation, trade, sustainability, and exploitation of medicinal plant species. *Development of Plant-Based Medicines: Conservation, Efficacy and Safety*, 1–15.
Luo, L., Wang, B., Jiang, J., Fitzgerald, M., Huang, Q., Yu, Z., Li, H., Zhang, J., Wei, J., Yang, C., 2021. Heavy metal contaminations in herbal medicines: Determination, comprehensive risk assessments, and solutions. *Frontiers in Pharmacology* 11, 595335.
Mahomoodally, M.F., Zengin, G., Sinan, K.I., Ak, G., Sadeer, N.B., Angeloni, S., Mustafa, A.M., Caprioli, G., Maggi, F., Cakilcioglu, U., 2021. Two medicinal plants (*Alkanna trichophila* and *Convolvulus galaticus*) from Turkey: Chemical characterization and biological perspectives. *Chemistry & Biodiversity* 18, e2100356.
Naz, R., Khan, M., Hafeez, A., Fazil, M., Khan, M., Ali, B., Javed, M., Imran, M., Shati, A., Alfaifi, M., 2022. Assessment of phytoremediation potential of native plant species naturally growing in heavy metal-polluted industrial soils. *Brazilian Journal of Biology* 84.
Nkansah, M.A., Hayford, S.T., Borquaye, L.S., Ephraim, J.H., 2016. Heavy metal contents of some medicinal herbs from Kumasi, Ghana. *Cogent Environmental Science* 2, 1234660.
Okaiyeto, K., Oguntibeju, O.O., 2021. African herbal medicines: Adverse effects and cytotoxic potentials with different therapeutic applications. *International Journal of Environmental Research and Public Health* 18, 5988.
Okem, A., Southway, C., Stirk, W., Street, R., Finnie, J., Van Staden, J., 2014. Heavy metal contamination in South African medicinal plants: A cause for concern. *South African Journal of Botany* 93, 125–130.
Okereafor, U., Makhatha, M., Mekuto, L., Uche-Okereafor, N., Sebola, T., Mavumengwana, V., 2020. Toxic metal implications on agricultural soils, plants, animals, aquatic life, and human health. *International Journal of Environmental Research and Public Health* 17, 2204.
Ozturk, R.B., Zengin, G., Sinan, K.I., Montesano, D., Zheleva-Dimitrova, D., Gevrenova, R., Uba, A.I., Çakılcıoğlu, U., Kaplan, A., Jugreet, S., 2022. Which extraction solvents and methods are more effective in terms of chemical composition and biological activity of Alcea fasciculiflora from Turkey? *Molecules* 27, 5011.
Plum, L.M., Rink, L., Haase, H., 2010. The essential toxin: Impact of zinc on human health. *International Journal of Environmental Research and Public Health* 7, 1342–1365.
Prasad, A.S., 2008. Zinc in human health: Effect of zinc on immune cells. *Molecular Medicine* 14, 353–357.
Sharma, P., Dubey, R.S., 2005. Lead toxicity in plants. *Brazilian Journal of Plant Physiology* 17, 35–52.
Sharma, R., Street, J., 1980. Public health aspects of toxic heavy metals in animal feeds. *Journal of the American Veterinary Medical Association* 177, 149–153.
Street, R., 2012. Heavy metals in medicinal plant products – an African perspective. *South African Journal of Botany* 82, 67–74.
Vaikosen, E.N., Alade, G.O., 2017. Determination of heavy metals in medicinal plants from the wild and cultivated garden in Wilberforce Island, Niger Delta region, Nigeria. *Journal of Pharmacy & Pharmacognosy Research* 5, 129–143.
Venkateshwar, C., Rao, Y., Rao, S., Piska, R., 2004. Toxic level heavy metal contamination of some medicinal plants of Apocynaceae. *Pollution Research* 23, 229–231.
Vodouhe, F.G., Dossou-Yovo, H.O., Chadaré, F.J., Gélinas, N., Assogbadjo, A.E., Coulibaly, O., 2016. Valuing the potential of non-timber forest products in the financial valuation of savannah formation in the Sudanian region. *Universal Journal of Agricultural Research* 4, 183–197.

8 Interaction between Fungi and Bacteria to Diminish the Effect of Salt Stress in Plants by Enhancing Enzymatic and Non-Enzymatic Antioxidants

Vahid Poozesh and Atefe Amirahmadi

8.1 INTRODUCTION

Plants face various biotic and abiotic stresses during their life cycle, including salinity, dryness, cold, flooding, heat, deficiency or excess of minerals, and exposure to toxins (Kumar *et al.*, 2013). The salinity of soil is considered a global environmental challenge that threatens cultivated areas and crop productivity. It has been found that 20% of all cultivated land and 33% of irrigated agricultural land are affected by salinity (Shrivastava and Kumar, 2015). It is estimated that by 2050, 50% of all arable land will be affected by primary and secondary salinity (Butcher *et al.*, 2016). Loss of arable land and declining yields are in direct conflict with the needs of a growing world population. Restoring saline lands is an essential part of nutrition in the world (Yamaguchi and Blumwald, 2005). In soils rich with nutrients, soil processes are no longer carried out properly due to the fact excessive soil salinity appreciably inhibits microbial ecological characteristics (activity), which is in particular mirrored in microbial participation in elemental transformation. In general, microbial activity in soils is not affected by static salt because soil salinity is dependent on temperature changes (evaporation) and rainfall, which can fluctuate greatly.

Stresses can have adverse effects on growth and development, leading to reduced productivity in various major food crops (Kesawat *et al.*, 2021, 2023). Plants are mostly subjected to multifactorial stresses that can exert cumulative effects on plants due to embedded signalling pathways (Munns and Tester, 2008; Volkov and Beilby, 2017). Meanwhile, salinity stress is one of the most severe abiotic stresses that control and regulate complex phenotypic variations. Disrupting the metabolic activities of plants leads to electrolyte imbalance and thus affects their growth and productivity all over the world (Munns and Tester, 2008; Volkov and Beilby, 2017). Salinity stress affects many parts of cropland (Hasanuzzaman *et al.*, 2020). Salinity affects plants' growth and development through the following pathways: (1) the effect of osmotic, (2) the effect of ionic, (3) the imbalance of hormonal, (4) nutrient imbalance, and (5) the production of ROS (Sachdev *et al.*, 2021; Singhal *et al.*, 2021).

ROS encompass hydrogen peroxide (H_2O_2), hydroxyl radical ($HO\bullet$), alkoxyl radicals ($2.5-5.2$ M NaC), singlet oxygen ($^1O2$), superoxide radical ($O_2^-\bullet$), and peroxy radicals ($ROO\bullet$) (Table 8.1). They are produced under regular growth conditions (Mehla *et al.*, 2017, Hasanuzzaman *et al.*, 2019). Nevertheless, severe abiotic stress leads to excessive production of ROS (Gill and Tuteja, 2010), which can modify cellular and molecular compositions, for example, when enzymes, proteins, lipids, carbohydrates, and DNA are oxidized, causing cell death (Singh *et al.*, 2019). To prevent oxidative damage, plants carefully control ROS production through various enzymatic and

DOI: 10.1201/b23155-8

TABLE 8.1
Different Types of ROS ($^1O_2$, HO•, O2, and H_2O_2) and Their Place of Production, Mechanisms of Action, and Scavenging Systems

Types of ROS	Type of Radical	Production Sites	System of Scavenging	Action Mechanisms
Hydrogen peroxide (H_2O_2)	Non-radical	Apoplast, chloroplasts, cytosol, mitochondria, peroxisomes	Ascorbic acid, ascorbate peroxidase enzyme, catalase enzyme, glutathione antioxidant, guaiacol peroxidase, peroxiredoxin enzymes	It reacts with DNA. It oxidizes proteins and forms HO• and $O2^{\bullet-}$. It mainly reacts with proteins by attacking methionine and cysteine residues. It reacts with heme proteins.
Hydroxyl radical (HO•)	Free radical	Chloroplasts, mitochondria, plasma membranes	Ascorbate, flavonoids, proline, sugars	It reacts with all biological molecules, including proteins, lipids, RNA, and DNA.
Singlet oxygen ($^1O_2$)	Non-radical	Chloroplasts, mitochondria, nucleus, plasma membranes	α-tocopherol and carotenoids	It oxidizes proteins containing residues of cysteine, methionine, tryptophan, tyrosine, and histidine, and also oxidizes polyunsaturated fatty acids like methylene interrupted polyenes, and DNA guanine residues.
Superoxide ($O_2^{\bullet-}$)	Free radical	Apoplasts, chloroplasts, electron transfer chains, mitochondria, peroxisomes, organelles	Ascorbic acid, flavonoids, superoxide dismutase enzymes	It becomes H_2O_2 and reacts with proteins containing double bonds such as iron–sulphur proteins through the iron atom.

Source: Adapted from Kesawat *et al.*, 2023

non-enzymatic antioxidants. The enzyme antioxidant defence system of plants includes ascorbate peroxidase (APX), catalase (CAT), dehydroascorbate reductase (DHAR), glutathione reductase (GR), monodehydroascorbate reductase (MDHAR), peroxidase (POD), and superoxide dismutase (SOD). The non-enzymatic antioxidant system includes ascorbate (AsA), carotenoids, flavonoids, stilbenes, tocopherols, and other vitamins. These systems neutralize oxidative stress either by restoring the level of endogenous antioxidants or by directly detoxifying overproduced ROS. The combination of these factors leads to an increase in the level of tolerance to oxidative stress (Raja *et al.*, 2017; Mittler, 2017). Irregular production of ROS has adverse effects on plant health. An explanation and understanding of ROS molecular basis signalling and the related downstream pathways would aid in reducing stress in plants.

To have healthy, profitable, and biologically salt-tolerant crops, it is important to use natural salt-tolerant plants and/or to exploit the rhizosphere and plant-related endophyte microorganisms to increase stress tolerance endogenous and exogenous plant salinity (Slama *et al.*, 2019). These helpful microbes promote host plant tolerance to soil salinity, amplify soil fertility, promote plant growth, and take away extra salt in their host plant life (Slama *et al.*, 2021). Mechanisms used by these microorganisms to set off tolerance include osmotic balance, synthesis of compatible solutes, production of exopolysaccharides, a lipid bilayer of Gram-negative bacteria, bacterial consortia interactions, and enhancement/modification genetics of secondary metabolites by regulating gene expression (Aung *et al.*, 2018). Therefore, to deal with stresses, especially salinity stress in crop production, it is necessary to provide a sustainable solution. In this regard, it is necessary to investigate and study the diversity of microorganisms to increase our knowledge and understanding of their physiological and functional characteristics and meet their potential. The knowledge gained from this research will improve our comprehension of the salinity effects on soil microorganisms and their relationships with plants and may supply guidance for the rehabilitation of saline soils.

8.2 SALINE SOIL MICROORGANISMS

Since microbial diversity performs a necessary position in maintaining microbial ecological multifunctionality (Jiao *et al.*, 2021), the impact of soil salinity on microbial diversity deserves unique attention. Analysis of the results obtained from surveys carried out in 111 areas of the world showed that the effect of salt stress on the composition of the microbial community is more predominant than factors such as pH, temperature, or other environmental physicochemical factors (Lozupone and Knight, 2007). Many studies confirmed that salinity decreases microbial activity and microbial biomass and changes microbial community structure (Rousk *et al.*, 2011; Setia *et al.*, 2011). In addition to the structure of the microbial community, the stress of salinity can also alter interactions among species (Hernandez *et al.*, 2021). There is growing proof that the function and consistency of bacterial communities are strongly influenced by this interplay (Rodriguez-Verdugo and Ackermann, 2021). By examining the richness of microbial community composition and molecular ecology networks, we can better understand microbial communities' succession because of soil salinity. Today, increasingly, agricultural and soil scientists emphasize the study and investigation of the activity of soil microorganisms to evaluate the health and stability of soil ecosystems (Guerra *et al.*, 2021). It was shown that microbial activity pursues a non-linear response to salinity, where slight salinity variations beyond the threshold will cause drastic oscillations in the activity of microorganisms (Jeppesen *et al.*, 2020). Some reports have shown that the direct negative effects of salinity on microbes might not be the sole cause of the inhibition of microbial activity (Rath and Rousk, 2015).

This inhibition may also be related to the lack of organic matter (OM), which leads to poor plant growth in saline soils. The addition of exogenous organic matter to saline soils can significantly enhance microbial activity (Wong *et al.*, 2009); however, there is little data on the growth reaction of microbial communities to carbon input in soils of saline. Microbial strategies to assimilate C to keep and grow will supply effective details on the physiological mechanisms of microorganisms' resistance to environmental stress. Generally, diversity is considered a favourable characteristic that reflects the capacity of the ecosystem to supply perfect functions (Naeem *et al.*, 2012). Several reports have demonstrated that bacterial richness is not always affected by diverse salinity amounts. Greater species richness and higher Shannon-Wiener index were observed at high salinity levels in a coastal area in central Italy (Canfora *et al.*, 2017). Additionally, surprisingly abundant and diverse bacterial communities with a wide range of species have been found in highly saline environments (Kalwasinska *et al.*, 2019). In hypothesis, the survival techniques of microbes and the assembly mechanisms of microbial communities are mostly determined by their historical biotic and abiotic situations (Stegen *et al.*, 2018).

It is thought that most species could originate from marine environments and therefore could have a potent adaptation to soil salinity. Therefore, severe salinity stress does not reduce soil bacterial diversity. Conversely, short-term increases in salinity can greatly decrease the richness of the microbial community, as demonstrated by an anaerobic membrane bioreactor with added NaCl (Chen *et al.*, 2019). Results from many prior studies show that soil salinity, expressed as EC and Na, significantly affects bacterial composition (Ibekwe *et al.*, 2017; Huang *et al.*, 2019). Such a change is attributed to the selection of salt applied to organisms rather than nutrient status (Hu *et al.*, 2022). In such environments, salt-tolerant organisms are more favoured, resulting in an evolution towards a more tolerant community (Rath and Rousk, 2015). Firmicutes are generally considered the particular indicators in high-salinity soils (Wang *et al.*, 2020). On the other hand, a significant bacterial richness has also been reported in highly saline soils (Dong *et al.*, 2022). Bacteria may share common strategies and strengthen their collaboration to minimize salt stress (Zheng *et al.*, 2017). There have been many reports that bacterial species, including the genera *Agrobacterium*, *Bacillus*, *Enterobacter*, *Klebsiella*, *Ochromobacter*, *Pseudomonas*, and *Streptomyces*, have been proposed on an extensive scale as effective biological inoculants in saline soils in agricultural areas (Sharma *et al.*, 2016; Singh and Jha, 2016; Sarkar *et al.*, 2018). Prior studies (Ansari *et al.*, 2019; Xie *et al.*, 2019) have shown a significant share of various bacterial species belonging to *Bacillus*, *Exiguobacterium*, and *Pseudomonas* (Kasana and Pandey, 2018) genera in plant growth promotion in growth-limiting conditions.

In areas with saline soils, though the low diversification of functional properties of a microbial community is no longer the highest quality for its steadiness and sustainability (Blüethgen *et al.*, 2016), increases in cell–cell dependency amongst microbial individuals can toughen their connections and neutralize this negative effect. It is important to remember that it takes a lot of energy to support and protect many of these links and to maintain such a complex network. In areas with very saline soils, respiratory activities are increased, also due to high cell-to-cell dependence, and a large amount of C is used, and as a result, the percentage of C that remains as biomass in the soil is greatly reduced.

8.3 INTERACTION BETWEEN MICROORGANISMS AND PLANTS IN SALINE SOILS

It can simply be said that salinity is a type of non-biological (chemical) factor that leads to the accumulation of soluble salt in the root zone (rhizosphere system). It is estimated that the world population will reach around 8 billion people by 2025. To prevent or minimize food deficiencies, saline soils must be restored and managed to meet the nutritional needs of growing human populations (Ladeiro, 2012). Microorganisms of soil make up much less than 0.5% (w/w) of soil mass; however, they have a key function in soil attributes and processes. Salinity can affect plants and microorganisms through osmotic stress or special ionic effects (Oren, 1999; Chhabra, 1996). Soluble salts reduce the osmotic potential of soil water to negative values, and as a result, they cause water to escape from plant cells (plasmolysis) and reduce soil microbial activities. On the other hand, a low osmotic potential (less negative) makes it more difficult for roots and microbes to absorb water from the soil (Oren, 1999). Plants and microorganisms can adapt to the low osmotic potential of their environment by accumulating compounds such as osmolytes; nevertheless, large amounts of energy are required for osmolyte synthesis, which can lead to reduced growth and development (Oren, 1999; Wichern *et al.*, 2006). On the other hand, special ions such as Na^+, Cl^-, and $HCO3^-$ at high concentrations can be toxic to many plants (Chhabra, 1996).

Until now, one of the ways to revitalize these soils was the use of different types of mineral and organic fertilizers (Qayyum *et al.*, 2016). At the same time, farmers and biotechnologists are in a constant struggle to develop salinity-tolerant cultivars from various routes, including natural selection, QTL mapping, marker-assisted selection, or genetic manipulation through the introduction of salinity-tolerant crops (Qadir *et al.*, 2017). Despite the positive results of applying such biological tools at the field level, for various reasons, this increase in stress tolerance among agricultural and economic products has not been completely satisfactory (Khare *et al.*, 2018). Recently, the exploitation and use of root-adherent plant growth-promoting rhizobacteria (PGPR) established in hypersaline conditions as an alternative biological approach with suitable adaptation to the environment to achieve better product productivity from salt-deteriorating land has been considered (Talaat, 2015). The enhancement of plant growth using these microbes is well documented (Barnawal *et al.*, 2012; Bharti *et al.*, 2014). An increase in the index of salt tolerance and a significant enhancement in percentage germination and seedling growth was reported in wheat after inoculation with *P. fluorescens* under salinity caused by NaCl (Safari *et al.*, 2018). A similar reaction to parameters of barley growth and yield (spike length, fertility index, and seed weight) was observed when it was treated with ACC deaminase-producing *P. fluorescens* under salinity stress (Azadikhah *et al.*, 2019). Many reports about *Bacillus* sp.'s potential have been documented. In a study, it was well shown that bacterial strains belonging to the Bacillus genus played an important role in reducing salinity stress in wheat due to their ability to generate EPS, ACCD, and IAA in laboratory conditions (Din *et al.*, 2019). In addition, reports showed that Bacillus strains resistant to salinity stress could help to increase KC and

Ca_2C absorption and increase the content of protein and nitrogen in rice seedlings grown under salt-stress conditions (Khattak *et al.*, 2019). It has been illustrated that plants inoculated with *B. pumilus* exhibited maximum fresh and dry root weight, accumulation of free proline and total soluble protein content in leaf tissues, and reduced activity of activity enzymatic antioxidants (Nawaz *et al.*, 2020).

A reduction in antioxidant activity is one of the clear hallmarks of PGPR-induced modulation of plant physiology, resulting in reduced ROS content (Singh and Jha, 2016). Some PGPR strains belonging to the genera *Azospirillum*, *Azotobacter*, *Bacillus*, *Enterobacteria*, *Ochrobactrum*, *Pseudomonas*, *Rhizobium*, *Serratia*, and *Stenotrophomonas* (Maleki *et al.*, 2011) have been widely reported and used as efficient candidates to be formulated as biofertilizers (green method) that can be a suitable alternative in the agriculture of saline areas (Egamberdieva *et al.*, 2019). Plant growth–promoting rhizobacteria (halophilic/halotollogenic) utilize important mechanisms by colonizing the rhizosphere of the plant to cope with severe environmental stresses and the resulting deleterious performance penalties. Also, these rhizobacteria can play an important role in activities such as the synthesis of new osmolytes to regulate cell osmosis, to regulate ion transporters, to maintain homeostasis, to reduce the toxic effects of Na^+ and Cl^- ions, and to activate the ROS-inhibiting defense system of plants to deal with the harmful effects of oxidative stress (Munns and Tester, 2008).

It has been hypothesized that plant growth–promoting rhizobacteria living in hypersaline ecological situations have the potential to regulate plant physiology by inducing systemic resistance to promote growth in soils degraded by salt. Halophilic and halo-tolerant microorganisms can survive in habitats of hyper-saline (Talaat, 2018). Important mechanisms that are responsible for their survivorship in stressful environments include de novo synthesis or osmotic buffer uptake and specialized ion transport systems such as NaC/HC countertransport (Egamberdieva *et al.*, 2019).

8.4 ADAPTATION STRATEGIES OF BACTERIA AND FUNGI IN SALINE ECOSYSTEMS

A salt-stressed situation negatively affects vital soil activities such as respiration, nitrification, microbial diversity, and mineralization residue decomposition (Schirawski and Perlin, 2018). In a dry and hot environment where low humidity and soil salinity are the most stressful factors for soil microbial variety, the activity and metabolism of the microorganisms decrease (Szabo *et al.*, 2016). Salinity affects almost every feature of plant physiology, morphology, and biochemistry and therefore causes significant loss of crop yield. A higher concentration of salt in soil restricts the uptake of water and critical nutrients by roots (Singh *et al.*, 2018).

Life under the stress of salinity has a requirement of high bioenergetics because the microflora needs to continue the osmotic equilibrium among the cytoplasm of the microbes and the surrounding environment (Figure 8.1). Microbes under salt stress situation survive by excluding the sodium ions from the cell inside; therefore, microorganisms need high energy, which is sufficient for osmoregulation (Oren, 2002). Salt-tolerant microbes are divided into four groups: non-halophilic <0.2 M NaCl, slight halophilic $0.2 - 0.5$ M NaC, moderate halophilic $0.5 - 2.5$ M NaC, and extreme halophiles $2.5 - 5.2$ M NaC (Kumawat *et al.*, 2022). The halo-tolerant microorganisms can endure high salt concentrations but grow best in media containing <0.2 M (1%) salt. This description of 'halo-tolerant' is generally accepted (Kushner, 1993; Gros *et al.*, 2003).

Saline soil consists of an abundance of halophilic microorganisms in the soil and most dominantly belongs to the genera of *Bacillus*, *Pseudomonas*, *Micrococcus*, and *Alcaligenes* (Chowdhury *et al.*, 2011). The accumulation of the ions that are essential for the metabolism of cells occurs in halo-tolerant microbes. The other mechanism of cell adaptation in salt stress conditions is the production of organic compounds that will neutralize the concentration gradient between the cell cytoplasm and soil solution (Wichern *et al.*, 2006). The microorganisms are exposed to the high-osmotic environment; rapid fluxes of the cell water out of the cell occur, resulting in a decrease in the turgor and the dehydration of the cytoplasm. Different types of adaptation have been achieved against the outflow of the cell water (Miao *et al.*, 2014).

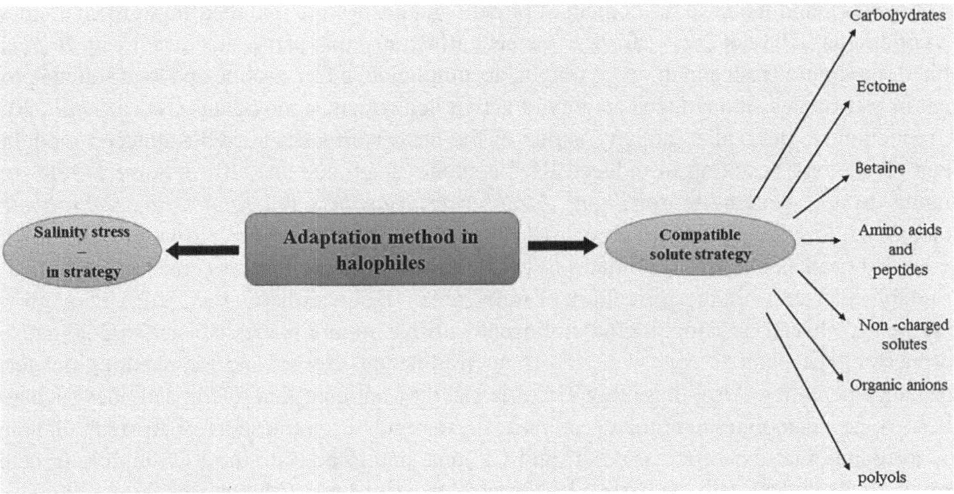

FIGURE 8.1 Various adaptation strategies in saline soils by halophilic bacteria (adapted from Kumawat *et al.*, 2022).

Halophilic bacteria are capable of balancing the osmotic pressure in the environment. They adapt to salinity by a different method, assisting the plant in surviving under salt-stress situations. To avoid excessive water loss in such an environment and preserve cellular structure, halophilic bacteria have evolved two techniques to deal with high salt concentrations (Morton *et al.*, 2019). The first strategy is the salt-in strategy, whereas the second is the compatible solute strategy.

Bacterial cells keep the internal and exterior environments osmotically equal by collecting a high concentration of KCl. This process is carried out by the cell through changes in various physiological metabolisms such as enzyme activity, cellular component production, and the shape and function of organelles (Wood, 2011; Wang *et al.*, 2018a).

The high-salt-in method protects halophiles from a saline environment by accumulating inorganic ions intracellularly to keep the salt concentrations in their environment balanced. Halophiles consist of the Cl^- pumps and transfer Cl^- from the environment into the cytoplasm. Most halophilic microorganisms protect the cell from high salt concentrations by the accumulation of compatible solutes such as organic (proline, betaine, ectoine, trehalose) and inorganic solutes $\left(K^+, Mg^{2+}, Na^+\right)$. The osmolytes (compatible solutes) are released in the cytoplasm by the bacterial cell itself or they are taken from the medium (Moghaddam *et al.*, 2016). Compatible solutes are low-molecular-weight compounds that are synthesized by bacteria to preserve cellular turgidity and electrolyte concentrations. The role of compatible solutes overlaps with that of potassium uptake in the maintenance of osmotic equilibrium (Ruppel *et al.*, 2013; Tanimura *et al.*, 2016).

Generally, bacteria maintain high intracellular potassium and low sodium concentrations for effective cellular enzyme activity, and some marine bacteria may keep a sodium gradient for energy metabolism. In addition, halophilic bacteria such as Halomonas and Halobacterium synthesize ectoine and L-glutamate, respectively, to survive in the salinity-stressed environment (Youssef *et al.*, 2014). Some halotolerant bacteria, including Bacillus, Aeromonas, Pseudomonas, and Zymomonas, use the polyols compounds such as sorbitol, glycerol, arabitol, and mannitol for osmoadaptation (Edbeib *et al.*, 2016).

Several polar and non-charged organic molecules have also been used as osmolytes to protect cells. Some sugar molecules such as trehalose have been detected in halotolerants and halophiles and have been used as compatible solutes to cope with desiccation, cold, heat, and a hypersaline environment. To adapt to high salt concentrations, cyanobacteria accumulate organic osmotic solutes. The most halophilic or halotolerant cyanobacteria usually accumulate glycine betaine to reach osmotic balance. The use of L-glutamate betaine (N-trimethyl-L-glutamate) has also been reported (Tanimura *et al.*, 2016; Singh and Singh, 2015). Several proteobacteria and marine cyanobacteria are known to accumulate sucrose as an osmolyte in salt-stressed environments. Some proteins such as proline, acetylated glutamine dipeptide, and carboxamine also act as osmolytes and protect the cell. They are mostly found in halophilic purple sulphur bacteria and marine phototrophic bacteria (Weralupitiya *et al.*, 2022).

On the other hand, several bacteria with particular mechanisms are present, which help to solubilize the iron nutrient and make it accessible to the plants. Halotolerant bacteria assist the plant in surviving under salt stress situations and increase iron accessibility through siderophore synthesis. Siderophores are low-molecular-weight compounds that chelate iron and transfer it into cells (Kuzyk *et al.*, 2021). Additionally, halophilic strains help convert insoluble phosphorus to soluble phosphorus and maintain soil P levels. Much research has been conducted on salt-tolerant strains capable of solubilizing and making phosphorus available (Razzaghi Komaresofla *et al.* 2019).

Many scientists have studied the effect of mycorrhiza on plants' adaptation to salt stress. Fungi seem more sensitive to the salinity situation and osmotic stress than other microorganisms.

Arbuscular mycorrhizal fungi (AMF) are fungi that have a symbiotic relationship with terrestrial plant roots. Mycorrhizal inoculation influences nutrient solubilization and mobilization, photosynthesis efficiency, and physiological and biochemical performance of plant growth and helps to reduce salt tolerance (Gopal *et al.*, 2012).

Mycorrhizal fungi play a critical role in membrane transport proteins and ion regulation, which govern the host plant's ion homeostasis. Consequently, it is clear that AMF's association with plants significantly enhances the concentration of macro- and micronutrients. The AMF relationship considerably increases antioxidant activities such as peroxidase, catalase, superoxide dismutase, and others.

Also, arbuscular mycorrhizal fungi play an important role in improving soil quality and health. Glomalin, a glycoprotein produced by AMF hyphae in the soil, aids in soil aggregation (Smith *et al.*, 2011). The positive effect of mycorrhizal fungus on some plants' growth under salt-stress conditions was related to an increase in proline synthesis and phosphorus absorption. In addition, it was shown that AMF improved the performance of photosystem II in plants under salinity stress by civilizing the consumption of photons and electron transport and reducing photo-inhibition (Wang *et al.*, 2018). Therefore, various plant species under salinity stress show different responses to AMF inoculation (Table 8.2).

TABLE 8.2
Plants' Response to Arbuscular Mycorrhizal Fungi Inoculation in Saline Environments

AMF	Host Plants	Kind of Plant Response	References
Glomus mosseae	False wheatgrass	Improve colonization rate; increase seedling weight, water content, and amount of phosphorus and nitrogen	Lin *et al.*, 2017
Rhizophagus irregularis	Tomato	Increase in leaf area, number of leaves, dry weight of root and shoot, and growth hormones	Khalloufi *et al.*, 2017
Glomus intraradices	Tomato	Cause improvement in chlorophyll content, dry matter, growth parameters, and ion uptake	Hajiboland *et al.*, 2010
Claroideoglomus etunicatum	Rice	Increase PSII quantum efficiency and photosynthetic rate	Porcel *et al.*, 2015
Claroideoglomus etunicatum	Aeluropus	Increase root, dry mass of shoot, free amino acids (FAAs), total soluble sugar (TSS)	Hajiboland *et al.*, 2015
Glomus fasciculate	Gum arabic tree	Improve root growth, increase the amount of dry mass of shoots and the content of phosphorus, zinc, and copper	Giri *et al.*, 2007
Glomus etunicatum, Glomum, intraradices, Glomus mosseae	Cucumber	Increase biomass amount, the synthesis of photosynthetic pigments, and antioxidant enzyme content	Hashem *et al.*, 2018

Source: Adapted from Kumawat *et al.*, 2022

8.5 ANTIOXIDATIVE ACTIVITY

Salt stress induces the creation of reactive oxygen species (ROS), which destroy various biomolecules such as lipids and proteins and cause plant death (del Río *et al.*, 2003). Salt stress interferes with carbon metabolism and thus fosters ROS generation (Figure 8.2). Stomatal conductance decreases during stress and lowers transpiration. Under salinity stress, each type of organelle employs different mechanisms of ROS creation. Conversely, cell metabolism generates ROS at low rates as a regular side product. Salinity stress often increases the generation of reactive oxygen species. This may lead to metabolic disorders, cellular damage, and untimely senescence (Acosta-Motos *et al.*, 2015; Miller *et al.*, 2010). To manage the oxidative stress resulting from ROS, higher plants have developed a complex scavenging system as well as enzymatic and non-enzymatic (antioxidants). Coordinated work of ROS-scavenging pathways from different cellular compartments in modulating the level of ROS in cells preventing cellular damage and controlling ROS signalling may play a key role in plant salt tolerance (Gill and Tuteja, 2010). There are many antioxidative enzymes (superoxide dismutase, catalase, and peroxidase) and non-enzymatic antioxidants (ascorbic acid, glutathione) that help inhibit ROS (Miller *et al.*, 2010; Gilroy *et al.*, 2014; Ferreira *et al.*, 2008).

Many studies have confirmed that salt-tolerant plant growth-promoting rhizobacteria (PGPR) are supplied by different mechanisms to plants. Some halotolerants, such as *Rhizobium leguminosarum*, are known to produce these enzymes (SOD, POX, CAT) and assist in the plant's survival under salt stress environments (Vaishnav *et al.*, 2016).

PGPR inoculation mitigates the harmful effects caused by oxidative stress through enzymatic and non-enzymatic mechanisms under saline-stressed situations. The significant function of PGPR is to increase key physiological processes in plants, including photosynthesis and mineral and water uptake (Ilangumaran and Smith, 2017). The pigment production and packaging of the DNA with proteins and chromatins present alternate sites for the attack of reactive oxygen species. Several non-enzymatic antioxidant compounds also prevent reactive oxygen species (Rahman *et al.*, 2021; Safdarian *et al.*, 2019). On the other hand, the enzymatic method produces different enzymes such as superoxide dismutase, glutathione peroxidases, catalases, and peroxiredoxins without generating further reactive species. Antioxidant enzymes alter harmful products into less harmful molecules or locate them and degrade them. These methods also preserve the suitable physiological levels of the reactive species such as ROS (Santoyo *et al.*, 2021). Synthesis and accumulation of osmolytes are one of the first responses of plants to combat osmotic and oxidative damage imposed by salinity stress. PGPR help plants accumulate compatible osmolytes in a saline situation to minimize osmotic stress, maintain high turgor pressure, and sustain ion equilibrium in the cytoplasm (Sunita *et al.*, 2020).

8.6 FUTURE CHALLENGES FOR SALT STRESS MIGRATION THROUGH HALOPHILIC MICROORGANISMS

All soils certainly contain a mixture of water-soluble salts, some of which are essential for plant growth. Because of hydrological and climatic events, in past eras, there have been large quantities of salt in sedimentary rocks, in surface and detrital waters, and in seas and oceans (Zaman *et al.*, 2018). From these deposits, through different mechanisms, the salts reach the soil. In some situations, the formation of a water evaporation front in the soil moves water by evaporation rather than by percolation; thus the salts remain in the soil and accumulate (Daliakopoulos *et al.*, 2016). Salinization represents a serious threat to agriculture because it can cause the modification of the delicate balance of ecological processes occurring in soil. Salinity affects almost every aspect of plant physiology, morphology, and biochemistry and therefore causes a major loss of crop yield. A higher concentration of salt in soil restricts the uptake of water and critical nutrients by plant roots (Kumar *et al.*, 2020; Zaman *et al.*, 2018).

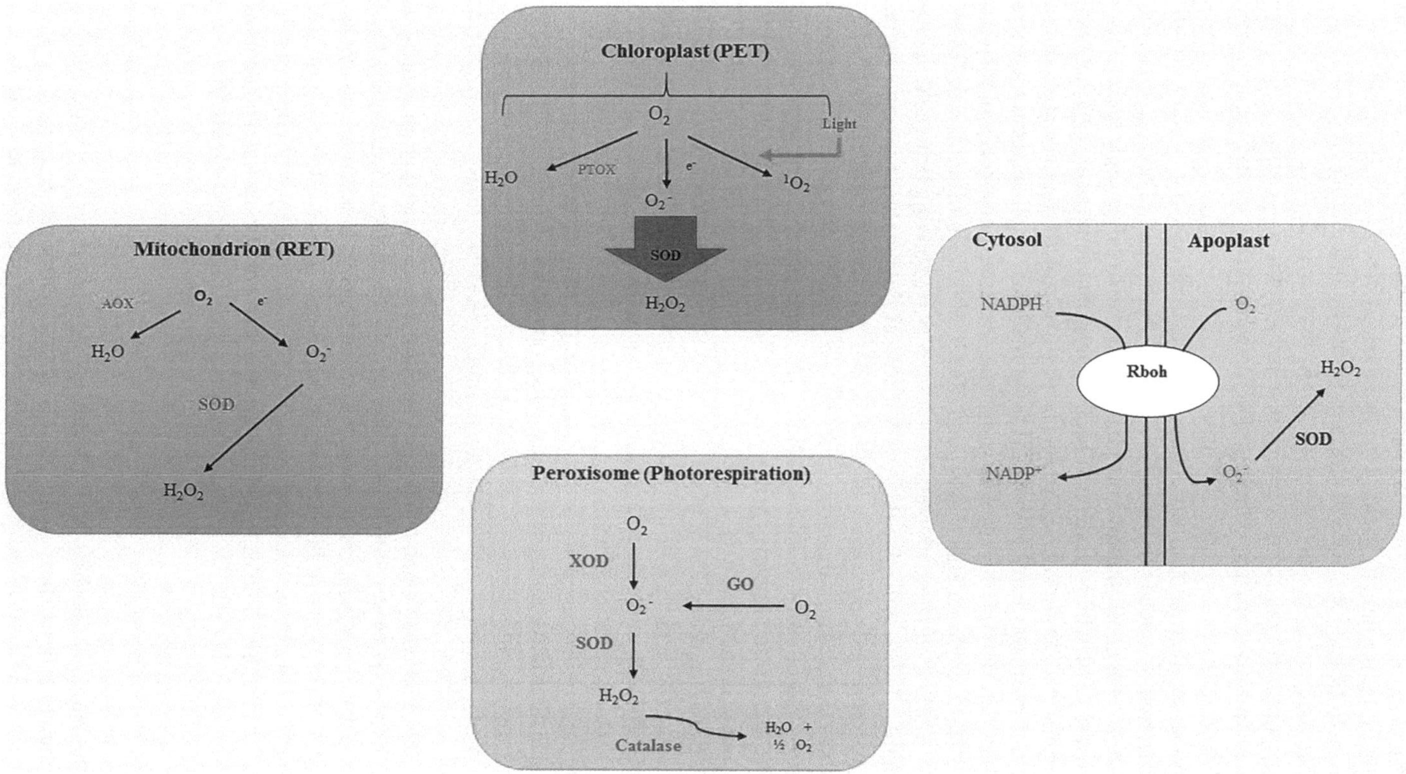

FIGURE 8.2 Several major sites of ROS production in plant cells under salt stress conditions include chloroplasts, mitochondria, peroxisomes, and plasma membranes. AOX (alternative oxidase); GO (glycolate oxidase); PET (electron transport of photosynthetic); PTOX (plastid terminal oxidase); RBOH (respiratory burst oxidase homolog); RET (electron transport of respiratory); SOD (superoxide dismutase); XOD (xanthine oxidase) (adapted from Hossain and Dietz, 2016).

The known halophilic plant growth-promoting microbes need to be applied in agriculture to improve crop yields under a stress environment. High salinity decreases the number of organisms where just halophilic or halotolerant ones can survive in such a hypersaline ecosystem (Ventosa *et al.*, 2015). The development of biological products based on valuable halophiles can expand the variety of options for maintaining the healthy yield of crops in salt-affected soils. In recent years, a new approach has been developed to improve salt stress in plants by inoculating crop seeds and seedlings with salt-tolerant plant growth-promoting microbes (Kumawat *et al.*, 2022). Therefore, to provide a sustainable solution for agriculture and cope with salinity stress, it is critical to discover the variety of microbes to understand their physiological and functional features and harness their potential (Zribi *et al.*, 2012).

In addition, studies on the detailed mechanism of mycorrhizal fungi-associated plant growth under salt stress are lacking, and this needs to be explored. As a mechanism of soil-plant natural ecosystems, symbiosis with AM fungi can improve the plant's response to salinity and have beneficial effects on plant growth and yield, which makes AM fungi appropriate candidates to bio-ameliorate salinized soils (Evelin *et al.*, 2009; Giri *et al.*, 2013). The positive growth response observed in mycorrhized plants under salt stress was linked to an AM fungi-mediated improvement in nutrient acquisition, particularly phosphorus. This effect was significant because phosphorus absorption (Pi) is typically reduced in saline environments (Shokri and Maadi, 2009; Smith and Smith 2015).

The hopeful approach toward tackling the problem of soil salinity by utilizing beneficial microorganisms including halophilic PGPR will play the greatest role in the agricultural economy, as they supply a cheap and environmental approach to alleviate salt stress (Kumawat *et al.*, 2022). The expansion of biological products based on valuable microorganisms can extend the range of options for maintaining the healthy yield of crops in saline habitats. In recent years, a new approach has been developed to improve salt stress in plants by treating crop seeds and seedlings with PGPR. The great opportunity for salt tolerance research is its ability to be combined with halophilic PGPR (Pérez-García *et al.*, 2023; Kumawat *et al.*, 2022). PGPR improves seed germination, as it facilitates the production of metabolites; the synthesis of antibiotics, enzymes, and phytohormones (auxin, cytokinin, and gibberellic acid); and other related activities. By enhancing phosphate solubilization in the soil and promoting root colonization (Swarna Lakshmi *et al.*, 2020), which increases nutrient uptake capacity and protection against plant pathogens, plant growth-promoting rhizobacteria (PGPR) present a sustainable alternative for agricultural production (Nazir *et al.*, 2020). Therefore, microbial formulation and application technology are critical for the development of commercial salt-tolerant bioformulation that is useful under salt-stress situations (Pérez-García *et al.*, 2023; Kumar *et al.*, 2020).

The increasing rate and development of areas under salinity stress have created an insecurity of food demands in several countries. However, despite the considerable research that has already been done in this area, there is a lack of knowledge about some important aspects. In addition, the mechanisms of growth promotion in the presence of PGPB under salinity stress still need to be determined in more detail up to the gene level. Additional research should focus on the operation of effective salt-tolerant PGPB in salt-affected agricultural fields to endorse the development of bacterial inoculants as commercial biofertilizers for improving salinity tolerance and crop yield (Kumar *et al.*, 2020).

8.7 CONCLUSION

Plants and microorganisms have relationships and interactions through a very complex mechanism. To ensure a good mutual relationship, a network of signals, hormones, enzymes, volatile compounds, genes, and metabolites controls these interactions. We tried to explain as much as possible the mechanisms involved in salt reduction by plants and bacteria. Changes in soil microbial community composition were largely influenced by high soil salinity; also, salinity stress strengthened and expanded the dependence between species. Nevertheless, microbial diversity response to increasing salinity levels in agricultural soils (limited salinization) is still not fully understood.

Based on previous studies, it was concluded that PGPR could have beneficial effects on the physical and chemical characteristics of inoculated plants. This leads to the reduction of damage caused by salinity stress by improving water relations, increasing the accumulation of compatible solutes, stimulating the absorption of elements such as potassium and calcium, and reducing antioxidant activity. These changes in the state of cell metabolism can ultimately lead to improved growth and performance of varieties resistant and sensitive to salt stress. In addition, due to climate change, genetic manipulation of signalling pathways will help to develop better stress-resistant cultivars and minimize stress-induced losses. Strong salt-resistant bacteria isolated from halo-tolerant plants can be used in agriculture to ameliorate the resistance of agricultural plants to increased salt concentration in soil and water. This biological modification is simple, economical, and eco-friendly. Nevertheless, more attention has been paid to the benefits provided by endophytic microorganisms to better plant tolerance to saline conditions.

REFERENCES

Acosta-Motos, J.R., Diaz-Vivancos, P., Àlvarez, S., Fernández-García, N., Sanchez-Blanco, M.J., Hernández, J.A. 2015. Physiological and biochemical mechanisms of the ornamental *Eugenia myrtifolia* L. plants for coping with NaCl stress and recovery. *Planta*. 242: 829–846. doi: 10.1007/s00425-015-2315-3.

Ansari, F.A., Ahmad, I., Pichtel, J. 2019. Growth stimulation and alleviation of salinity stress to wheat by the biofilm-forming *Bacillus pumilus* strain FAB10. *Appl. Soil Ecol.* 143: 45–54. doi: 10.1016/j.apsoil.2019.05.023.

Aung, K., Jiang, Y., He, S.Y. 2018. The role of water in plant-microbe interactions. *Plant J.* 93(4): 771–780. doi: 10.1111/tpj.13795.

Azadikhah, M., Jamali, F., Nooryazdan, H.R., Bayat, F. 2019. Growth promotion and yield enhancement of barley cultivars using ACC deaminase-producing *Pseudomonas fluorescens* strains under salt stress. *Span. J. Agric. Res.* 17: e0801. doi: 10.5424/sjar/2019171-13828.

Barnawal, D., Bharti, N., Maji, D., Chanotiya, C. S., Kalra, A. 2012. 1-Aminocyclopropane-1-carboxylic acid (ACC) deaminase-containing rhizobacteria protect Ocimum sanctum plants during waterlogging stress via reduced ethylene generation. *Plant Physiol. Biochem.* 58: 227–235. doi: 10.1016/j.plaphy.2012.07.008.

Bharti, N., Barnawal, D., Awasthi, A., Yadav, A., Kalra, A. 2014. Plant growth-promoting rhizobacteria alleviate salinity-induced negative effects on growth, oil content, and physiological status in *Mentha arvensis*. *Acta Physiol. Plant.* 36: 45–60. doi: 10.1007/s11738-013-1385-8.

Blüethgen, N., Simons, N.K., Jung, K., Prati, D., Renner, S.C., Boch, S., Fischer, M., Hoelzel, N., Klaus, V.H., Kleinebecker, T., *et al.* 2016. Land use imperils plant and animal community stability through changes in asynchrony rather than diversity. *Nat. Commun.* 7: 10697. doi: 10.1038/ncomms10697.

Butcher, K., Wick, A.F., DeSutter, T., Chatterjee, A., Harmon, J. 2016. Soil salinity: A threat to global food security. *Agron. J.* 108: 2189–2200. doi: 10.2134/agronj2016.06.0368.

Canfora, L., Salvati, L., Benedetti, A., Francaviglia, R. 2017. Is soil microbial diversity affected by soil and groundwater salinity? Evidence from a coastal system in central Italy. *Environ. Monit. Assess.* 189(7): 319. doi: 10.1007/s10661-017-6040-1.

Chen, L., Hu, Q., Zhang, X., Chen, Z., Wang, Y., Liu, S. 2019. Effects of salinity on the biological performance of anaerobic membrane bioreactor. *J. Environ. Manag.* 238: 263–273. doi: 10.1016/j.jenvman.2019.03.012.

Chhabra, R. 1996. *Soil Salinity and Water Quality*. Rotterdam, The Netherlands: Balkema.

Chowdhury, N., Marschner, P., Burns, R.G. 2011. Soil microbial activity and community composition: Impact of changes in matric and osmotic potential. *Soil Biol. Biochem.* 43: 1229–1236.

Daliakopoulos, I.N., Tsanis, I.K., Koutroulis, A., Kourgialas, N.N., Varouchakis, A.E., Karatzas, G.P., Ritsema, C.J. 2016. The threat of soil salinity: A European scale review. *Sci. Total Environ.* 573: 727–739. doi: 10.1016/j.scitotenv.2016.08.177.

del Río, L.A., Corpas, F.J., Sandalio, L.M., Palma, J.M., Barroso, J.B. 2003. Plant peroxisomes, reactive oxygen metabolism, and nitric oxide. *IUBMB Life*. 55(2): 71–81. doi: 10.1080/1521654031000094694.

Din, B.U., Sarfraz, S., Xia, Y., Kamran, M.A., Javed, M.T., Sultan, T., Hussain Munis, M.F., Chaudhary, H.J. 2019. Mechanistic elucidation of germination potential and growth of wheat inoculated with exopolysaccharide and ACC-deaminase producing Bacillus strains under induced salinity stress. Ecotoxicol. *Environ. Saf.* 183: 109466. doi: 10.1016/j.ecoenv.2019.109466.

Dong, Y., Zhang, J., Chen, R., Zhong, L., Lin, X., Feng, Y. 2022. Microbial Community Composition and Activity in Saline Soils of Coastal Agro–Ecosystems. *Microorganisms*. 10(4): 835. doi: 10.3390/microorganisms10040835.

Edbeib, M.F., Wahab, R.A., Huyop, F. 2016. Halophiles: Biology, adaptation, and their role in decontamination of hypersaline environments. *World J. Microbiol. Biotechnol.* 32(8): 135. doi: 10.1007/s11274-016-2081-9.

Egamberdieva, D., Wirth, S., Bellingrath-Kimura, S.D., Mishra, J., Arora, N.K. 2019. Salt-tolerant plant growth promoting rhizobacteria for enhancing crop productivity of saline soils. *Front. Microbiol.* 10: 2791. doi: 10.3389/fmicb.2019.02791.

Evelin, H., Kapoor, R., Giri, B. 2009. Arbuscular mycorrhizal fungi in alleviation of salt stress: A review. *Ann. Bot.* 104: 1263–1280. doi: 10.1093/aob/mcp251.

Ferreira, A.L., Arrabaca, J.D., Vaz-Pinto, V., Lima-Costa, M.E. 2008. Induction of alternative oxidase chain under salt stress conditions. *Biol. Plant.* 52: 66–71. doi: 10.1007/s10535-008-0009-4.

Gill, S.S., Tuteja, N. 2010. Reactive oxygen species and antioxidant machinery in abiotic stress tolerance in crop plants. *Plant Physiol. Biochem.* 48: 909–930. doi: 10.1016/j.plaphy.2010.08.016.

Gilroy, S., Suzuki, N., Miller, G., Choi, W.G., Toyota, M., Devireddy, A.R., Mittler, R. 2014. A tidal wave of signals: Calcium and ROS at the forefront of rapid systemic signaling. *Trends Plant Sci.* 19(10): 623–630. doi: 10.1016/j.tplants.2014.06.013.

Giri, B., Kapoor, R., Mukerji, K.G. 2007. Improved tolerance of *Acacia nilotica* to salt stress by *arbuscular mycorrhiza, Glomus fasciculatum* may be partly related to elevated K/Na ratios in root and shoot tissues. *Microb. Ecol.* 54(4): 753–760. doi: 10.1007/s00248-007-9239-9.

Giri, B., Kapoor, R., Mukerji, K.G. 2013. Influence of arbuscular mycorrhizal fungi and salinity on growth, biomass, and mineral nutrition of *Acacia auriculiformis*. *Biol. Fertil. Soils.* 38: 170–175. doi: 10.1007/s00374-003-0636.

Gopal, S., Chandrasekaran, M., Shagol, C., Kim, K., Sa, T. 2012. Spore-associated bacteria (SAB) of arbuscular mycorrhizal fungi (AMF) and plant growth-promoting rhizobacteria (PGPR) increase nutrient uptake and plant growth under stress conditions. *Korean J. Soil Sci. Fertil.* 45(4): 582–592. doi: 10.7745/KJSSF.2012.45.4.582.

Gros, R., Poly, F., Jocteur-Monrozier, L., Faivre, P. 2003. Plant and soil microbial community responses to solid waste leachates diffusion on grassland. *Plant Soil.* 255: 445–455.

Guerra, C.A., Bardgett, R.D., Caon, L., Crowther, T.W., Delgado-Baquerizo, M., Montanarella, L., Navarro, L.M., Orgiazzi, A., Singh, B.K., Tedersoo, L., *et al.* 2021. Tracking, targeting, and conserving soil biodiversity. *Science.* 371: 239–241. doi: 10.1126/science.abd7926.

Hajiboland, R., Aliasgharzadeh, N., Laiegh, S.F., Poschenrieder, C. 2010. Colonization with arbuscular mycorrhizal fungi improves salinity tolerance of tomato (*Solanum lycopersicum* L.) plants. *Plant Soil.* 331: 313–327. doi: 10.1007/s11104-009-0255-z.

Hajiboland, R., Dashtebani, F., Aliasgharzad, N. 2015. Physiological responses of halophytic C4 grass *Aeluropus littoralis* to salinity and *arbuscular mycorrhizal* fungi colonization. *Photosynthetica.* 53(4): 572–584. doi: 10.1007/s11099-015-0131-4.

Hasanuzzaman, M., Bhuyan, M.B., Anee, T.I., Parvin, K., Nahar, K., Mahmud, J.A., Fujita, M. 2019. Regulation of ascorbate-glutathione pathway in mitigating oxidative damage in plants under abiotic stress. *Antioxidants.* 8: 384. doi: 10.3390/antiox8090384.

Hasanuzzaman, M., Bhuyan, M.B., Zulfiqar, F., Raza, A., Mohsin, S.M., Mahmud, J.A., Fujita, M., Fotopoulos, V. 2020. Reactive oxygen species and antioxidant defense in plants under abiotic stress: Revisiting the crucial role of a universal defense regulator. *Antioxidants.* 9(8): 681. doi: 10.3390/antiox9080681.

Hashem, A., Alqarawi, A.A., Radhakrishnan, R., Al-Arjani, A.B.F., Aldehaish, H.A., Egamberdieva, D., Abd-Allah, E.F. 2018. Arbuscular mycorrhizal fungi regulate the oxidative system, hormones, and ionic equilibrium to trigger salt stress tolerance in *Cucumis sativus* L. *Saudi J. Biol. Sci.* 25(6): 1102–1114. doi: 10.1016/j.sjbs.2018.03.009.

Hernandez, D.J., David, A.S., Menges, E.S., Searcy, C.A., Afkhami, M.E. 2021. Environmental stress destabilizes microbial networks. *ISME J.* 15: 1722–1734. doi: 10.1038/s41396-020-00882-x.

Hossain, M.S., Dietz, K.J. 2016. Tuning of redox regulatory mechanisms, reactive oxygen species and redox homeostasis under salinity stress. *Front Plant Sci.* 7: 548. doi: 10.3389/fpls.2016.00548.

Hu, Y., Jiang, X., Shao, K., Tang, X., Qin, B., Gao, G. 2022. Convergence and stability responses of bacterial communities to salinization in arid and semiarid areas: Implications for global climate change in lake ecosystems. *Front. Microbiol.* 12: 741645. doi: 10.3389/fmicb.2021.741645.

Huang, J., Yang, J., Jiang, H., Wu, G., Xie, Z., Dong, H. 2019. Surviving onshore soil microbial communities differ among the Qing – Tibetan lakes with different salinity. *FEMS Microbiol. Ecol.* 95(11): fiz156. doi: 10.1093/femsec/fiz156.

Ibekwe, A.M., Ors, S., Ferreira, J.F.S., Liu, X., Suarez, D.L. 2017. Seasonal-induced changes in spinach rhizosphere microbial community structure with varying salinity and drought. *Sci. Total Environ.* 579: 1485–1495. doi: 10.1016/j.scitotenv.2016.11.151.

Ilangumaran, G., Smith, D.L. 2017. Plant growth promoting rhizobacteria in amelioration of salinity stress: A systems biology perspective. *Front. Plant Sci.* 8: 1768. doi: 10.3389/fpls.2017.01768.

Jeppesen, E., Beklioglu, M., Ozkan, K., Akyurek, Z. 2020. Salinization increase due to climate change will have substantial negative effects on inland waters: A call for multifaceted research at the local and global scale. *Innovation.* 1: 100030. doi: 10.1016/j.xinn.2020.100030.

Jiao, S., Peng, Z., Qi, J., Gao, J., Wei, G. 2021. Linking bacterial – fungal relationships to microbial diversity and soil nutrient cycling. *Msystems.* 6: e01052–20. doi: 10.1128/mSystems.01052-20.

Kalwasinska, A., Deja-Sikora, E., Szabo, A., Felfoldi, T., Kosobucki, P., Brzezinska, M.S., Walczak, M. 2019. Salino – alkaline lime of anthropogenic origin a reservoir of diverse microbial communities. *Sci. Total Environ.* 655: 842–854. doi: 10.1016/j.scitotenv.2018.11.246.

Kasana, R.C., Pandey, C. 2018. Exiguobacterium: An overview of a versatile genus with potential in industry and agriculture. *Critical.* 38: 141–156. doi: 10.1080/07388551.2017.1312273.

Kesawat, M.S., Kherawat, B.S., Singh, A., Dey, P., Kabi, M., Debnath, D., Saha, D., Khandual, A., Rout, S., Ali, A. 2021. Genome-wide identification and characterization of the brassinazole-resistant (BZR) gene family and its expression in the various developmental stages and stress conditions in wheat (*Triticum aestivum* L.). *Int. J. Mol. Sci.* 22(16): 8743. doi: 10.3390/ijms22168743.

Kesawat, M.S., Satheesh, N., Kherawat, B.S., Kumar, A., Kim, H.U., Chung, S.M., Kumar, M. 2023. Regulation of reactive oxygen species during salt stress in plants and their crosstalk with other signaling molecules-current perspectives and future directions. *Plants (Basel).* 12(4): 864. doi: 10.3390/plants12040864.

Khalloufi, M., Martínez-Andújar, C., Lachaâl, M., Karray-Bouraoui, N., Pérez-Alfocea, F., Albacete, A. 2017. The interaction between foliar GA3 application and arbuscular mycorrhizal fungi inoculation improves growth in salinized tomato (*Solanum lycopersicum* L.) plants by modifying the hormonal balance. *J. Plant Physiol.* 214: 134–144. doi: 10.1016/j.jplph.2017.04.012.

Khare, E., Mishra, J., Arora, N.K. 2018. Multifaceted interactions between endophytes and plant: Developments and prospects. *Front. Microbiol.* 9: 2732. doi: 10.3389/fmicb.2018.02732.

Khattak, M.R., Jan, S.U., Malook, I., Khattak, S.R., Akhtar, N., Khan, S., Jamil, M. 2019. Effects of halophilic bacteria on biochemical characteristics of rice cultivars under salinity stress conditions. In *Saline Soil-based Agriculture by Halotolerant Microorganisms*, eds. M. Kumar, H. Etesami, and V. Kumar. Cham: Springer, 161–174. doi: 10.1007/978-981-13-8335-9_7.

Kumar, A., Singh, S., Gaurav, A.K., Srivastava, S., Verma, J.P. 2020. Plant growth-promoting bacteria: Biological tools for the mitigation of salinity stress in plants. *Front. Microbiol.* 11: 1216. doi: 10.3389/fmicb.2020.01216.

Kumar, K., Kumar, M., Kim, S.R., Ryu, H., Cho, Y.G. 2013. Insights into the genomics of salt stress response in rice. *Rice (NY).* 6(1): 27. doi: 10.1186/1939-8433-6-27.

Kumawat, C., Kumar, A., Parshad, J., Sharma, S.S., Patra, A., Dogra, P., Yadav, G.K., Dadhich, S.K., Verma, R., Kumawat, G.L. 2022. Microbial diversity and adaptation under salt-affected soils: A review. *Sustainability.* 14: 9280. https://doi.org/10.3390/su14159280.

Kushner, D.J. 1993. Growth and nutrition of halophilic bacteria. In *The Biology of Halophilic Bacteria*, eds. R.H. Vreeland, and L.I. Hochstein. Boca Raton, FL, USA: CRC Press, 87–89.

Kuzyk, S.B., Hughes, E., Yurkov, V. 2021. Discovery of siderophore and metallophore production in the aerobic anoxygenic phototrophs. *Microorganisms.* 9(5): 959. doi: 10.3390/microorganisms9050959.

Ladeiro, B. 2012. Saline agriculture in the 21st century: Using salt contaminated resources to cope food requirements. *J. Bot.* 2012: 1–7. doi: 10.1155/2012/310705.

Lin, J., Wang, Y., Sun, S., Mu, C., Yan, X. 2017. Effects of arbuscular mycorrhizal fungi on the growth, photosynthesis, and photosynthetic pigments of *Leymus chinensis* seedlings under salt-alkali stress and nitrogen deposition. *Sci. Total Environ.* 576: 234–241. doi: 10.1016/j.scitotenv.2016.10.091.

Lozupone, C.A., Knight, R. 2007. Global patterns in bacterial diversity. *Proc. Natl. Acad. Sci. USA.* 104: 11436–11440. doi: 10.1073/pnas.0611525104.

Maleki, M., Mokhtarnejad, L., Mostafaee, S. 2011. Screening of rhizobacteria for biological control of cucumber root and crown rot caused by *Phytophthora drechsleri*. *Plant Pathol. J.* 27: 78–84. doi: 10.5423/PPJ.2011.27.1.078.

Mehla, N., Sindhi, V., Josula, D., Bisht, P., Wani, S.H. 2017. An introduction to antioxidants and their roles in plant stress tolerance. In *Reactive Oxygen Species and Antioxidant Systems in Plants: Role and Regulation under Abiotic Stress*. Berlin/Heidelberg, Germany: Springer, 1–23.

Miao, C., Jia, F., Wan, Y., Zhang, W., Lin, M., Jin, W. 2014. Halomonas huangheensis sp. nov. a moderately halophilic bacterium isolated from a saline – alkali soil. *Int. J. Syst. Evol. Microbiol.* 64(Pt 3): 915–920. doi: 10.1099/ijs.0.056556-0.

Miller, G., Susuki, N., Ciftci-Yilmaz, S., Mittler, R. 2010. Reactive oxygen species homeostasis and signaling during drought and salinity stresses. *Plant Cell Environ.* 33(4): 453–467. doi: 10.1111/j.1365-3040.2009.02041.x.

Mittler, R. 2017. ROS are good. *Trends Plant Sci.* 22: 11–19. doi: 10.1016/j.tplants.2016.08.002.

Moghaddam, J.A., Boehringer, N., Burdziak, A., Kunte, H.J., Galinski, E.A., Schäberle, T.F. 2016. Different strategies of osmoadaptation in the closely related marine myxobacteria *Enhygromyxa salina* SWB007 and *Plesiocystis pacifica* SIR-1. *Microbiology.* 162(4): 651–661. doi: 10.1099/mic.0.000250.

Morton, M.J., Awlia, M., Al-Tamimi, N., Saade, S., Pailles, Y., Negrão, S., Tester, M. 2019. Salt stress under the scalpel – dissecting the genetics of salt tolerance. *Plant J.* 97(1): 148–163. doi: 10.1111/tpj.14189.

Munns, R. Tester, M. 2008. mechanisms of salinity tolerance. *Annu. Rev. Plant Biol.* 59: 651.

Naeem, S., Duffy, J.E., Zavaleta, E. 2012. The functions of biological diversity in an age of extinction. *Science.* 336: 1401–1406. doi: 10.1126/science.1215855.

Nawaz, A., Shahbaz, M., Asadullah, I.A., Marghoob, M.U., Imtiaz, M., Mubeen, F. 2020. Potential of salt tolerant PGPR in growth and yield augmentation of wheat (*Triticum aestivum* L.) under saline conditions. *Front. Microbiol.* 11: 2019. doi: 10.3389/fmicb.2020.02019.

Nazir, U., Zargar, M., Baba, Z., Mir, S., Mohiddin, F., Bhat, N. 2020. Isolation and characterization of plant growth promoting rhizobacteria associated with pea rhizosphere in North Himalayan region. *Int. J. Chem. Stud.* 8: 1131–1135.

Oren, A. 1999. Bioenergetic aspects of halophilism. *Microbiol. Mol. Biol. Rev.* 63: 334–340.

Oren, A. 2002. Molecular ecology of extremely halophilic archaea and bacteria. *FEMS Microbiol. Ecol.* 39(1): 1–7. doi: 10.1111/j.1574-6941.2002.tb00900.x.

Pérez-García, L.-A., Sáenz-Mata, J., Fortis-Hernández, M., Navarro-Muñoz, C.E., Palacio-Rodríguez, R., Preciado-Rangel, P. 2023. Plant-growth-promoting rhizobacteria improve germination and bioactive compounds in cucumber seedlings. *Agronomy.* 13(2): 315. doi: 10.3390/agronomy13020315.

Porcel, R., Redondo-Gómez, S., Mateos-Naranjo, E., Aroca, R., Garcia, R., Ruiz-Lozano, J.M. 2015. Arbuscular mycorrhizal symbiosis ameliorates the optimum quantum yield of photosystem II and reduces non-photochemical quenching in rice plants subjected to salt stress. *J. Plant Physiol.* 185: 75–83. doi: 10.1016/j.jplph.2015.07.006.

Qadir, M., Oster, J., Schubert, S., Noble, A., Sahrawat, K. 2017. Phytoremediation of sodic and saline-sodic soils. *Adv. Agron.* 96: 197–247. doi: 10.1016/s0065-2113(07)96006-x.

Qayyum, F., Saifullah, M.S., Akhtar, J. 2016. Salt-affected soils: Sources genesis and management. In *Soil Science Concepts and Applications*, eds. M. Sabir, J. Akhtar, and K.R. Hakeem. Faisalabad: University of Agriculture Faisalabad.

Rahman, M.M., Mostofa, M.G., Keya, S.S., Siddiqui, M.N., Ansary, M.M.U., Das, A.K., Rahman, M.A., Tran, L.S. 2021. Adaptive mechanisms of halophytes and their potential in improving salinity tolerance in plants. *Int. J. Mol. Sci.* 22: 10733. doi: 10.3390/ijms221910733.

Raja, V., Majeed, U., Kang, H., Andrabi, K.I., John, R. 2017. Abiotic stress: Interplay between ROS, hormones, and MAPKs. *Environ. Exp. Bot.* 137: 142–157. doi: 10.1016/j.envexpbot.2017.02.010.

Rath, K.M., Rousk, J. 2015. Salt effects on the soil microbial decomposer community and their role in organic carbon cycling: A review. *Soil Biol. Biochem.* 81: 108–123. doi: 10.1016/j.soilbio.2014.11.001.

Razzaghi Komaresofla, B., Alikhani, H.A., Etesami, H., Khoshkholgh-Sima, N.A. 2019. Improved growth and salinity tolerance of the halophyte Salicornia sp. by co-inoculation with endophytic and rhizosphere bacteria. *Appl. Soil Ecol.* 138: 160–170. doi: 10.1016/j.apsoil.2019.02.022.

Rodriguez-Verdugo, A., Ackermann, M. 2021. Rapid evolution destabilizes species interactions in a fluctuating environment. *ISME J.* 15: 450–460. doi: 10.1038/s41396-020-00787-9.

Rousk, J., Elyaagubi, F.K., Jones, D.L., Godbold, D.L. 2011. Bacterial salt tolerance is unrelated to soil salinity across an arid agroecosystem salinity gradient. *Soil. Biol Biochem.* 43: 1881–1887. doi: 10.1016/j.soilbio.2011.05.007.

Ruppel, S., Franken, P., Witzel, K. 2013. Properties of the halophyte microbiome and their implications for plant salt tolerance. *Funct. Plant Biol.* 40: 940–951. doi: 10.1071/FP12355.

Sachdev, S., Ansari, S.A., Ansari, M.I., Fujita, M., Hasanuzzaman, M. 2021. Abiotic stress and reactive oxygen species: Generation, signaling, and defense mechanisms. *Antioxidants.* 10: 277. doi: 10.3390/antiox10020277.

Safari, D., Jamali, F., Nooryazdan, H.R., Bayat, F. 2018. Evaluation of ACC deaminase-producing Pseudomonas fluorescens strains for their effects on seed germination and early growth of wheat under salt stress. *Aust. J. Crop Sci.* 12: 413–421. doi: 10.21475/ajcs.18.12.03.pne801.

Safdarian, M., Askari, H., Shariati, J.V., Nematzadeh, G. 2019. Transcriptional responses of wheat roots inoculated with *Arthrobacter nitroguajacolicus* to salt stress. *Sci. Rep.* 9: 1792. doi: 10.1038/s41598-018-38398-2.

Santoyo, G., Urtis-Flores, C.A., Loeza-Lara, P.D., Orozco-Mosqueda, M., Glick, B.R. 2021. Rhizosphere colonization determinants by plant growth-promoting rhizobacteria (PGPR). *Biology*. 10(6): 475. doi: 10.3390/biology10060475.

Sarkar, A., Ghosh, P. K., Pramanik, K., Mitra, S., Soren, T., Pandey, S., Mondal, M.H., Maiti, T.K. 2018. A halotolerant Enterobacter sp. displaying ACC deaminase activity promotes rice seedling growth under salt stress. *Microbiol. Res.* 169: 20–32. doi: 10.1016/j.resmic.2017.08.005.

Schirawski, J., Perlin, M.H. 2018. Plant-microbe interaction 2017-the good, the bad, and the diverse. *Int. J. Mol. Sci.* 19(5): 1374. doi: 10.3390/ijms19051374.

Setia, R., Marschner, P., Baldock, J., Chittleborough, D., Verma, V. 2011. Relationships between carbon dioxide emission and soil properties in salt-affected landscapes. *Soil. Biol. Biochem.* 43: 667–674. doi: 10.1016/j.soilbio.2010.12.004.

Sharma, S., Kulkarni, J., Jha, B. 2016. Halotolerant rhizobacteria promote growth and enhance salinity tolerance in peanuts. *Front. Microbiol.* 7: 1600. doi: 10.3389/fmicb.2016.0160.

Shokri, S., Maadi, B. 2009. Effects of arbuscular mycorrhizal fungus on the mineral nutrition and yield of *Trifolium alexandrinum* plants under salinity stress. *J. Agron.* 8: 79–83. doi: 10.3923/ja.2009.79.83.

Shrivastava, P., Kumar, R. 2015. Soil salinity: A serious environmental issue and plant growth promoting bacteria as one of the tools for its alleviation. *Saudi J. Biol. Sci.* 22(2): 123–131. doi: 10.1016/j.sjbs.2014.12.001.

Singh, A., Kumar, A., Yadav, S., Singh, I.K. 2019. Reactive oxygen species-mediated signaling during abiotic stress. *Plant Gene*. 18: 100173. doi: 10.1016/j.plgene.2019.100173.

Singh, B.K., Trivedi, P., Singh, S., Macdonald, C.A., Verma, J.P. 2018. Emerging microbiome technologies for sustainable increase in farm productivity and environmental security. *Microbiol. Aust.* 39: 17–23. doi: 10.1071/MA18006.

Singh, R.P., Jha, P.N. 2016. The multifarious PGPR Serratia marcescens CDP-13 augments induced systemic resistance and enhanced salinity tolerance of wheat (*Triticum aestivum* L.). *PLoS One*. 11: e0155026. doi: 10.1371/journal.pone.0155026.

Singh, V., Singh, D.V. 2015. Cyanobacteria modulated changes and their impact on bioremediation of saline-alkaline soils. *Bangladesh J. Bot.* 44: 653–658. doi: 10.3329/BJB.V44I4.38646.

Singhal, R.K., Saha, D., Skalicky, M., Mishra, U.N., Chauhan, J., Behera, L.P., Lenka, D., Chand, S., Kumar, V., Dey, P. 2021. Crucial cell signaling compounds crosstalk and integrative multi-omics techniques for salinity stress tolerance in plants. *Front. Plant Sci.* 12: 670369. doi: 10.3389/fpls.2021.670369.

Slama, H.B., Cherif-Silini, H., Bouket, A.C., Silini, A., Alenezi, F.N., Luptakova, L., Vallat, A., Belbahri, L. 2021. Biotechnology and bioinformatics of endophytes in biocontrol, bioremediation, and plant growth promotion. In *Endophytes: Mineral Nutrient Management* (Volume 3). Cham, Switzerland: Springer, 181–205.

Slama, H.B., Triki, M., Chenari Bouket, A., Mefteh, F., Alenezi, F., Luptakova, L., Cherif-Silini, H., Vallat, A., Oszako, T., Gharsallah, N., Belbahri, L. 2019. Microorganisms screening of the high-rhizosphere competent *Limoniastrum monopetalum* culturable endophyte microbiota allows the recovery of multifaceted and versatile biocontrol agents. *Microorganisms*. 7(8): 249. doi: 10.3390/microorganisms7080249.

Smith, F.A., Smith, S.E. 2015. How harmonious are arbuscular mycorrhizal symbioses? Inconsistent concepts reflect different mindsets as well as results. *New Phytol.* 205: 1381–1384. doi: 10.1111/nph.13202.

Smith, S.E., Jakobsen, I., Grønlund, M., Smith, F.A. 2011. Roles of arbuscular mycorrhizas in plant phosphorus nutrition: Interactions between pathways of phosphorus uptake in arbuscular mycorrhizal roots have important implications for understanding and manipulating plant phosphorus acquisition. *Plant Physiol.* 156(3): 1050–1057. doi: 10.1104/pp.111.174581.

Stegen, J.C., Bottos, E.M., Jansson, J.K. 2018. A unified conceptual framework for prediction and control of microbiomes. *Curr. Opin. Microbiol.* 44: 20–27. doi: 10.1016/j.mib.2018.06.002.

Sunita, K., Mishra, I., Mishra, J., Prakash, J., Arora, N.K. 2020. Secondary metabolites from halotolerant plant growth promoting rhizobacteria for ameliorating salinity stress in plants. *Front. Microbiol.* 11: 567768. doi: 10.3389/fmicb.2020.567768.

Swarna Lakshmi, K., Yadav, V., Tyagi, D., Dhar, D.W., Kannepalli, A., Kumar, S. 2020. Significance of plant growth promoting rhizobacteria in grain legumes: Growth promotion and crop production. *Plants (Basel).* 9(11): 1596. doi: 10.3390/plants9111596.

Szabo, S., Hossain, M.S., Adger, W.N., Matthews, Z., Ahmed, S., Lázár, A.N., Ahmad, S. 2016. Soil salinity, household wealth and food insecurity in tropical deltas: Evidence from south-west coast of Bangladesh. *Sustain. Sci.* 11(3): 411–421. doi: 10.1007/s11625-015-0337-1.

Talaat, N.B. 2015. Effective microorganisms improve growth performance and modulate the ROS-scavenging system in common bean (*Phaseolus vulgaris* L.) plants exposed to salinity stress. *J. Plant Growth Regul.* 34: 35–46. doi: 10.1007/s00344-014-9440-2.

Talaat, N.B. 2018. Exploring halotolerant rhizomicrobes as a pool of potent genes for engineering salt stress tolerance in crops. *Sali. Responses Toleran. Plant.* 2: 49–76. doi: 10.1007/978-3-319-90318-7_3.

Tanimura, K., Matsumoto, T., Nakayama, H., Tanaka, T., Kondo, A. 2016. Improvement of ectoine productivity by using sugar transporter-overexpressing Halomonas elongate. *Enzyme Microb. Technol.* 89: 63–68. doi: 10.1016/j.enzmictec.2016.03.006.

Vaishnav, A., Kumari, S., Jain, S., Varma, A., Tuteja, N., Choudhary, D.K. 2016. PGPR-mediated expression of salt tolerance gene in soybean through volatiles under sodium nitroprusside. *J. Basic Microbiol.* 56(11): 1274–1288. doi: 10.1002/jobm.201600188.

Ventosa, A., de la Haba, R.R., Sánchez-Porro, C., Papke, R.T. 2015. Microbial diversity of hypersaline environments: A metagenomic approach. *Curr. Opin. Microbiol.* 25: 80–87. doi: 10.1016/j.mib.2015.05.002.

Volkov, V., Beilby, M.J. 2017. Editorial: Salinity tolerance in plants: Mechanisms and regulation of ion transport. *Front. Plant Sci.* 8: 1795. doi: 10.3389/fpls.2017.01795.

Wang, S., Sun, L., Ling, N., Zhu, C., Chi, F., Li, W., Hao, X., Zhang, W., Bian, J., Chen, L., Wei, D. 2020. Exploring soil factors determining composition and structure of the bacterial communities in saline-alkali soils of Songnen Plain. *Front. Microbiol.* 10: 2902. doi: 10.3389/fmicb.2019.02902.

Wang, W., Wu, Z., He, Y., Huang, Y., Li, X., Ye, B.C. 2018a. Plant growth promotion and alleviation of salinity stress in Capsicum annuum L. by Bacillus isolated from saline soil in Xinjiang. *Ecotox. Environ. Safe.* 164: 520–529. doi: 10.1016/j.ecoenv.2018.08.070.

Wang, Y., Wang, M., Li, Y., Wu, A., Huang, J. 2018b. Effects of arbuscular mycorrhizal fungi on growth and nitrogen uptake of Chrysanthemum morifolium under salt stress. *PLoS One.* 13: e0196408. doi: 10.1371/journal.pone.0196408.

Weralupitiya, C., Wanigatunge, R.P., Gunawardana, D., Vithanage, M., Magana-Arachchi, D. 2022. Cyanotoxins uptake and accumulation in crops: Phytotoxicity and implications on human health. *Toxicon.* 211: 21–35. doi: 10.1016/j.toxicon.2022.03.003.

Wichern, J., Wichern, F., Joergensen, R.G. 2006. Impact of salinity on soil microbial communities and the decomposition of maize in acidic soils. *Geoderma.* 137: 100–108. doi: 10.1016/j.geoderma.2006.08.001.

Wong, V.N.L., Dalal, R.C., Greene, R.S.B. 2009. Carbon dynamics of sodic and saline soils following gypsum and organic material additions: A laboratory incubation. *Appl. Soil Ecol.* 41: 29–40. doi: 10.1016/j.apsoil.2008.08.006.

Wood, J.M. 2011. Bacterial osmoregulation: A paradigm for the study of cellular homeostasis. *Annu. Rev. Microbiol.* 65: 215–238. doi: 10.1146/annurev-micro-090110-102815.

Xie, Z., Chu, Y., Zhang, W., Lang, D., Zhang, X. 2019. Bacillus pumilus alleviates drought stress and increases metabolite accumulation in *Glycyrrhiza uralensis* Fisch. *Environ. Exp.* 158: 99–106. doi: 10.1016/j.envexpbot.2018.11.021.

Yamaguchi, T., Blumwald, E. 2005. Developing salt-tolerant crop plants: Challenges and opportunities. *Trends Plant Sci.* 10: 615–620. doi: 10.1016/j.tplants.2005.10.002.

Youssef, N.H., Savage-Ashlock, K.N., McCully, A.L., Luedtke, B., Shaw, E.I., Hoff, W.D., Elshahed, M.S. 2014. Trehalose/2-sulfotrehalose biosynthesis and glycine-betaine uptake are widely spread mechanisms for osmoadaptation in the Halobacteriales. *ISME J.* 8(3): 636–649. doi: 10.1038/ismej.2013.165.

Zaman, M., Shahid, S.A., Heng, L. 2018. *Guideline for Salinity Assessment, Mitigation and Adaptation Using Nuclear and Related Techniques.* Switzerland AG: Springer International Publishing, 164. doi: 10.1007/978-3-319-96190-3.

Zheng, W., Xue, D., Li, X., Deng, Y., Rui, J., Feng, K., Wang, Z.-L. 2017. The responses and adaptations of microbial communities to salinity in farmland soils: A molecular ecological network analysis. *Appl. Soil Ecol.* 120: 239–246. doi: 10.1016/j.apsoil.2017.08.019.

Zribi, O.T., Labidi, N., Slama, I., Debez, A., Ksouri, R., Rabhi, M., Smaoui, A., Abdelly, C. 2012. Alleviation of phosphorus deficiency stress by moderate salinity in the halophyte *Hordeum maritimum L. Plant Growth Regul.* 66: 75–85. doi: 10.1007/s10725-011-9631-9.

9 Ameliorative Role of Endophytes in Mitigating Biotic and Abiotic Stresses in Medicinal Plants

Muhammad Shahbaz Ali, Farrukh Azeem, Muhammad Shahid, Faisal Mahmood, Tanvir Shahzad, Yasir Bilal, Muhammad Bilal, Muhammad Talha Khokhar, and Sabir Hussain

9.1 INTRODUCTION

Climatic instability and other ecological parameters have a detrimental effect on the life of plants and animals in an ecosystem (Chaudhry and Sidhu, 2022; Kumar and Nautiyal, 2022). Plants, being immobilized, are more susceptible to environmental stresses. Different biotic and abiotic stress like nutrient deficiency, variable temperature, drought, oxidative stress, light intensity, viruses, fungi, bacteria, insects, nematodes, and weeds have negative effects on the development, growth, and productivity of plants (Flora *et al.*, 2013; Jia *et al.*, 2016; Ogbe *et al.*, 2020; Verma *et al.*, 2021a; Zhang *et al.*, 2022; Kumar and Nautiyal, 2022; Hanaka *et al.*, 2022). Moreover, a substantial number of valuable medicinal plants are also influenced by the previously mentioned environmental constraints (Sahoo *et al.*, 2010; Flora *et al.*, 2013; Namdar *et al.*, 2019; Ogbe *et al.*, 2020; Wu *et al.*, 2021). These medicinal plants have gained recognition in developing countries because of their role in generating revenue for residents and serving as resources in pharmaceutical industries (Sahoo *et al.*, 2010; Roy and Roy, 2016; Ogbe *et al.*, 2020). However, these environmental stresses lead to the vulnerability of these valuable medicinal plants. Nonetheless, medicinal plants have evolved diverse mechanisms to counteract these environmental stresses. One notable mechanism involves the synthesis of secondary metabolites, which have lower molecular mass and play a pivotal part in a plant's response to the environment. These secondary metabolites help the plant to counter environmental stresses through adaptation, accommodation, and defence (Wink, 2003; Jia *et al.*, 2016; Ogbe *et al.*, 2020; Verma *et al.*, 2021b; Wu *et al.*, 2021). Secondary metabolites such as terpenoids, steroids, lignans, alkaloids, flavonoids, phenolics, anthocyanins, and quinones are also responsible for the therapeutic properties and are involved in defence against different abiotic and biotic stresses (Ogbe *et al.*, 2020).

Moreover, previous literature reported that not only the host plants are involved in mitigating the environmental stresses but also some unidentified organisms, namely endophytes that dwell in the plants' tissues, are also engaged in the alleviation of different stresses by enhancing tolerance and resistance in plants (Yuan *et al.*, 2016; Jia *et al.*, 2016; Ogbe *et al.*, 2020). These endophytes are considered a significant part of the plant ecosystem and reside in healthy plants' tissues. They can develop a special association with the host plants, including medicinal plants that provide the formation of secondary metabolites and better-quantity and quality crude products (Firáková *et al.*, 2007; Jia *et al.*, 2016). Plants and endophytes are commonly associated with each other by mutualism, antagonism, and neutralism.

DOI: 10.1201/b23155-9

These beneficial associations may confer benefits like acceleration of growth, development propagation, alteration of gene expression, increased resistance to diseases, and many other biotic agents (Rodriguez *et al.*, 2009; Jia *et al.*, 2016; Gupta *et al.*, 2018; Gupta and Chaturvedi, 2019; Ek-Ramos *et al.*, 2019; Eid *et al.*, 2019; Ogbe *et al.*, 2020; Gupta and Chaturvedi, 2020; Wu *et al.*, 2021). Moreover, plants infected with endophytes showed more potential for germination as compared to non-infected plants during unfavourable environmental conditions. In addition, the production and accumulation of various bioactive compounds like secondary metabolites are also triggered by these mutualistic relationships that cause the production of higher-quality crude drugs in medicinal plants (Jia *et al.*, 2016). Acting as biological growth regulators, endophytes produce siderophores, which serve as plant growth regulators, for instance, auxins, cytokinin, abscisic acid, gibberellins, and ethylene, which are needed for the regulation of seed germination and growth of the plant (Malinowski and Belesky, 1999; Firáková *et al.*, 2007; Ogbe *et al.*, 2020). In response, host plants provide shelter and nutrients to endophytic microbes (Ogbe *et al.*, 2020).

Moreover, endophytes serve as an essential part in mitigating abiotic and biotic stresses in medicinal plants by actively participating in the synthesis of antioxidants, antibiotics, antifungal compounds, secondary metabolites, and phenolic compounds (Jennings *et al.*, 1998; Zhou *et al.*, 2009; Pang and Wang, 2010; Torres *et al.*, 2012; Lata *et al.*, 2018). They are biological factors that help in controlling plants' active metabolism under stressed conditions (Rodriguez *et al.*, 2012; Ogbe *et al.*, 2020). Consequently, previous studies have shown that plants infected with endophytes exhibit greater resilience against environmental stresses as compared to endophyte-free plants (Waller *et al.*, 2005; Saikkonen *et al.*, 2010; Ogbe *et al.*, 2020). Furthermore, some endophytes are reported as important agents to mitigate multiple environmental stresses (Piccoli and Bottini, 2013; Hume *et al.*, 2016; Zhang *et al.*, 2019). Indeed, recently advanced technologies have brought light to the remarkable potential of endophytes in alleviating various abiotic stresses, for instance, nutrient deficiencies, oxidative stress, temperature fluctuations, and drought. Endophytes have also demonstrated their effectiveness in countering biotic stresses caused by pathogenic attacks and nematodes. During abiotic stresses, endophytes enhance the formation of antioxidants, leading to the biosynthesis of antistress biomolecules that aid in bolstering plant defences under adverse environmental conditions (Herrera-Carillo *et al.*, 2009; Torres *et al.*, 2009; Saraf *et al.*, 2010; Hameed *et al.*, 2014; Gond *et al.*, 2015; Ogbe *et al.*, 2020; Kumar and Nautiyal, 2022). However, in the case of biotic stresses, endophytes protect their hosts by secreting a variety of chemicals that inhibit or eliminate biological agents. They also bolster the immunity of plants, including medicinal plants, against pathogens by providing an additional layer of defence (Saikkonen *et al.*, 1998; Zhang *et al.*, 2006; Kusari *et al.*, 2012; Ogbe et al., 2020; Wu *et al.*, 2021).

This chapter elaborates on endophytes as biological agents to mitigate adverse environmental stresses. It also includes a detailed mechanism involving the mitigation of different stresses in medicinal plants. This chapter also explores the different types of abiotic and biotic stresses that influence plant growth and survival under unfavourable environmental conditions. Future perspectives and challenges relating to plant resilience are also discussed in this chapter.

9.2 ENDOPHYTES AND THEIR PROPERTIES

German botanist Fredrich Link stated that endophytes as a group of partially parasitic organisms (fungi, archaea, and bacteria) colonize both vascular and non-vascular healthy plants' tissues during their life cycle without resulting in any harm (Sun and Guo, 2012; Wani *et al.*, 2015; Meshram *et al.*, 2016; Aguiar-Pulido *et al.*, 2016; Gupta *et al.*, 2020; Ogbe *et al.*, 2020). However, some endophytes exhibit disease in host plants during senescence (Rodriguez and Redman, 2008; Wani *et al.*, 2015; Ogbe *et al.*, 2020). Moreover, some endophytes are organ specific, and it may be due to the capability to confront stipulated physiological and microecological conditions of those organs (Carroll, 1988).

Based on morphology, taxonomy, and mode of transmission, endophytes are categorized into systematic and non-systematic endophytes. Systematic endophytes inhibit the plant tissue, don't cause any disease in them, and also sustain symbiotic associations with their host (Wani *et al.*, 2015), whereas non-systematic endophytes complete their life process in the tissues of the plant and do not cause any disease under normal conditions; however, they become infectious to host plants under stress conditions (Wani *et al.*, 2015; Ogbe *et al.*, 2020). Regarding the mode of transmission, both endophytes and pathogens share a comparable approach while entering host plants (Sieber, 2007), but the relationship between endophytes and their host plants is mostly mutualistic (Carroll, 1988). For example, Hardoim *et al.* (2015) described the associations among endophytes and host plants as categorized by opportunistic, facultative, and obligative endophytes. Obligate endophytes reside in their host cells, as they do not have a free-living stage, and, for their existence, they rely on the host's metabolism (Hardoim *et al.*, 2015). Mycorrhizal fungi and a few genera such as *Epichloe* and *Balansia* are examples of obligate endophytes (Parniske, 2008). Similarly, Bashan and de-Bashan (2010) revealed that opportunistic endophytes exist exterior to plants and sporadically move into the tissues of plants. A few groups of fungi like *Trichoderma* and *Hypocrea* as well as a few bacteria like *Azospirillum* and *Pseudomonas* are examples of opportunistic endophytes. Facultative endophytes are intermediate and mainly heterogeneous and dwell outside their particular host body for a specific portion of their circulation, for instance, *Rhizobia* sp., *Beauveria bassiana*, and *Trichoderma* species which relate completely to legumes (Card *et al.*, 2016). These are generally in a very sophisticated intimate relationship with the microbial assortment present in the neighbourhood and specifically in their microbiome (Hardoim *et al.*, 2015; Ogbe *et al.*, 2020).

9.2.1 Properties

Endophytes are significant in improving the growth and survival of medicinal plants by activating several enzymes, altering gene expression, improving the accretion of secondary metabolites, and strengthening resistance (Malinowski and Belesky, 1999; Firáková *et al.*, 2007; Jia *et al.*, 2016). For instance, studies presented an increase in hormonal formation of cytokinins, indole-3-acetic acid, and indole-3-acetonitrile for plants' growth and fitness (Gravel *et al.*, 2007; Jia *et al.*, 2016; Fouda *et al.*, 2021). Furthermore, a remarkable increase in the growth affiliated with higher light utilization efficacy and photosynthetic carbon uptake was demonstrated by inoculating plant seedlings with the *Paenibacillus ehimensis* strain (Zhao *et al.*, 2015; Wu *et al.*, 2021). Additionally, endophytes also assist plants by providing nutritional elements like P and N (Zhang *et al.*, 2006; Hartley and Gange, 2009; Vaishnav *et al.*, 2019). For example, the growth and seed germination of the *Gastrodia elata* host plant could be promoted through *Mycena dendrobii*, which is involved in the secretion of indoleacetic acid, which is a growth promotor hormone (Guo and Wang, 2001). Likewise, nitrogen was directly transferred by *Metarhizium robertsii* from insects to host plants through hyphae (Behie *et al.*, 2012). Similarly, Abd_Allah *et al.* (2018) reported that, in *Cicer arietinum*, the concentrations of K, N, Mg, and Ca were enhanced in addition to the facilitation of synthesis of photosynthetic pigments after inoculation with *Bacillus subtilis* (Wu *et al.*, 2021). Moreover, the significance of endophytes was also revealed for triggering numerous gene and enzyme expression, which can enhance the strength and development of host plants (Vendan *et al.*, 2010; Devi *et al.*, 2016; Hassan, 2017; Ibrahim *et al.*, 2019). For instance, stimulation of the expression of nitrate reductase enzymes and starch-degrading enzymes (glucan-water dikinase) may lead to an enhancement in the growth of tobacco roots (Sherameti *et al.*, 2005). Various other advantages of plant growth that are credited to endophytes are enhancement in solubilization activity, phosphate siderophore production, stomatal regulation, osmotic adjustment, and improvement of root morphology (Compant *et al.*, 2005; Chen *et al.*, 2016; Zhu, 2019; Wu *et al.*, 2021). Regarding induced immunity in host plants, it is also observed that endophytes can play a part in chemical defence through bioactive secondary metabolite production (Ji *et al.*, 2009; Jia *et al.*, 2016; Wu *et al.*, 2021). Moreover, two distinct theories have been presented for the plant-endophyte coevolution hypothesis: acquired immune systems and the mosaic effect (Kusari *et al.*, 2012). The former theory stated that endophytes have a remarkable part in modifying host plants with acquired immune systems. The latter theory suggested that in various organs, endophytes secrete mixed chemicals, which are particularly linked to the protection of their host plants, therefore creating variants in the efficiency of such organs hostile to pathogens and herbivores (Carroll, 1991; Kusari *et al.*, 2012). They play a role in enhancing the germination of seeds of a range of host plant species, and undoubtedly such symbiotic relationships can be crucial for the lucrative growth of a few species (Verma *et al.*, 2019). The importance of such implications is that they can increase the potential for seed growth during unfavourable environments as already observed in a few endangered and unique medicinal species including *Dendrobium chrysanthum* and *Dendrobium nobile* which are very tough to germinate in normal circumstances. However, these seeds can potentially germinate in the presence of endophytic fungi associated with the *Mycena* genus, which has aided in the artificial culture of such medicinal plants (Chen and Guo, 2005). Researchers also reported the assembly of secondary metabolites by endophytes that are primitively generated by plants. It may help in improving the formation and assembly of bioactive compounds of host plants for better-quality crude drugs (Liu *et al.*, 2015; Zhou *et al.*, 2016; Yang *et al.*, 2019; Yin *et al.*, 2019). Furthermore, such metabolites can be generated by both host plants and endophytic fungi (Shwab and Keller, 2008). It may also be beneficial for plants such as medicinal plants to improve their development and resistance to biotic or abiotic stresses (Jia *et al.*, 2016). These secondary metabolites have anti-inflammatory, analgesic, antimalarial, antipyretic, and anti-tumour properties, which provide resistance to plants against pathogens. For instance, it is also demonstrated that plants of the family *Taxaceae* concerning endophytes are capable of producing taxol with antitumour activity. Likewise, various secondary metabolites, including cinchonine, cinchonidine, quinidine, quinine, and cinchona alkaloid, exhibited anti-inflammatory, analgesic, antimalarial, and antipyretic properties (Maehara *et al.*, 2012). In addition, it is noted that productive and excellent-quality medicines are gained from medicinal plants that showed their association with fungal endophytes. Specific endophytic fungi in medicinal plants can be involved in the generation of particular bioactive compounds needed for humans. For example, *Huperziaserrata*, which is a medicinal plant found in tropical regions, can generate Huperzine-A compounds intended to be provoked by endophytic fungi species (Zhou *et al.*, 2009; Wang *et al.*, 2009). This is the reason that in traditional Chinese medicine, doctors usually opt for the usage of specific habitats or locations where the chemical type and content of specific compounds can be predicted.

9.3 MAJOR BIOTIC AND ABIOTIC STRESSES IN MEDICINAL PLANTS

Both biotic and abiotic elements of the environment are impacted by climate change (Chaudhry and Sidhu, 2022). The plant's physical environment is influenced by a variety of different aspects. Animals can employ various strategies to reduce the impact of differences, but plants can't do so because of their immobility. To survive in the changing external environment, plants completely rely on their internal mechanisms. As a result of extremely unfavourable ecological conditions, there is a chance that different stresses will damage the plant's health and growth (Wu *et al.*, 2021). Such stresses are categorized into two main types, abiotic and biotic stresses. The major biotic and abiotic stresses affecting medicinal plants are summarized in Figure 9.1. Abiotic stress is described as those non-living factors that adversely affect plants in a particular agro-ecological area. Pollutant toxicity, saltiness, drought, light, extreme temperatures (low and high), reactive oxygen species (ROS), mineral unavailability, and nutrient deficiency are some abiotic stresses that vastly affect plant development, growth, quality, and yield of produce (Wani *et al.*, 2008; Gosal *et al.*, 2009; Zhang *et al.*, 2022). Moreover, abiotic stressors mostly affect the ratio of ions like Na^+ and K^+ in the cytoplasm of plant cells. Wang and Frei (2011) stated that exposure to environmental and abiotic stress may cause a variety of physiological changes as well as changes in the chemical makeup of plants.

Likewise, biotic factors also have detrimental effects on plants' growth and survival. The main cause of pre- and post-harvest crop losses is biotic stresses. Organisms, in particular bacteria, fungi, nematodes, arachnids, viruses, weeds, and insects, induce biotic stress in plants, including medicinal plants. These organisms get their nourishment from their host, or neighbourhood in the instance

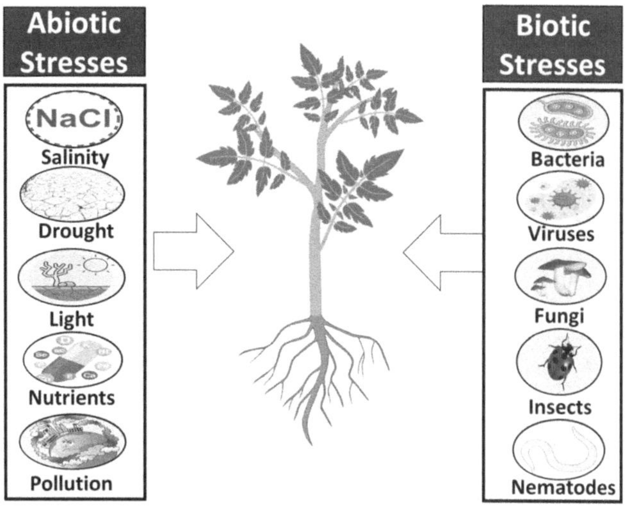

FIGURE 9.1 Different biotic and abiotic stresses on medicinal plants.

of weeds, which kill plants. Biotic stresses are quite different from abiotic stresses. They may lead to a lower leaf surface area and leaf-munching, which has an impact on the photosynthetic rate and physiology of a normal plant (Papp-Rupar *et al.*, 2022; Kumar and Nautiyal, 2022). Plants have a dearth of adaptive immune systems. Yet they can counteract biotic stress by inducing intricate procedures within the cells (Hanaka *et al.*, 2022). The genetic material of a plant innately conforms to the defensive mechanism that performs against biotic stress, and the genes resistant to biotic stress are present in the genome of the plant. Crop plants are subjected to various biotic stresses depending on the soil and environment, and these stresses can be countered by the crop plants. Numerous biotic agents, for instance, viruses, fungi, bacteria, and leaf insect feeding, reduce leaf surface area, which lowers the photosynthesis rate and development of plants. A feasible agricultural system and the ability of endophytic microorganisms to cope with a variety of challenges have been linked (Jana *et al.*, 2022).

9.4 ROLE OF ENDOPHYTES IN MITIGATING THE STRESSES IN MEDICINAL PLANTS

Biotic and abiotic stresses enhance the vulnerability of plants, including medicinal plants, to distinct kinds of tissue damage. These parameters hinder the productivity, survival, and growth of these plants. In this stress condition, endophytes are reported as effective biological factors that play a beneficial part in the development of healthy metabolism under stress conditions (Rodriguez *et al.*, 2012). The major roles of endophytes in mitigating abiotic and biotic stresses in medicinal plants are presented in Figure 9.2.

Moreover, this association between plants and endophytes has been outlined to play a significant part in the reluctance of medicinal plants to stresses by production of antioxidants, antibiotic and antifungal compounds, phenolic compounds, and secondary metabolites (Jennings *et al.*, 1998; Zhou *et al.*, 2003; Pang and Wang, 2010; Torres *et al.*, 2012; Lata *et al.*, 2018; Singh and Dubey, 2018; Ghiasvand *et al.*, 2020). Therefore, it was observed that endophyte-infected plants are generally healthier and are less vulnerable to detrimental effects of stresses in contrast to endophyte-free hosts (Waller *et al.*, 2005; Saikkonen *et al.*, 2010). This evidence showed the potential of endophytes to alleviate abiotic and biotic stress in plants. Table 9.1 shows the endophyte strains that have a beneficial impact on medicinal plant growth and help alleviate both biotic and abiotic stresses.

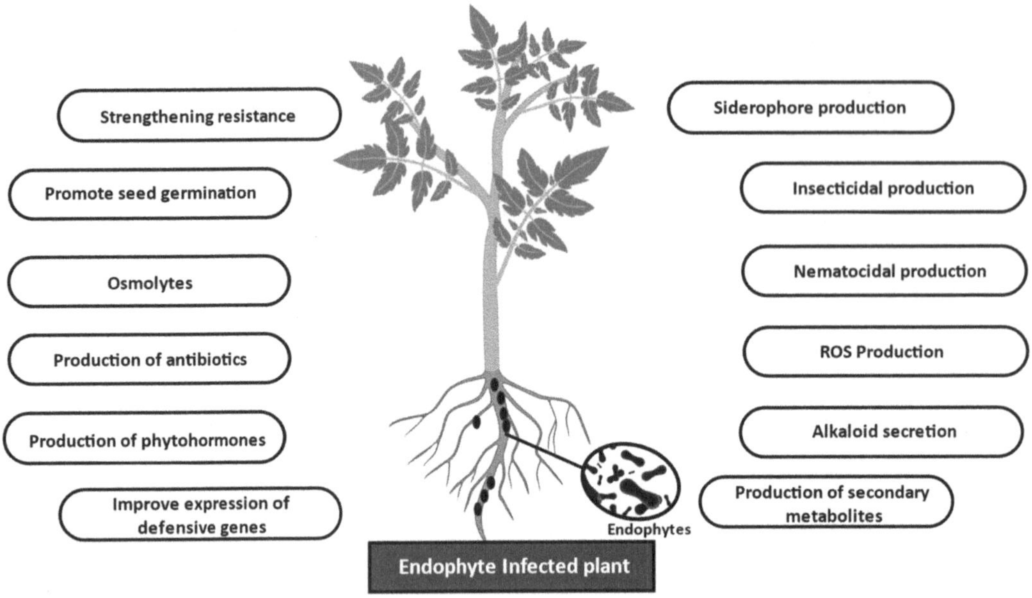

FIGURE 9.2 Beneficial role of endophytes in the alleviation of stresses in medicinal plants.

TABLE 9.1
Endophytes and Their Beneficial Role in Medicinal Plant Growth and Mitigating Stresses

Endophytic Bacteria	Host Plant	Comment	References
Micrococcus luteus *Lysinibacillus fusiformis*	*Panax ginseng*	These bacterial endophytes played a vital role in promoting plant growth and enhancing seedling biomass of Panax ginseng.	Vendan *et al.*, 2010
Bacillus cereus *Bacillus subtilis*	*Teucrium polium*	These endophytes played an important role in increasing root weights, length, and root site.	Hassan, 2017
Microbacterium *Burkholderia*	*Coptis teeta*	These endophytes produced berberine, which belongs to a class of alkaloids that exhibited anti-tumour, anti-inflammatory, and blood sugar–lowering activities.	Liu *et al.*, 2020
Alternaria sp. *Fusarium oxysporum*	*Catharanthus roseus*	These endophytes produced the vinblastine secondary metabolite, which showed antitumour effects.	Zhang *et al.*, 1998
Bacillus *Paenibacillus spp.*	*Curcuma longa*	It increased root number, root length, and shoot length and therefore acted as a plant growth promoter.	Aswathy *et al.*, 2013
Bacillus amyloliquefaciens	*Panax notoginseng*	These bacterial endophytes enhanced the plant resistance to phytopathogens; therefore, host plants were protected from phytopathogen infection.	Ma *et al.*, 2013
Serratia marcescens	*Achyranthes aspera*	It also served as plant growth promoter and increased the shoot length, leaf area, and fresh weight of roots and shoots.	Devi *et al.*, 2016
Pseudonocardia sp.	*Artemisia annua*	It played a significant role in increasing the artemisinin content.	Li *et al.*, 2012
Stenotrophomonas maltophilia *Bacillus* sp.	*Panax ginseng.*	These endophytes suppressed pathogen mycelial growth.	Hong *et al.*, 2018
Bacillus sp.	*Curcuma longa*	These endophytes played a significant part in inducing the host disease resistance.	Jayakumar *et al.*, 2019

(Continued)

TABLE 9.1 (*Continued*)
Endophytes and Their Beneficial Role in Medicinal Plant Growth and Mitigating Stresses

Endophytic Bacteria	Host Plant	Comment	References
Bacillus siamensis	*Coriandrum sativum*	These endophytes served as plant growth promoters and enhanced the shoot length, root length, and its dry weight.	Ibrahim *et al.*, 2019
Cohnella sp. *Paenibacillus* sp. *Pantoea* sp.	*Centella asiatica*	These bacterial endophytes increased plant resistance to phytopathogens by stimulating plant defence systems.	Rakotoniriana *et al.*, 2013
Kytococcus schroeteri	*Ephedra foliata*	It is observed that these endophytes produced camptothecin that belongs to the class alkaloids and showed cancer fighting activity.	Ghiasvand *et al.*, 2020
Bacillus amyloliquefaciens	*Ginkgo biloba*	It played a vital role in production of antibiotics, as well as producing systemic resistance.	Yang *et al.*, 2015
Phyllobacterium myrsinacearum	*Epimedium brevicornu*	It is observed that these endophytes depressed pathogen development via enhancing host plant tolerance to phytopathogens.	He *et al.*, 2009
Achromobacter xylosoxidans	*Catharanthus roseus*	These endophytes increased the rate of germination and root weight in a salty environment by enhancing host plant abiotic stress tolerance.	Karthikeyan *et al.*, 2012
Bacillus subtilis	*Ligusticum chuanxiong*	It is involved in production of ligustrazine, which played a significant role in treatment of various vascular diseases.	Yin *et al.*, 2019
Pseudomonas fluorescens	*Atractylodes macrocephala*	It helped in promoting plant metabolite accumulation by increasing sesquiterpenoid production.	Yang *et al.*, 2019
Citrobacter putida	*Euphorbia milii*	It played a remarkable role in removing airborne benzene.	Khaksar *et al.*, 2016
Streptomyces sp.	*Allium tuberosum*	It is involved in production of 6-prenylindole that belongs to the class alkaloids, which exhibited the bio-property of antifungal activity.	Singh and Dubey, 2018
Pseudomonas fluorescens	*Atractylodes lancea*	These endophytes served as promoters for plant metabolite accumulation and accumulated essential oil.	Zhou *et al.*, 2016
Entrophospora infrequens	*Nothapodytes foetida*	These fungal endophytes are responsible for production of camptothecin, which showed antitumour effects.	Rehman *et al.*, 2008
Paenibacillus polymyxa	*Panax ginseng*	These bacterial endophytes induce ginsenoside production.	Gao *et al.*, 2015
Microbacterium sp.	*Catharanthus roseus*	It is reported that these endophytes produced vindoline, which helped in treating acute leukaemia and Hodgkin's disease.	Anjum and Chandra, 2019
Pseudomonas fluorescens	*Atractylodes lancea*	It is reported that these endophytes synthesized indole-3-acetic acid, which belongs to the class alkaloids that enhance plant root growth, and carbohydrate products were boosted.	Zhou *et al.*, 2018
Bacillus atrophaeus *Bacillus mojavensis*	*Glycyrrhiza uralensis*	It is involved in production of phthalic acid and benzoic acid, which belong to the class organic acids that presented antibacterial and antifungal activity.	Mohamad *et al.*, 2018
Streptomyces sp.	*Boesenbergia rotunda*	It is involved in production of fisetin, 2',7-dihydroxy-4',5'-dimethoxyisoflavone and 7-methoxy-3,30,40,6-tetrahydroxyflavone, which belong to the class flavonoids that showed cancer-fighting activity.	Taechowisan *et al.*, 2014
Burkholderia sp.	*Panax ginseng*	These endophytes produced ginsenoside Rg3, which belongs to the class saponins and showed cancer fighting activity.	Fu *et al.*, 2017

(*Continued*)

TABLE 9.1 (*Continued*)
Endophytes and Their Beneficial Role in Medicinal Plant Growth and Mitigating Stresses

Endophytic Bacteria	Host Plant	Comment	References
Microbispora sp.	*Clinacanthus siamensis Bremek.*	Production of linfuranone A is reported by these endophytes.	Indananda *et al.*, 2013
Bacillus subtilis	*Thymus vulgaris*	These bacterial endophytes produced Tetracosane, bis (2-ethylhexyl) phthalate, dibutyl phthalate, and 1,3-dimethyl-, p-xylene, which showed antimicrobial activity.	Mohamad *et al.*, 2018
Epulorhiza sp. *Mycena* sp.	*Dendrobium nobile, D. chrysanthum*	These endophytes were observed to be helpful in enhancing plant nutrients absorption; therefore, host seed germination was promoted.	Chen and Guo, 2005
Penicillium chrysogenum	*Lycopodium serratum*	These fungal endophytes produced huperzine A metabolite, which functioned as anticholinesterase.	Zhou *et al.*, 2009
Agrobacterium sp.	*Panax ginseng*	It is observed that these endophytes produced ginsenoside Rh2 that belongs to the class saponins and showed anticancer activity.	Yan *et al.*, 2019
Streptomyces sp.	*Maytenus hookeri*	It is reported that these endophytes produced naphthomycins and grignard lactone A that belong to the class polyketones, which exhibited microbial fighting activity.	Yang *et al.*, 2018
Mycena osmundicola *Mycena orchidicola* *M. anoectochili*	*Gastrodia elata*	These fungal endophytes produced plant hormones that play an important role to promote germination in seeds of host plants.	Guo and Wang, 2001
Phyllosticta citricarpa	*Citrus medica*	It is reported for production of the paclitaxel secondary metabolite, which exhibited antitumour effects.	Kumaran *et al.*, 2008
Fusarium solani	*Camptotheca acuminata*	These endophytes of fungi are also in charge of producing camptothecin, a secondary metabolite with anticancer properties.	Kusari *et al.*, 2009
Pestalotiopsis terminaliae	*Terminalia arjuna*	It is reported for production of paclitaxel, a secondary metabolite which exhibited antitumour effects.	Gangadevi and Muthumary, 2009
Aspergillus nidulans *A. oryzae*	*Ginkgo biloba*	These endophytes produced quercetin, a secondary metabolite, which performed inflammatory fighting functions.	Qiu *et al.*, 2010
Streptomyces aureofaciens	*Zingiber officinale*	These endophytes have been reported to produced 5,7-dimethoxy-4-phenylcoumarin that showed antimicrobial activities against various pathogens.	Taechowisan *et al.*, 2007
Nocardiopsis sp.	*Zingiber officinale*	It is reported that these endophytes produced 1–heptacosanol that showed antimicrobial activity.	Sabu *et al.*, 2017
Papulaspora sp. *Ozonium* sp. *Ectostroma* sp. *Tubercularia* sp.	*Taxus chinensis* var. *mairei*	These endophytes produced paclitaxel, a metabolite that performed antitumour functions.	Zhou *et al.*, 2007 Guo *et al.*, 2009Wu *et al.*, 2013
Sclerotium sp.	*Atracty lancea*	It played an important role in increasing leaf metabolic capability of the host and cell protection from desiccation.	Chen *et al.*, 2008
Acremonium sp. *Shiraia* sp.	*Huperzia serrata*	These endophytes produced huperzine A, a metabolite that acted as anticholinesterase.	Li *et al.*, 2007
Blastomyces sp. *Botrytis* sp.	*Phlegmariurus cryptomerianus*	It is observed that these endophytes produced huperzine A that is a secondary metabolite that functioned as anticholinesterase.	Ju *et al.*, 2009

(*Continued*)

TABLE 9.1 (*Continued*)
Endophytes and Their Beneficial Role in Medicinal Plant Growth and Mitigating Stresses

Endophytic Bacteria	Host Plant	Comment	References
Phomopsis sp. *Penicillium* sp. *Fomitopsis* sp. *Arthrinium* sp.	*Cinchona ledgeriana*	It is reported that these endophytes produced cinchona alkaloids that are secondary metabolites, including cinchonine, cinchonidine, quinidine, and quinine, alkaloids that have anti-inflammatory, analgesic, antimalarial, and antipyretic functions.	Maehara *et al.*, 2012
Ceratobasidium sp.	*Rehmannia glutinosa*	These fungal endophytes produced indoleacetic acid.	Chen B. *et al.*, 2011
Taxomyces andreanae	*Taxus brevifolia*	It is responsible for production of paclitaxel, a secondary metabolite which showed antitumour effects.	Stierle *et al.*, 1995
Alternaria sp.	*Phellodendron amurense*	It plays a vital role in production of berberine that is a secondary metabolite for antibacterial activity.	Duan *et al.*, 2009
Fusarium solani *Metarhizium anisopliae* *Mucor rouxianus*	*Taxus chinensis*	These endophytes are reported to produce paclitaxel, a secondary metabolite that exhibited antitumour bioproperties.	Deng *et al.*, 2009 Liu *et al.*, 2009
Fusarium solani	*Taxus celebica*	It is observed that these endophytes produced paclitaxel, a secondary metabolite that has antitumour properties.	Chakravarthi *et al.*, 2008
Pestalotiopsis pauciseta	*Cardiospermum helicacabum*	It has also been used for the formation of paclitaxel secondary metabolites with anticancer activity.	Gangadevi *et al.*, 2008
Pithomyces sp.	*Taxus sumatrana*	It is reported that these endophytes produced paclitaxel, a secondary metabolite that has antitumour bioproperties.	Strobel *et al.*, 1996
Botryodiplodia theobroma *Pestalotia bicilia*	*Taxus baccata*	These endophytes produced paclitaxel, which showed antitumour effects in host plants.	Venkatachalam *et al.*, 2008
Chaetomium globosum	*Hypericum perforatum*	These endophytes are reported to produce hypericin, which is a secondary metabolite that performed an anti-depressant function in host.	Kusari *et al.*, 2008
Trichoderma atroviride D16	*Salvia miltiorrhiza*	It is reported for the production of tanshinone I and tanshinone IIA that are secondary metabolites that exhibited anti-inflammatory and antibacterial properties.	Ming *et al.*, 2012
Sordariomycete sp.	*Eucommia ulmoides*	It is observed that these endophytes produced chlorogenic acid, a secondary metabolite that exhibited antitumour effects.	Chen *et al.*, 2010
Fusarium oxysporum	*Ginkgo biloba*	It is observed that these endophytes produced ginkgolide B, which showed anti-inflammatory, antiallergic, and antishock activity.	Cui *et al.*, 2012
Arbuseular mycorrhiza *Penicillium griseofulvum*	*Glycyrrhiza uralensis*	These endophytes alleviated drought and salt stress by increasing protective enzyme activity and osmotica content.	Wang L. *et al.*, 2009
Arbuseular mycorrhiza	*Salvia miltiorrhiza*	Drought stress was lowered by these endophytes through increasing the nutrient absorption and altering metabolic activities in the host.	Meng and He, 2011
Leucocoprinus gongylophorus	*Cordia alliodora*	These endophytes killed the insects by producing few substances hostile to ants' fungal symbiont.	Bittleston *et al.*, 2011

(*Continued*)

TABLE 9.1 (*Continued*)
Endophytes and Their Beneficial Role in Medicinal Plant Growth and Mitigating Stresses

Endophytic Bacteria	Host Plant	Comment	References
Penicillium sp. *Aspergillus* sp.	*Ginkgo biloba*	It is reported that these endophytes produced 5-hydroxy-2-[(2S, 3R, 4S, 5S, 6-R)-3,4,5-trihydroxy-6-(hydroxymethyl) oxan-2-yl] oxybenzoate. Bilobalide acted as anti-inflammatory, antioxidant, neuroprotective and antidepressant.	Fei *et al.*, 2016
150 foliar fungal endophytes	*Picea rubens*	*Choristoneura fumiferana* insects were killed by the toxic chemicals produced by these endophytes.	Sumarah *et al.*, 2010
Bacillus subtilis *Myxormia* sp.	*Angelica sinensis*	These are involved in the production of some chemicals that can protect plants from the pathogenic effects of *Fusarium solani* and *Fusarium oxysporum*.	Yang *et al.*, 2012
Penicillium fumicalsuri, Aspergillus niger	*Azadirachta indica*	It is reported that these endophytes showed antifungal effects in *Azadirachta indica*, which belongs to the Meliaceae family.	Shaharoona *et al.*, 2006 Sharma *et al.*, 2020
Fusarium oxysporium	*Curcuma longa*	Antifungal function was performed by these endophytes in host plant *Curcuma longa* that belongs to the Zingiberaceae family.	Shaharoona *et al.*, 2006 Sharma *et al.*, 2020
Cryptosporiopsis cf. quercina	*Triptergyium wilfordii*	It is reported that these endophytes killed *Pyricularia oryzae* pathogenic fungi by producing pathogens that are poisoned by cryptocandin and cryptocin.	Strobel *et al.*, 1999
Bacillus subtilis	*Aloe vera*	It is observed that endophytes showed antibacterial effects in the aloe vera plant, which belongs to the Liliaceae family.	Shaharoona *et al.*, 2006 Sharma *et al.*, 2020
Escherichia coli *Proteus vulgaris*	*Curcuma longa*	Antibacterial function was performed by these endophytes in host plant *Curcuma longa* that belongs to the Zingiberaceae family.	Shaharoona *et al.*, 2006 Sharma *et al.*, 2020
Bacillus amyloliquefaciens	*Scutellaria baicalensis*	It is reported that these bacterial endophytes showed wide-spectrum antifungal and antibacterial activities in *Scutellaria baicalensis* that belongs to the Lamiaceae family.	Shaharoona *et al.*, 2006 Sharma *et al.*, 2020
Trichothecium roseum	*Maytenus hookeri*	It is reported that these endophytes killed pathogenic fungi by producing trichothecin, which is toxic to pathogens.	Zhang *et al.*, 2010
Taphylococcus aureus	*Azadirachta indica*	These endophytes showed antimicrobial activity in host plant *Azadirachta indica*, which belongs to the Meliaceae family.	Shaharoona *et al.*, 2006 Sharma *et al.*, 2020
Chaetomium globosum L18	*Curcuma wenyujin*	Pathogenic fungi are destroyed by pathogen-killing compounds generated by these endophytes.	Wang *et al.*, 2012
Bacillus thuringiensis	*Andrographis paniculata*	These endophytes showed anti-infective properties in host plant *Andrographis paniculata* that belongs to the Acanthaceae family.	Shaharoona *et al.*, 2006 Sharma *et al.*, 2020
Pseudonocardia sp.	*Artemisia annua*	The antimalarial compound artemisinin was produced by these endophytes in host plant *Artemisia annua*, which belongs to the Asteraceae family.	Shaharoona *et al.*, 2006 Sharma *et al.*, 2020
Gilmaniella sp. *AL12.*	*Atractylodes lancea*	These endophytes destroy pathogenic fungi by generating jasmonic acid, which triggers immune responses.	Ren and Dai, 2012

(*Continued*)

TABLE 9.1 (*Continued*)
Endophytes and Their Beneficial Role in Medicinal Plant Growth and Mitigating Stresses

Endophytic Bacteria	Host Plant	Comment	References
Pseudomonas sp.	*Zingiber officinale*	It is observed that endophytes showed antimicrobial effects in the aloe vera plant, which belongs to the *Liliaceae* family.	Shaharoona *et al.*, 2006 Sharma *et al.*, 2020
Phialemonium curvatum	*Passiflora edulis*	It is involved in production of solaniol, uridine, and 3-indole acetic acid, which exhibited antioxidant and antifungal activity and brine shrimp lethality as well.	Rathnayake *et al.*, 2018
Bipolaris sorokiniana	*Costus speciosus*	It is reported that these fungal endophytes produced helminthosporol and helminthosporal acid that led to toxicity against brine shrimp.	Qader *et al.*, 2017
Lasiodiplodia theobromae	*Morinda citrifolia*	It is reported that these endophytes synthesized taxol that led to cytotoxicity against human breast cancer.	Pandi *et al.*, 2011
Alternaria macrospora	*Berberis aristata DC*	These endophytes performed antioxidant and antibacterial activity.	Sharma *et al.*, 2018
Alternaria alternata	*Azadirachta indica*	It is involved in production of dichloronitromethane that showed antimicrobial effects.	Chatterjee *et al.*, 2019
Rosellinia sanctaecruciana	*Albizia lebbeck*	It is observed that these endophytes produced jammosporin A that led to cytotoxicity activity.	Sharma *et al.*, 2018
Streptomyces parvulus	*Abutilon indicum*	It is reported that these endophytes produced fengycin, actinomycin X 0b, and actinomycin D, which exhibited antibiotic effects.	Chandrakar and Gupta, 2019
Colletotrichum gloeosporioides	*Centella asiatica*	These fungal endophytes produced asiaticoside, which stimulated new cell growth, sped up wound healing, and showed anti-inflammatory activity.	Gupta *et al.*, 2018
Colletotrichum sp.	*Tinospora cordifolia*	It is observed that these endophytes produced allopurinol that actively functions against hyperuricemia.	Kapoor and Saxena, 2018
Penicillium citrinum	*Tragia involucrata*	It is reported that these endophytes produced flavonoid and phenolic content, which led to cytotoxic activity against cancer cells.	Danagoudar *et al.*, 2018
Cochliobolus sp.	*Aerva lanata*	It is involved in production of flavonoid and phenolic content that acts as anti-inflammatory, antidiabetic, and antioxidant.	Shoba and Mythili, 2018
Bacillus thuringiensis	*Picrorhiza kurroa*	These endophytes produced polymyxin, gramicidin S, and bacitracin that showed antibacterial activity against a variety of human diseases.	Raina *et al.*, 2018
Paraconiothyrium sp.	*Zingiber officinale*	It is reported that these endophytes produced dantron, which showed antimicrobial activity.	Anisha *et al.*, 2018
Pestalotiopsis species	*Terminalia arjuna*	It showed antihypertensive and antioxidant activity in the host plant.	Tejesvi *et al.*, 2008
Aspergillus terreus	*Achyranthes aspera*	Production of terrein compound is reported by these fungal endophytes that exhibited anticancer effects in the host.	Goutam *et al.*, 2017
Enterobacter sp.	*Sida cordifolia*	It is reported that these endophytes showed antibacterial activity against different microbes like *Salmonella typhi* and *Klebsiella pneumonia* and also acted as immunomodulating agents.	Rajan *et al.*, 2016

(*Continued*)

TABLE 9.1 (*Continued*)
Endophytes and Their Beneficial Role in Medicinal Plant Growth and Mitigating Stresses

Endophytic Bacteria	Host Plant	Comment	References
Trichoderma lixii	*Bacopa monnieri*	These endophytes produced a tribacopin AV compound that showed antifungal efficacy against *Candida albicans.*	Katoch *et al.*, 2019
Chaetomium arcuatum	*Semecarpus anacardium*	It is involved in production of 6-hydroxymethyleugenin and eugenetin compounds, which led to cytotoxicity of cancerous cells.	Kumar *et al.*, 2017
Ralstonia sp.	*Aloe vera*	In the host plant, these endophytes demonstrated antibacterial activity against *Escherichia coli, Staphylococcus aureus*, and *Pseudomonas aeruginosa.*	Sinha *et al.*, 2015
Aspergillus flavus	*Aegle marmelos*	It is observed that these endophytes produced a bioflavonoid rutin compound that showed antibacterial efficacy against common fungal, bacterial, and human diseases.	Patil *et al.*, 2015
Colletotrichum gloeosporioides	*Justicia gendarussa*	These endophytes are reported to produce taxol, which showed anticancer activity.	Gangadevi and Muthumary, 2008
Colletotrichum gloeosporioides	*Huperzia serrata*	It is involved in production of huperzine A that aided in treatment of Alzheimer's disease.	Shu *et al.*, 2014
Xylaria sp.	*Ginkgo biloba L*	7-amino-4-methylcoumarin compounds are produced by these endophytes that showed antifungal and antibacterial effects.	Liu *et al.*, 2008
Fusarium	*Dysosma versipellis*	It is involved in the production of podophyllotoxin that showed anticancer and antimicrobial activity.	Tan *et al.*, 2018
Colletotrichum *Fusarium* *Phoma* *Penicillium*	*Pereskia bleo*	These endophytes are reported to produce asparaginase, which showed anticancer activity.	Chow and Ting, 2015
Alternaria sp.	*Erythrina variegata*	It is involved in the production of altersolanol that showed antiangiogenic activity.	Pompeng *et al.*, 2013
Aspergillus sp.	*Swietenia macrophylla*	It is reported that these fungal endophytes produced 2-(2,4,5-trichlorophenoxy) propanoic acid that showed antibacterial activity against human pathogens.	Leong *et al.*, 2018
Fusarium solani	*Cassia alata*	It is involved in the production of azaanthraquinones and napthaquinones that showed cytotoxicity against brine shrimp and also showed antioxidant and antimicrobial properties.	Khan *et al.*, 2018
Pestalotiopsis microspora	*Artocarpus heterophyllus*	It is involved in the production of hydroxypestalotin, which showed cytotoxicity against murine leukaemia.	Riga *et al.*, 2019
Trichoderma sp., *Aspergillus sydowii*	*Sonneratia alba*	These endophytes exhibited antimicrobial and cytotoxic activities.	Handayani *et al.*, 2018
Lecythophora sp.	*Alyxia reinwardtii*	It is involved in the production of a lecythomycin compound, which showed antifungal activity against *Candida kruzei* and *Aspergillus fumigatus.*	Sugijanto *et al.*, 2011

9.4.1 Role of Endophytes in Mitigating Biotic Stresses in Medicinal Plants

Biotic stressors deprive the host of nutrients, spread disease, diminish vigour, and in extreme situations cause host mortality. Worldwide, biotic stressors cause pre- and post-harvest yield losses. Sessile plants are unable to escape from these environmental cues. Plants do not have an adaptive immune system, yet they can develop complex mechanisms to withstand biotic stressors. On the other hand, the special association between endophytes and host plants, including medicinal plants, can protect the plant from different biotic stresses (Kumar and Nautiyal, 2022, Papp-Rupar *et al.*, 2022). Two theories were proposed to show the association between host plants and endophytes. The mosaic effect theory states that endophytes shield their hosts by secreting a variety of chemicals like secondary metabolites in various plant organs that are relevant to their species (Carroll, 1991; Wu *et al.*, 2021), which results in a difference in how effective these organs are at warding off herbivores and pathogens (Tan and Zou, 2001; Zhang *et al.*, 2006; Kusari *et al.*, 2012; Ogbe et al., 2020; Wu *et al.*, 2021). According to another theory, host plants are supported by endophytes with acquired immune systems (Wu *et al.*, 2021).

9.4.1.1 Pathogenic Microorganisms

Pathogenic organisms on crops lead to many plant-related diseases that are the main causes of economic harm in agriculture. It may be prevented by the presence of endophytic microbes. It was observed that endophytes play a remarkable part in antifungal and antibacterial activities by preventing phytopathogen growth and aiding the production of bioactive compounds (Huang *et al.* 2008; Rodriguez *et al.*, 2009; Ancheeva *et al.*, 2020; Ogbe *et al.*, 2020; Jayaram *et al.*, 2021; Kumar and Nautiyal, 2022). Many reports have stated the role of both bacterial and endophytic fungi in alleviating stress on plants, including medicinal plants. Endophytic fungi play a dominant part in the accumulation and production of bioactive secondary metabolites that help in reducing reactive oxygen species damage, activating the host defence system, and improving genes that protect host plants, including medicinal plants, from pathogenic attack (Rodriguez and Redman, 2008; Sasan and Bidochka, 2013; Jaber and Salem, 2014; Jaber, 2015; Zhang *et al.*, 2019; Ogbe *et al.*, 2020; Ancheeva *et al.*, 2020; Jayaram *et al.*, 2021; Kumar and Nautiyal, 2022). Endophytic fungi are also responsible for the induction of biocontrol agents like phytoalexins that showed resistance against pathogenic attack (Gao *et al.*, 2010; Kumara *et al.*, 2014; Zhang *et al.*, 2019). For example, Ting *et al.* (2012) reported that the growth of the pathogen *Fusarium oxysporum* can be restricted in the presence of *Penicillium citrinum* by enzymatic activities. Similarly, bacterial endophytes have also been reported in various studies that showed their potential to alleviate pathogenic stress (Rosenblueth and Martinez-Romero, 2006; Ryan *et al.*, 2008; Munif *et al.*, 2013; Adam *et al.*, 2014; Upreti and Thomas, 2015; Zhang *et al.*, 2019). Moreover, the endophytes have also been found to be concerned with the production of antibiotics and suit of lipopeptides that can restrain bacterial and fungal pathogens (Scherwinski *et al.* 2008; Adesina *et al.* 2009; Müller *et al.* 2009; Köberl *et al.*, 2013; Gond *et al.*, 2015; Upreti and Thomas, 2015; Zhang *et al.*, 2019). For example; Wang *et al.* (2016) showed the potential of bacterial endophytes against *Aspergillus parasiticusor aflatoxins* present in Chinese peanuts. Similar to non-medicinal plants, medicinal plants also benefit from an association of endophytes because it is involved in the synthesis of numerous antibiotics which directly impede the growth of pathogens (Lacava and Azevedo, 2013; Wu *et al.*, 2021). For example, Hong *et al.* (2018) reported that endophytic bacteria like *Panax ginseng* in medicinal plants showed antimicrobial activity against *Cylindrocarpon destructans* and *Botrytis cinerea*. Similarly, Zhao *et al.* (2015) revealed the potential of *Paenibacillus* in *Lonicera japonica* medicinal plants to exhibit a hindering effect against phytopathogenic fungi that act as an ideal organism for investigating different aspects of host–pathogen relationships. Moreover, Yang *et al.* (2015) reported that *Bacillus amyloliquefaciens* Fy11 appears to suppress phytophthora blight on pepper indirectly by promoting induced systemic resistance (ISR). In addition, endophytes like *Bacillus amyloliquefaciens*, *Stenotrophomonas maltophilia*, *Bacillus amyloliquefaciens*, and *Phyllobacterium myrsinacearum* also play a significant part in resisting the attack of pathogens in different medicinal plants, including *Panax notoginseng*, *Panax ginseng*, *Ginkgo biloba*, and *Epimedium brevicornu* (Rakotoniriana *et al.*, 2013; Ma *et al.*, 2013; Yang *et al.*, 2015; Hong *et al.*, 2018; Jayakumar *et al.*, 2019).

9.4.1.2 Insects

Insect and pest infestations may lead to preharvest losses of the crop ($13-16\%$) and post-harvest crop storage loss ($50-60\%$). They are responsible for almost 20–36% loss of crop production costs, a loss of \$470 billion annually (Savary *et al.*, 2019; Kumar and Nautiyal, 2022). However, numerous studies have reported that the presence of entomopathogenic fungi reduces plant diseases by reducing the pest development rate, larva survival, and reproduction rate and inhibiting insect feeding (Russo *et al.*, 2019; Aguilera-Sammaritano *et al.*, 2021; Kumar and Nautiyal, 2022). Moreover, endophyte-infected plants exhibit more resistance toward oxidative stresses caused by insect infestation (Chen *et al.*, 2022). For example, Aguilera-Sammaritano *et al.* (2021), examined the mortality of different phases of *Lobesia botrana* insect pests, including larva (75%), pupae, and adult (85%), in *Vitis vinifera* L. under the activity of entomopathogenic fungi *Metarhizium anisopliae.* Similar to endophytic fungi, the efficiency of endophytic bacteria regarding the regulation of the growth of insect pests is also described in the literature. For example, Mishra *et al.* (2017), reported a 100% death rate of the larva of the Bihar hairy caterpillar in the presence of the endophyte *Bacillus thuringiensis.* This endophytic association with plants is also applicable to medicinal plants to support their survival and growth. For example, an endophytic fungus of the genus *Neotyphodium* is involved in the production of alkaloids that can resist insect attack. With the higher strain endurance, host plants that are infected by endophytic fungi could subdue indigenous plants without any fungal infection and subsequently become intrusive (Tofern *et al.*, 1999; Clement *et al.*, 2005; Jia *et al.*, 2016). Moreover, *Trichoderma viride, Cladosporium cladosporioides, Chaetomium cochliodes, L. cf. psalliotae, Lecanicillium dimorphum, Beauveria bassiana,* and *Leucocoprinus gongylophorus* were also reported as effective endophytes in the alleviation of biotic stress due to insects in various medicinal plants like *Phoenix dactylifera, Cirsium arvense,* and *Cordia alliodora* (Bittleston *et al.*, 2011; Hartley and Gange, 2009).

9.4.1.3 Nematodes

Endophytes serve as biocontrol agents to resist pathogenic nematodes in host plants (Rosenblueth and Martinez-Romero, 2006; Ryan *et al.*, 2008; Munif *et al.*, 2013; Adam *et al.*, 2014; Upreti and Thomas, 2015; Kumar and Nautiyal, 2022). The presence of pathogenic nematodes adversely affected root development, plant growth, and photosynthesis rate. However, this effect can be alleviated by the use of endophytes (Kumar and Nautiyal, 2022). It was observed that endophytic organisms like *Bacillus cereus* (BCM 2), *Bacillus wiedmannii, Beauveria bassiana, Serratia liquefaciens, Pantoea agglomerans* (MN34), *Bacillus cereus* (SZ5), *Bacillus altitudinis* (CCM 7), *Paecilomyces lilacinus, Streptomyces* sp. *(CBG9)*, and *Pseudomonas putida* (MN12) have a beneficial effect in retarding the growth and egg masses of the nematode *Meloidogyne incognita* in different host plants (Muhae-ud-Din *et al.*, 2018; Youssef *et al.*, 2020; Moslehi *et al.*, 2021). Similarly, Maulidia *et al.* (2020) reported that the mortality rate of *Meloidogyne incognita* was 95.46% under the action of *Bacillus thuringiensis* strain AK08, which may cause the production of cholest-5-en-3-ol (3. beta.)-carbonochloridate that showed nematocidal activities. Similarly, Marques-Pereira *et al.* (2022) reported the production of serratamolide-like lipopeptides by Serratia stress with a mortality of 77% of a pinewood nematode, *Bursaphelenchus xylophilus* (Marques-Pereira *et al.*, 2022). Moreover, Farhat *et al.* (2022) reported 100% juvenile mortality of *Meloidogyne javanica* in the presence of the endophyte *Fusarium solani.* Similarly, Sharma and Pandey (2009) reported that fungi *Arthrobotrys oligospora, Paecilomyces lilacinus,* and *Trichoderma harzianum* are effective in the alleviation of stress due to *Meloidogyne incognita* in *Withania somnifera.*

9.4.2 Role of Endophytes in Mitigating Abiotic Stresses in Medicinal Plants

In addition to biotic stresses, abiotic stresses like drought, salt, water, and temperature negatively influence the development and growth of plants, including medicinal plants (Miller *et al.*, 2010; Mastouri *et al.*, 2010; Yang *et al.*, 2010; Jha *et al.*, 2011; Khan *et al.*, 2012; Spiers *et al.*, 2012; Zhang *et al.*, 2012, 2019; Murphy *et al.*, 2014; Naveed *et al.*, 2014; Vargas *et al.*, 2014; Cherif *et al.*, 2015; Nagata *et al.*, 2015). Therefore, the association between plants and endophytes is a dominating factor that plays a critical part in relieving abiotic stress by promoting plant growth, producing a variety of antioxidant compounds, biosynthesizing antistress biomolecules, and assisting plants in developing a strong defence system under stress conditions (Herrera-Carillo *et al.*, 2009; Torres *et al.*, 2009; Saraf *et al.*, 2011; Hameed *et al.*, 2014; Gond *et al.*, 2015; Ogbe *et al.*, 2020; Kumar and Nautiyal, 2022). The role of both endophytic fungi and bacteria was also reported in the literature regarding abiotic stress alleviation (Waqas *et al.*, 2012; Hashem *et al.*, 2016; Zhang *et al.*, 2019).

9.4.2.1 Drought Stress

An abiotic factor, drought, greatly influences plants' development, growth, and yield. It may be because of the limited supply of water and the high transpiration rate in plants. In temperate regions, generally water stress happens during midafternoon hours and negatively influences the rate of plant growth (Zhang *et al.*, 2019; Takahashi *et al.*, 2020; Kumar and Nautiyal, 2022). It may decrease seed germination, plant growth, and photosynthetic rate and increase the concentration of reactive oxygen species. (Wang *et al.*, 2003; Waqas *et al.*, 2012; Zia *et al.*, 2021; Kumar and Nautiyal, 2022). During this condition, the population of plants successfully adopts different strategies of stress tolerance because of the role played by endophytes (Cherif *et al.* 2015; Zhang *et al.*, 2019; Kumar and Nautiyal, 2022). Bacterial endophytes significantly take part in the production of growth promotors to improve growth under drought conditions. Several bacterial endophytes synthesized different bioactive compounds like spermidine, 1-aminocyclopropane-1-carboxylate deaminase, indole acetic acid, exopolysaccharide, and colanic acid siderophore which serve a remarkable function in alleviation of drought strain in host plants, including medicinal plants (Namwongsa *et al.*, 2019; Zhang *et al.*, 2020; Kumar and Nautiyal, 2022). For example; *Bacillus methylotrophicus* strain 5.18, *Micrococcus luteus* strain 4.43, and *Bacillus aquimaris* strain 3.13 are involved in the production of 1-aminocyclopropane-1-carboxylate (ACC) deaminase, Indole Acetic Acid (IAA), and exopolysaccharide, respectively. All these secondary metabolites are produced to enhance drought tolerance (Namwongsa *et al.*, 2019). Similarly, Cohen *et al.* (2015) showed that the reaction of *Arabidopsis thaliana* to drought was ameliorated by *Azospirillum brasilense* by increasing the abscisic acid (ABA) levels. Cohen *et al.* (2015) revealed the role of *Azospirillum brasilense* in the generation of abscisic acid during drought stress on *Arabidopsis thaliana*. Similarly, Xu *et al.* (2017) also stated that *Piriformospora indica* mediated plant safeguards against the harmful impacts of drought stress in maize plants by improving the accumulation of proline, production of antioxidant enzyme activity, and expression of drought-related genes. Similarly, Naveed *et al.* (2014) reported the role of *Enterobacter* sp. and *Burkholderia phytofirmans* in the alleviation of drought strain by catalysing the production of enzymes such as superoxide dismutase, catalase, and peroxidase. Moreover, an endophytic organism, *Pseudomonas* (LTGT-11–2Z) that was identified in the desert medicinal plant *Alhagi sparsifolia*, was found to be effective in enhancing drought tolerance in plants during water scarcity (Zhang *et al.*, 2020). From the analysis, it was observed that plants infected with endophytes showed higher root and shoot biomass, high efficiency to uptake water, and high photosynthetic rate as compared to plants free from endophytes. Similarly, Meng and He (2011) reported the role of *Arbuseular mycorrhiza* in the alleviation of drought stress by increasing nutrient absorption for *Salvia miltiorrhiza*.

9.4.2.2 Salinity Stress

In addition to drought, excess concentration of salt is also a devastating factor that affects the growth and metabolism of plants and also increases plants' susceptibility to soil-borne pathogens (Joe *et al.*, 2012; Zhao *et al.*, 2016; Zhang *et al.*, 2019; Wu *et al.*, 2021; Kumar and Nautiyal, 2022). In this regard, the association between endophytes and plants has a notable part in relieving salinity strain because of its involvement in the reduction of reactive oxygen species damage and ethylene level and acceleration of detoxifying enzymes' action and photosynthetic rate (Joe *et al.*, 2012; Zhao *et al.*, 2016; Zhang *et al.*, 2019; Siddiqui *et al.*, 2022; Kumar and Nautiyal, 2022).

Moreover, endophytes serve a significant function in the production of biomolecules like tocopherol, glutathione, ascorbate, mono-dihydro-ascorbate reductase (MDHAR), thiol-reliant peroxidase (APX), glutathione reductase (GR), and catalase (CAT) that are involved in removal of ROS by the direct enzymatic action (Kumar and Nautiyal, 2022). The role of two endophytes, *Pseudomonas fluorescens* and *Pseudomonas migulae*, was stated by Ali *et al.* (2014). They observed that both endophytes were significantly involved in the production of 1-aminocyclopropane-1-carboxylate (ACC) deaminase which increased tomato plants' ability to resist a high level of salt. Similarly, *Piriformospora indica* was also considered an effective endophyte to relieve salt stress in rice and barley by improving photosynthetic pigment content and detoxifying enzyme activity (Waller *et al.*, 2005; Jogawat *et al.*, 2013). Similar to non-medicinal plants, medicinal plants are also influenced by their association with endophytes. For example; Karthikeyan *et al.* (2012), reported the production of antioxidant enzymes including superoxide dismutase, catalase, and ascorbate peroxidase in a saline environment in *Achromobacter xylosoxidans*'s presence in *Catharanthus roseus*. Similarly, *Arbuscular mycorrhiza* and *Penicillium griseofulvum* are also involved in the alleviation of stress due to salinity by promoting protective enzyme activity and osmotica content (Wang *et al.*, 2009).

9.4.2.3 Temperature Stress

Temperature is also a notable parameter that affects plant growth and survival, including medicinal plants (Ding *et al.*, 2011; Valluru *et al.*, 2012; Wang *et al.*, 2016; Zhang *et al.*, 2019), due to extreme climatic variation and temperature stress intensity causing a great decline in the yield. In this drastic condition, the symbiotic interaction by endophytes serves a remarkable function in alleviating temperature stress because it enhances carbohydrate metabolism, osmotic regulation, antioxidant defence system, and related gene expression (Zhang *et al.*, 2019). For instance, Wang *et al.* (2016) specified that the tolerance level of tomato plants can be increased by the presence of endophytes *Serratia* sp., *Bacillus subtilis*, and *Bacillus cereus*, because they are involved in promoting soluble sugar, activating stress-related genes, and improving the antioxidant protection system. Similarly, the role of *Burkholderia phytofirmans* PsJN was demonstrated in making Vitis (a medicinal herb) adapt to cold by the regulation of carbohydrate metabolism, which can be one of the significant impacts of lessening chilling-induced harm (Zhang *et al.*, 2019).

9.4.2.4 Oxidative Stress

Abiotic stresses, including oxidative stress, results in the formation of reactive oxygen species in plant bodies, and long-term exposure to ROS cause cellular damage, lipid peroxidation, oxidative degradation of nucleic acid, and eventually death of cells (Wang *et al.*, 2003; Waqas *et al.*, 2012; Verma *et al.*, 2021b). However, the association of plants with endophytes diminished lipid peroxidation and the production of ROS and accelerated the production of antioxidant defence enzyme activities (Zhang *et al.*, 2019; Kumar and Nautiyal, 2022; Ismail *et al.*, 2022). Moreover, endophytes also play a part in the production of different metabolites like ascorbate peroxidase, peroxidase (POD), superoxide dismutase (SOD), and enzymatic and nonenzymatic antioxidants and alteration of expression of stress tolerance genes that play a significant part in alleviating the concentration of ROS and enhance their survival and productivity (Kumar and Nautiyal, 2022). For example, Abd_Allah *et al.* (2018), reported that *Cicer arietinum* under salinity exhibits more resistance toward oxidative stress because of improved ROS-scavenging antioxidant enzymes (catalase, peroxidase, superoxide, and dismutase). Similarly, Mastouri *et al.* (2010) showed that *Trichoderma harzianum* reduced the amount of lipid peroxidation in *Lycopersicum esculentum* in saline conditions. Like non-medicinal plants, studies also revealed the reduction of oxidative stress in medicinal plants in endophytes' presence. For instance, Khan *et al.* (2016) described that in a medicinal plant, *Populas* sp., oxidative stress can be alleviated by increasing the amount of indole acetic acid in the proximity of an endophytic consortium composed of *Burkholderia vietnamiensis, Rhizobium tropici, Rahnella* sp., *Rhodotorula graminis,* and *Acinetobacter calcoaceticus* when inoculated in a medicinal plant, *Populas* sp.

9.4.2.5 Nutrient Stress

Essential nutrients like nitrogen, phosphate, and potassium influence plant growth and yield. Moreover, the availability of micro- and macro-nutrients mainly depends on the association of the host with the microflora involved in extracting nutrients (Kumar and Nautiyal, 2022). Similarly, endophytes play an important role in adjusting the root morphology, rhizosphere exudates, and fixation of atmospheric nitrogen into ammonium (NH_4^+) and increasing the mobility of essential nutrients like NPK (Muangthong *et al.*, 2015; Verma *et al.*, 2021a; Kumar and Nautiyal, 2022). In addition to that, in case of insufficient nutrient resource, endophytes play an important role in developing root system architecture through the improvement of biosynthesis and phytohormone signalling in host roots (Harman and Uphoff, 2019; Kumar and Nautiyal, 2022). Similarly, root growth and root hairs rely on auxin, produced by endophytes. Therefore, it is revealed that plants with endophytes exhibit more stress tolerance to nutrient deficiency because endophytes improve root system architecture through plant hormone (ethylene, gibberellin, cytokinins, and auxin) signalling (Passari *et al.*, 2016). For example, Singh and Dubey (2018), revealed that in the wheat plant, *Arthrobacter* sp. (DS-179) and *B. subtilis* (DS-178) played an important role in the assimilation and mobilization of zinc, and *Enterococcus hirae* (DS-163) and *Arthrobacter sulfonivorans* (DS-68) helped in the uptake of iron (Singh and Dubey, 2018). During the presence of endophytes DS-163, DS-68, DS-179, and DS-178, the wheat plant showed almost a 1–7-fold increase in Fe and Zn assimilation and output increase in comparison to uninoculated control. The positive impact of endophytes on the betterment of crops and the alleviation of nutrient stresses is significant (Kumars and Nautiyal, 2022). Li et al. (2012) demonstrated the beneficial impact of endophytes on the growth of *Artemisia annua*, a medicinal plant known for the synthesis of the antimalarial compound artemisinin. The researchers isolated endophytic bacteria from *Artemisia annua* roots and found that these bacteria significantly enhanced the solubilization of phosphorus, leading to increased plant biomass and artemisinin content. This example highlights how endophytes can contribute to the nutrient status and medicinal value of plants under nutrient stress. Similarly, Li et al. (2012) stated the function of endophytes in enhancing P solubilization in the medicinal plant *Artemisia annua*, which led to increased plant biomass and artemisinin content. Moreover, Khan *et al.* (2016) also investigated the significance of endophytic fungi in enhancing enzymatic activity that led to organic matter decomposition in the soil, releasing nitrogen and phosphorus in the soil to fulfil nutrient deficiency in *Withania somnifera*.

9.4.2.6 Metals and Other Pollutant Stress

Because of rapid industrialization and geo-chemical activities, heavy metal contamination greatly affects the survival of plants, including medicinal plants. Therefore, the function of endophytes in mitigating metal tensity cannot be neglected (Zhang *et al.*, 2012; Waqas *et al.*, 2014). Antioxidant enzymatic activities, accumulation of heavy metal ions, changing translocation, and photosynthesis are also enhanced by endophytes (Zhang *et al.*, 2019). Moreover, the phenomena of metal sequestration or chelation in endophytic fungi also acts as the main player in increasing the tolerances of host plants (Weyens *et al.*, 2009). Likewise, Shi *et al.* (2016) demonstrated the presence of *Pseudomonas aeruginosa* could increase the accumulation, translocation, and stress tolerance of cadmium in plants, including medicinal plants, from roots to shoots. Jin *et al.* (2018) demonstrated that *Acrocalymma vagum* may reduce tobacco's levels of cadmium (Cd), arsenic (As), and lead (Pb), as well as heavy metal ions that are transported to shoots from roots.

Aside from heavy metal stress, the friendly relationship between plants and endophytes also strengthens plants against several other pollutants. Barac *et al.* (2004) described that the presence of *Burkholderia cepacia* G4 endophyte increases the capability of plants to tolerate high toluene levels and also helps in lesser transpiration of toluene in the atmosphere. Similarly, Newman and Reynolds (2005) concluded the possible benefits of employing endophytic microbes for enhancing the remediation of xenobiotics. The most significant benefit of employing the endophytes is that it is relatively easier to manipulate their genes through genetic engineering for a better functioning of processes including the remediation of xenobiotics affecting the growth of plants. (Newman and Reynolds, 2005; Wu *et al.*, 2021). Moreover, Govarthanan *et al.* (2016) reported the function of the endophytic bacteria *Paenibacillus* sp. in the mitigation of heavy metal stress in a medicinal plant, *Tridax procumbens*. Similarly, *Agrobacterium* spp. (Xu *et al.*, 2016) and *Citrobacter putida* (Khaksar *et al.*, 2016) also played an important role in relieving xenobiotic distress in medicinal plants, including *Pteris vittate* and *Euphorbia milii*, respectively.

9.4.3 Role of Endophytes in Concurrent Mitigation of Multiple Stresses in Medicinal Plants

Different reports suggested that endophytes play a significant part in mitigating multiple stresses in plants by promoting growth, increasing resilience to stress, and promoting secondary metabolite production. Moreover, endophytes can mitigate a range of abiotic stress concurrently which make plants normally grow (Zhang *et al.*, 2019). For example, rhizobacteria, a plant growth stimulant that was studied by Piccoli and Bottini (2013), can reduce plant tension caused by high temperatures, droughts, heavy metals, and solar ultraviolet-B radiation (280–315 nm). Hume *et al.* (2016) showed that by boosting the amount of anti-herbivore chemicals derived from endophytes, Epichlo endophytes may help perennial ryegrass and tall fescue in tolerating circumstances like metal toxicity, mineral shortages, salt, low light, and water scarcity. Similarly, Glick (2012) and Glick *et al.* (1998) demonstrated that Rhizobacteria that are predominantly plant growth promoters and contain ACC deaminase can efficiently combat growth inhibition caused by stress conditions such as low temperature, raised levels of organic pollutants and metals, drought, and flooding. Moreover, Mastouri *et al.* (2010) also revealed an association between tomato and *Trichoderma* spp. that could increase stress resilience by managing the harm induced by reactive oxygen species during abiotic stresses. Moreover, endophytic fungus *Chaetomium* Ch1001 also showed the potential to alleviate the combined stress of both insects and nematodes in the medicinal plant *Cucumis sativus*, because it is responsible for generating abscisic acid that affected both *Meloidogyne incognita* and root-knot (Yan *et al.*, 2019). Likewise, Wang *et al.* (2009) also revealed that endophytic fungi *Penicillium griseofulvum* and *Arbuscular mycorrhiza* play an important role in alleviation of the combined effect of drought and salt stress on *Glycyrrhiza uralensis* (a medicinal plant).

9.4.4 Mechanisms for Alleviation of Medicinal Plant Stresses by Endophytes

The association between plants and endophytes may result in some beneficial adaptations, including the inauguration of the host metabolism through endophytes, initiation of endophyte metabolism by the host, certain routes shared by endophyte and host, and the metabolization of secondary metabolites from the host plant (Ludwig-Muller, 2015). Such relationships exert a substantial influence on the resistance of medicinal plants to biotic and abiotic stresses by transcription factors, promoter elements, generation of antioxidants, phytohormone generation, and the expression and stimulation of stress-responsive genes of the host (Lata *et al.*, 2018; Ogbe *et al.*, 2020). Therefore, it is observed that endophyte-infected plants exhibit better health and less susceptibility to different stress (Waller *et al.*, 2005; Saikkonen *et al.*, 2010). Moreover, the affiliation between medicinal host plant metabolism and bacterial endophytes is also responsible for the biotransformation of compounds and synthesis of secondary metabolites that are very significant in soothing stresses (Tiwari *et al.*, 2010; Brader *et al.*, 2014; Wu *et al.*, 2021). For example, the endophyte *Bacillus altitudinis* KX230132.1 serves as an efficient trigger in a medicinal herb, ginseng, in accelerating the concentration of ginsenoside, which causes the biochemical transformation of active ingredients in medicinal plants (Song *et al.*, 2017; Wu *et al.*, 2021). Similarly, in the biotransformation of the major ginsenoside Rb1 into the minor ginsenoside Rg3, *Burkholderia* sp. GE 17–7 was demonstrated to act as an antitumour compound (Fu *et al.*, 2017; Wu *et al.*, 2021). Plants associated with bacterial endophytes were also used for the production of essential secondary compounds like sesquiterpenes, polyketones, flavonoids, and saponins (Ek-Ramos *et al.*, 2019; Wu *et al.*, 2021). Additional studies showed analogous properties like anti-HIV, anti-inflammatory, anticancer, and antimicrobial activities between endophytic bacteria and host plants, including medicinal plants (Rustamova *et al.*, 2020; Wu *et al.*, 2021). Similar to bacterial endophytes, endophytic fungi are also crucial for plants, including medicinal plants, in bioactive secondary metabolite accumulation that affects drug quality and quantity (Chen *et al.*, 2016; Jia *et al.*, 2016). These secondary metabolites included podophyllotoxin (Eyberger *et al.*, 2006), deoxy podophyllotoxin (Kusari *et al.*, 2009), camptothecin (Kusari *et al.*, 2011), azadirachtin (Kusari *et al.*, 2012), and hypericin and emodin (Kusari *et al.*, 2008, Kusari *et al.*, 2009). Importantly, endophytic fungi like *T. andreanae* (Stierle *et al.*, 1995), *Metarhizium anisopliae* (Liu *et al.*, 2009), *Tubercularia* sp. strain TF5 (Wang *et al.*, 2006), *Eimatoantlerium nepalense* (Bashyal, 1999), and *Seimatoantlerium tepuiense* are involved in the production of the best anticancer compound, taxol. Moreover, some endophytic fungi are vital for enhancing secondary metabolite accumulation that plays a crucial role in anti-inflammatory, analgesic, antimalarial, antimalarial, antipyretic, or antitumour medicinal treatments. For example, Wang *et al.* (2006) reported that *Coetotrichum gloesporioides* played a dominating role in artemisinin synthesis in hairy-root cultures of *Artemisia annua*. Additionally, research also showed that endophytes help host plants manage the quantity of ROS and protect them from their detrimental impacts by generating antioxidants and secondary metabolites that have previously been recognized in medicinal plants (Priti *et al.*, 2009; Ogbe *et al.*, 2020). Secondary metabolites and antioxidants produced by endophytes work to eliminate ROS that may evade plant defence (Torres *et al.*, 2012). Similarly, endophyte-infected plants were found to have increased concentrations of phenolic compounds and antioxidants compared to endophyte-free plants under various stresses (Ogbe *et al.*, 2020). A response is triggered in plant tissues due to reactive oxygen species produced by endophytes' presence that leads to high concentrations of proline, methionine flavonoids, and other phenols which help improve the plants' resistance capacity under different stressors (Baltruschat *et al.*, 2008; Herrera-Carillo *et al.*, 2009). The endophyte–host interaction facilitates rapid ROS along with antioxidant generation for systemic acquired resistance responses and hypersensitivity (Tanaka *et al.*, 2006). In this process, MAP kinase (MAPK) genes play a significant role in regulating oxidant synthesis while managing signalling pathways and synthesizing antioxidants (Asai *et al.*, 2002). Scavenging enzymes like SOD, CAT, and APX contribute to neutralizing oxidative stress resulting from either host or respondent-produced ROS. Such mechanisms play crucial roles in regulating oxidant burdens produced by various biotic factors restricting them from causing cellular damage (Eid *et al.*, 2019; Ogbe *et al.*, 2020). It has been also proposed that the primary chemical auxin mediates the plant cells' responses to ROS (Torres *et al.*, 2012). Recent research has described the importance of auxins – believed to be produced by endophytes in regulating ROS homeostasis inside plant cells (Pasternak *et al.*, 2005). This serves as an essential mechanism that reduces vulnerability to oxidative stress and hence lowers cellular sensitivity levels (Torres *et al.*, 2012). Tyburski *et al.* (2009) showed that the formation of secondary metabolites was promoted in tomato culture by auxins. Moreover, other hormones, such as gibberellins, brassinosteroids, and abscisic acid, can help plants withstand oxidative stress by increasing the amount of ROS species in their tissues (Nie *et al.*, 2013; Ogbe *et al.*, 2020).

9.5 CONCLUSIONS AND FUTURE PERSPECTIVES

The role of endophytes as stress-mitigating organisms that aid medicinal plants in alleviating both abiotic and biotic stresses is highlighted in this chapter. Additionally, it summarizes the mechanisms performed by various endophytes in stress alleviation, though some of these mechanisms are not fully understood. One valuable application is the involvement of endophytes in producing bioactive secondary metabolites, which not only assist plants in stress alleviation but also contribute significantly to plant survival. Such compounds serve a significant function in plant growth, seed germination, and photosynthesis acceleration and exhibit various antitumour activities, providing protection against biotic stresses like viruses and pathogens. However, these findings are currently limited to research, and further efforts are needed to implement this knowledge in the field to promote better plant and crop growth.

Looking ahead, there is a growing need to develop and utilize advantageous endophyte stress for cultivating and producing medicinal plants, thus enhancing the quality of pharmaceutical products. Additionally, creating a plant breeding library that leverages endophytes to improve seed germination could be vital in preserving medicinal plant species facing the threat of extinction.

REFERENCES

Abd_Allah, E. F., Alqarawi, A. A., Hashem, A., Radhakrishnan, R., Al-Huqail, A. A., Al-Otibi, F. O. N., . . . Egamberdieva, D. (2018). Endophytic bacterium Bacillus subtilis (BERA 71) improves salt tolerance in chickpea plants by regulating the plant defense mechanisms. *Journal of Plant Interactions*, 13(1), 37–44.

Adam, M., Heuer, H., & Hallmann, J. (2014). Bacterial antagonists of fungal pathogens also control root-knot nematodes by inducing systemic resistance in tomato plants. *PloS One*, 9(2), e90402.

Adesina, M. F., Grosch, R., Lembke, A., Vatchev, T. D., & Smalla, K. (2009). In vitro antagonists of Rhizoctonia solani tested on lettuce: Rhizosphere competence, biocontrol efficiency, and rhizosphere microbial community response. *FEMS Microbiology Ecology*, 69(1), 62–74.

Aguiar-Pulido, V., Huang, W., Suarez-Ulloa, V., Cickovski, T., Mathee, K., & Narasimhan, G. (2016). Metagenomics, metatranscriptomics, and metabolomics approaches for microbiome analysis: Supplementary issue: Bioinformatics methods and applications for big metagenomics data. *Evolutionary Bioinformatics*, 12, EBO-S36436.

Aguilera-Sammaritano, J., Caballero, J., Deymié, M., Rosa, M., Vazquez, F., Pappano, D., Lechner, B., & González-Teuber, M. (2021). Dual effects of entomopathogenic fungi on control of the pest Lobesia botrana and the pathogenic fungus Eutypella microtheca on grapevine. *Biological Research*, 54, 1–10.

Ali, S., Charles, T. C., & Glick, B. R. (2014). Amelioration of high salinity stress damage by plant growth-promoting bacterial endophytes that contain ACC deaminase. *Plant Physiology and Biochemistry*, 80, 160–167.

Ancheeva, E., Daletos, G., & Proksch, P. (2020). Bioactive secondary metabolites from endophytic fungi. *Current Medicinal Chemistry*, 27(11), 1836–1854.

Anisha, C., Sachidanandan, P., & Radhakrishnan, E. K. (2018). Endophytic Paraconiothyrium sp. from Zingiber officinale Rosc. displays broad-spectrum antimicrobial activity by the production of danthron. *Current Microbiology*, 75, 343–352.

Anjum, N., & Chandra, R. (2019). Endophytic bacteria of Catharanthus roseus as an alternative source of vindoline and application of response surface methodology to enhance its production. *Archives of Biological Sciences*, 71(1), 27–38.

Asai, T., Tena, G., Plotnikova, J., Willmann, M. R., Chiu, W.-L., Gomez-Gomez, L., Boller, T., Ausubel, F. M., & Sheen, J. (2002). MAP kinase signaling cascade in Arabidopsis innate immunity. *Nature*, 415(6875), 977–983.

Aswathy, A. J., Jasim, B., Jyothis, M., & Radhakrishnan, E. K. (2013). Identification of two strains of Paenibacillus sp. as indole 3 acetic acid-producing rhizome-associated endophytic bacteria from Curcuma longa. *3 Biotech*, 3, 219–224.

Baltruschat, H., Fodor, J., Harrach, B. D., Niemczyk, E., Barna, B., Gullner, G., Janeczko, A., Kogel, K., Schäfer, P., & Schwarczinger, I. (2008). Salt tolerance of barley induced by the root endophyte Piriformospora indica is associated with a strong increase in antioxidants. *New Phytologist*, 180(2), 501–510.

Barac, T., Taghavi, S., Borremans, B., Provoost, A., Oeyen, L., Colpaert, J. V, Vangronsveld, J., & Van Der Lelie, D. (2004). Engineered endophytic bacteria improve the phytoremediation of water-soluble, volatile, organic pollutants. *Nature Biotechnology*, 22(5), 583–588.

Bashyal, B. (1999). Seimatoantlerium nepalense, an endophytic taxol-producing coelomycete from Himalayan yew (*Taxus wallachiana*). *Mycotaxon*, 72, 33–42.

Behie, S. W., Zelisko, P. M., & Bidochka, M. J. (2012). Endophytic insect-parasitic fungi translocate nitrogen directly from insects to plants. *Science*, 336(6088), 1576–1577.

Bittleston, L. S., Brockmann, F., Wcislo, W., & Van Bael, S. A. (2011). Endophytic fungi reduce leaf-cutting and damage to seedlings. *Biology Letters*, 7(1), 30–32.

Brader, G., Compant, S., Mitter, B., Trognitz, F., & Sessitsch, A. (2014). Metabolic potential of endophytic bacteria. *Current Opinion in Biotechnology*, 27, 30–37.

Card, S., Johnson, L., Teasdale, S., & Caradus, J. (2016). Deciphering endophyte behavior: The link between endophyte biology and efficacious biological control agents. *FEMS Microbiology Ecology*, 92(8), fiw114.

Carroll, G. C. (1988). Fungal endophytes in stems and leaves: From latent pathogen to mutualistic symbiont. *Ecology*, 69(1), 2–9.

Carroll, G. C. (1991). Beyond pest deterrence – alternative strategies and hidden costs of endophytic mutualisms in vascular plants. In *Microbial ecology of leaves* (pp. 358–375). Springer.

Chakravarthi, B., Das, P., Surendranath, K., Karande, A. A., & Jayabaskaran, C. (2008). Production of paclitaxel by Fusarium solani isolated from Taxus celebica. *Journal of Biosciences*, 33, 259–267.

Chandrakar, S., & Gupta, A. K. (2019). Studies on the production of broad-spectrum antimicrobial compound polypeptide (actinomycins) and lipopeptide (fengycin) from Streptomyces sp. K-R1 associated with the root of Abutilon indicum against multidrug-resistant human pathogens. *International Journal of Peptide Research and Therapeutics*, 25, 779–798.

Chatterjee, S., Ghosh, R., & Mandal, N. C. (2019). Production of bioactive compounds with bactericidal and antioxidant potential by endophytic fungus Alternaria alternata AE1 isolated from Azadirachta indica A. Juss. *PLoS One*, 14(4), e0214744.

Chaudhry, S., & Sidhu, G. P. S. (2022). Climate change regulated abiotic stress mechanisms in plants: A comprehensive review. *Plant Cell Reports*, 41(1), 1–31.

Chen, B., Wang, M., Hu, Y., Lin, Z., Yu, R., & Huang, L. (2011). Preliminary study on promoting effects of endophytic fungi on the growth of Rehmannia glutinosa. *China Journal of Chinese Materia Medica*, 36(9), 1137–1140.

Chen, C.-Y., Huang, P.-H., Yeh, K.-W., & Wang, S.-J. (2022). Colonization of Piriformospora indica enhances insect herbivore resistance of rice plants through jasmonic acid and antioxidant-mediated defense mechanisms. *Journal of Plant Interactions*, 17(1), 9–18.

Chen, C.-Y., Xin, K., Li, M., Li, X., Cheng, J., Zhang, L., & Shen, X. (2016). Paenibacillus sinopodophylli sp. nov., a siderophore-producing endophytic bacterium isolated from roots of Sinopodophyllum hexandrum (Royle) Ying. *International Journal of Systematic and Evolutionary Microbiology*, 66(12), 4993–4999.

Chen, J. X., Dai, C. C., Li, X., Tian, L. S., & Xie, H. (2008). Endophytic fungi screening from Atracty lancea and inoculating into the host plantlet. *Guihaia*, 28(2), 256–260.

Chen, X. M., & Guo, S. X. (2005). Effects of four species of endophytic fungi on the growth and polysaccharide and alkaloid contents of Dendrobium nobile. *Zhongguo Zhong Yao Za Zhi= Zhongguo Zhongyao Zazhi= China Journal of Chinese Materia Medica*, 30(4), 253–257.

Chen, X. M., Sang, X., Li, S., Zhang, S., & Bai, L. (2010). Studies on chlorogenic acid-producing endophytic fungi isolated from Eucommia ulmoides Oliver. *Journal of Industrial Microbiology and Biotechnology*, 37(5), 447–454.

Cherif, H., Marasco, R., Rolli, E., Ferjani, R., Fusi, M., Soussi, A., Mapelli, F., Blilou, I., Borin, S., & Boudabous, A. (2015). Oasis desert farming selects environment-specific date palm root endophytic communities and cultivable bacteria that promote resistance to drought. *Environmental Microbiology Reports*, 7(4), 668–678.

Clement, S. L., Elberson, L. R., Bosque-Perez, N. A., & Schotzko, D. J. (2005). Detrimental and neutral effects of wild barley – Neotyphodium fungal endophyte associations on insect survival. *Entomologia Experimentalis et Applicata*, 114(2), 119–125.

Cohen, A. C., Bottini, R., Pontin, M., Berli, F. J., Moreno, D., Boccanlandro, H., . . . Piccoli, P. N. (2015). Azospirillum brasilense ameliorates the response of Arabidopsis thaliana to drought mainly via enhancement of ABA levels. *Physiologia Plantarum*, 153(1), 79–90.

Compant, S., Reiter, B., Sessitsch, A., Nowak, J., Clément, C., & Ait Barka, E. (2005). Endophytic colonization of Vitis vinifera L. by plant growth-promoting bacterium Burkholderia sp. strain PsJN. *Applied and Environmental Microbiology*, 71(4), 1685–1693.

Cui, Y., Yi, D., Bai, X., Sun, B., Zhao, Y., & Zhang, Y. (2012). Ginkgolide B produced endophytic fungus (Fusarium oxysporum) isolated from Ginkgo biloba. *Fitoterapia*, 83(5), 913–920.

Danagoudar, A., Joshi, C. G., Ravi, S. K., Kumar, H. G. R., & Ramesh, B. N. (2018). Antioxidant and cytotoxic potential of endophytic fungi isolated from medicinal plant Tragia involucrata L. *Pharmacognosy Research*, 10(2).

Deng, B. W., Liu, K. H., Chen, W. Q., Ding, X. W., & Xie, X. C. (2009). Fusarium solani, Tax-3, a new endophytic taxol-producing fungus from Taxus chinensis. *World Journal of Microbiology and Biotechnology*, 25, 139–143.

Devi, K. A., Pandey, P., & Sharma, G. D. (2016). Plant growth-promoting endophyte Serratia marcescens AL2–16 enhances the growth of Achyranthes aspera L., a medicinal plant. *HAYATI Journal of Biosciences*, 23(4), 173–180.

Ding, S., Huang, C. L., Sheng, H. M., Song, C. L., Li, Y. B., & An, L. Z. (2011). Effect of inoculation with the endophyte Clavibacter sp. strain Enf12 on chilling tolerance in Chorispora bungeana. *Physiologia Plantarum*, 141(2), 141–151.

Duan, L. I., Liwei, G., & Hong, Y. (2009). Isolation and identification of producing endophytic fungi of berberine from the plant Phellodendron amurense. *Journal of Anhui Agricultural Sciences* 22, 7.

Eid, A. M., Salim, S. S., Hassan, S. E.-D., Ismail, M. A., & Fouda, A. (2019). Role of endophytes in plant health and abiotic stress management. *Microbiome in Plant Health and Disease: Challenges and Opportunities*, 119–144.

Ek-Ramos, M. J., Gomez-Flores, R., Orozco-Flores, A. A., Rodríguez-Padilla, C., González-Ochoa, G., & Tamez-Guerra, P. (2019). Bioactive products from plant-endophytic gram-positive bacteria. *Frontiers in Microbiology*, 10, 463.

Eyberger, A. L., Dondapati, R., & Porter, J. R. (2006). Endophyte fungal isolates from Podophyllum peltatum produce podophyllotoxin. *Journal of Natural Products*, 69(8), 1121–1124.

Farhat, H., Urooj, F., Sohail, N., Ansari, M., & Ehteshamul-Haque, S. (2022). Evaluation of the nematicidal potential of endophytic fungi associated with healthy plants and GC-MS profiling of metabolites of endophytic Fusarium solani. *South African Journal of Botany*, 146, 146–161.

Firáková, S., Šturdíková, M., & Múčková, M. (2007). Bioactive secondary metabolites are produced by microorganisms associated with plants. *Biologia*, 62, 251–257.

Flora, S. J. S., Shrivastava, R., & Mittal, M. (2013). Chemistry and pharmacological properties of some natural and synthetic antioxidants for heavy metal toxicity. *Current Medicinal Chemistry*, 20(36), 4540–4574.

Fouda, A., Eid, A. M., Elsaied, A., El-Belely, E. F., Barghoth, M. G., Azab, E., Gobouri, A. A., & Hassan, S. E.-D. (2021). Plant growth-promoting endophytic bacterial community inhabiting the leaves of Pulicaria incisa (Lam.) DC inherent to arid regions. *Plants*, 10(1), 76.

Fu, Y., Yin, Z. H., & Yin, C. Y. (2017). Biotransformation of ginsenoside Rb1 to ginsenoside Rg3 by endophytic bacterium Burkholderia sp. GE 17-7 isolated from Panax ginseng. *Journal of Applied Microbiology*, 122(6), 1579–1585.

Gangadevi, V., Murugan, M., & Muthumary, J. (2008). Taxol determination from Pestalotiopsis pauciseta, a fungal endophyte of a medicinal plant. *Chinese Journal of Biotechnology*, 24(8), 1433–1438.

Gangadevi, V., & Muthumary, J. (2008). Isolation of Colletotrichum gloeosporioides, a novel endophytic taxol-producing fungus from the leaves of a medicinal plant, Justicia gendarussa. *Mycologia Balcanica*, 5(1), 1–4.

Gangadevi, V., & Muthumary, J. (2009). Taxol production by *Pestalotiopsis terminaliae*, an endophytic fungus of Terminalia arjuna (arjun tree). *Biotechnology and Applied Biochemistry*, 52(1), 9–15.

Gao, F., Dai, C., & Liu, X. (2010). Mechanisms of fungal endophytes in plant protection against pathogens. *African Journal of Microbiology Research*, 4(13), 1346–1351.

Gao, Y., Liu, Q., Zang, P., Li, X., Ji, Q., He, Z., Zhao, Y., Yang, H., Zhao, X., & Zhang, L. (2015). An endophytic bacterium isolated from Panax ginseng CA Meyer enhances growth, reduces morbidity, and stimulates ginsenoside biosynthesis. *Phytochemistry Letters*, 11, 132–138.

Ghiasvand, M., Makhdoumi, A., Matin, M. M., & Vaezi, J. (2020). Exploring the bioactive compounds from endophytic bacteria of a medicinal plant: *Ephedra foliata* (Ephedrales: Ephedraceae). *Advances in Traditional Medicine*, 20, 61–70.

Glick, B. R. (2012). Plant growth-promoting bacteria: Mechanisms and applications. *Scientifica*, 2012.

Glick, B. R., Penrose, D. M., & Li, J. (1998). A model for the lowering of plant ethylene concentrations by plant growth-promoting bacteria. *Journal of Theoretical Biology*, 190(1), 63–68.

Gond, S. K., Bergen, M. S., Torres, M. S., & White Jr, J. F. (2015). Endophytic Bacillus spp. produce antifungal lipopeptides and induce host defense gene expression in maize. *Microbiological Research*, 172, 79–87.

Gosal, S. S., Wani, S. H., & Kang, M. S. (2009). Biotechnology and drought tolerance. *Journal of Crop Improvement*, 23(1), 19–54.

Goutam, J., Sharma, G., Tiwari, V. K., Mishra, A., Kharwar, R. N., Ramaraj, V., & Koch, B. (2017). Isolation and characterization of "Terrein" an antimicrobial and antitumor compound from endophytic fungus Aspergillus terreus (JAS-2) associated with Achyranthus aspera Varanasi, India. *Frontiers in Microbiology*, 8, 1334.

Govarthanan, M., Mythili, R., Selvankumar, T., Kamala-Kannan, S., Rajasekar, A., & Chang, Y. C. (2016). Bioremediation of heavy metals using an endophytic bacterium Paenibacillus sp. RM isolated from the roots of Tridax procumbens. *3 Biotech*, 6, 1–7.

Gravel, V., Antoun, H., & Tweddell, R. J. (2007). Growth stimulation and fruit yield improvement of greenhouse tomato plants by inoculation with Pseudomonas putida or Trichoderma atroviride: Possible role of indole acetic acid (IAA). *Soil Biology and Biochemistry*, 39(8), 1968–1977.

Guo, S. X., and Wang, Q. Y. (2001). Character and action of good strain on stimulating seed germination of Gastrodia elata. Mycosystema 3, 408–412. doi: 10.3969/j.issn.1672-6472.2001.03.023

Guo, B. H., Wang, Y. C., Zhou, X. W., Hu, K., Tan, F., Miao, Z. Q., et al. (2009). An endophytic taxol-producing fungus BT2 isolated from Taxus chinensis var. mairei. Afr. J. Biotechnol. 5, 875–877.

Gupta, S., Bhatt, P., & Chaturvedi, P. (2018). Determination and quantification of asiaticoside in endophytic fungus from Centella asiatica (L.) Urban. *World Journal of Microbiology and Biotechnology*, 34, 1–10.

Gupta, S., & Chaturvedi, P. (2019). Enhancing secondary metabolite production in medicinal plants using endophytic elicitors: A case study of Centella asiatica (Apiaceae) and asiaticoside. *Endophytes for a Growing World*, 310–323.

Gupta, S., Chaturvedi, P., Kulkarni, M. G., & Van Staden, J. (2020). A critical review on exploiting the pharmaceutical potential of plant endophytic fungi. *Biotechnology Advances*, 39, 107462.

Hameed, A., Dilfuza, E., Abd-Allah, E. F., Hashem, A., Kumar, A., & Ahmad, P. (2014). Salinity stress and arbuscular mycorrhizal symbiosis in plants. *Use of Microbes for the Alleviation of Soil Stresses*, 1, 139–159.

Hanaka, A., Majewska, M., & Jaroszuk-Ściseł, J. (2022). Study of the influence of abiotic and biotic stress factors on horticultural plants. *Horticulturae*, 8(1), 6.

Handayani, D., Rivai, H., Mulyana, R., Suharti, N., Rasyid, R., & Hertiani, T. (2018). Antimicrobial and cytotoxic activities of endophytic fungi isolated from mangrove plant Sonneratia alba Sm. *Journal of Applied Pharmaceutical Science*, 8(2), 49–53.

Hardoim, P. R., Van Overbeek, L. S., Berg, G., Pirttilä, A. M., Compant, S., Campisano, A., Döring, M., & Sessitsch, A. (2015). The hidden world within plants: Ecological and evolutionary considerations for defining functioning of microbial endophytes. *Microbiology and Molecular Biology Reviews*, 79(3), 293–320.

Harman, G. E., & Uphoff, N. (2019). Symbiotic root-endophytic soil microbes improve crop productivity and provide environmental benefits. *Scientifica*, 2019.

Hartley, S. E., & Gange, A. C. (2009). Impacts of plant symbiotic fungi on insect herbivores: Mutualism in a multitrophic context. *Annual Review of Entomology*, 54, 323–342.

Hashem, A., Abd_Allah, E. F., Alqarawi, A. A., Al-Huqail, A. A., Wirth, S., & Egamberdieva, D. (2016). The interaction between arbuscular mycorrhizal fungi and endophytic bacteria enhances plant growth of Acacia gerrardii under salt stress. *Frontiers in Microbiology*, 7, 1089.

Hassan, S. E.-D. (2017). Plant growth-promoting activities for bacterial and fungal endophytes isolated from the medicinal plant of *Teucrium polium* L. *Journal of Advanced Research*, 8(6), 687–695.

He, R. L., Wang, G. P., Liu, X., Zhang, C. L., & Lin, F. C. (2009). Antagonistic bioactivity of an endophytic bacterium isolated from Epimedium brevicornu Maxim. *African Journal of Biotechnology*, 8(2).

Herrera-Carillo, Z., Torres, M. S., Singh, A. P., Vorsa, N., Gianfagna, T., Meyer, W., & White Jr, J. F. (2009). Phenolic, flavonoid, and antioxidant profiling for cool-season grasses with and without endophyte. *Proceedings of the 18th Annual Rutgers Turfgrass Symposium*, 12.

Hong, C. E., Jo, S. H., Jo, I. H., & Park, J. M. (2018). Diversity and antifungal activity of endophytic bacteria associated with Panax ginseng seedlings. *Plant Biotechnology Reports*, 12, 409–418.

Huang, Z., Cai, X., Shao, C., She, Z., Xia, X., Chen, Y., Yang, J., Zhou, S., & Lin, Y. (2008). Chemistry and weak antimicrobial activities of phomopsins produced by mangrove endophytic fungus Phomopsis sp. ZSU-H76. *Phytochemistry*, 69(7), 1604–1608.

Hume, D. E., Ryan, G. D., Gibert, A., Helander, M., Mirlohi, A., & Sabzalian, M. R. (2016). Epichloë fungal endophytes for grassland ecosystems. *Sustainable Agriculture Reviews*, 19, 233–305.

Ibrahim, E., Fouad, H., Zhang, M., Zhang, Y., Qiu, W., Yan, C., Li, B., Mo, J., & Chen, J. (2019). Biosynthesis of silver nanoparticles using endophytic bacteria and their role in inhibition of rice pathogenic bacteria and plant growth promotion. *RSC Advances*, 9(50), 29293–29299.

Indananda, C., Igarashi, Y., Ikeda, M., Oikawa, T., & Thamchaipenet, A. (2013). Linfuranone A, a new polyketide from plant-derived Microbispora sp. GMKU 363. *The Journal of Antibiotics*, 66(11), 675–677.

Ismail, I., Hamayun, M., Hussain, A., Iqbal, A., Khan, S. A., Ahmad, A., Gul, S., Kim, H.-Y., & Lee, I.-J. (2022). Stemphylium Solani stabilized the physicochemical characteristics of host plant species during stress. *Polish Journal of Environmental Studies*, 31(2).

Jaber, L. R. (2015). Grapevine leaf tissue colonization by the fungal entomopathogen Beauveria bassiana sl and its effect against downy mildew. *BioControl*, 60, 103–112.

Jaber, L. R., & Salem, N. (2014). Endophytic colonization of squash by the fungal entomopathogen Beauveria bassiana (Ascomycota: Hypocreales) for managing Zucchini yellow mosaic virus in cucurbits. *Biocontrol Science and Technology*, 24(10), 1096–1109.

Jana, S. K., Islam, M. M., & Mandal, S. (2022). Endophytic microbiota of rice and their collective impact on host fitness. *Current Microbiology*, 79(2), 37.

Jayakumar, A., Krishna, A., Mohan, M., Nair, I. C., & Radhakrishnan, E. K. (2019). Plant growth enhancement, disease resistance, and elemental modulatory effects of plant probiotic endophytic Bacillus sp. Fcl1. *Probiotics and Antimicrobial Proteins*, 11, 526–534.

Jayaram, H., Marigowda, V., & Thara Saraswathi, K. J. (2021). Secondary metabolite production and terpenoid biosynthesis in endophytic fungi Cladosporium cladosporioides isolated from wild *Cymbopogon martinii* (Roxb.) Wats. *Microbiology Research*, 12(4), 812–828.

Jennings, D. B., Ehrenshaft, M., Pharr, D. M., & Williamson, J. D. (1998). Roles for mannitol and mannitol dehydrogenase in active oxygen-mediated plant defense. *Proceedings of the National Academy of Sciences*, 95(25), 15129–15133.

Jha, Y., Subramanian, R. B., & Patel, S. (2011). A combination of endophytic and rhizospheric plant growth-promoting rhizobacteria in Oryza sativa shows higher accumulation of osmoprotectant against saline stress. *Acta Physiologiae Plantarum*, 33, 797–802.

Ji, H., Li, X., & Zhang, H. (2009). Natural products and drug discovery: Can thousands of years of ancient medical knowledge lead us to new and powerful drug combinations in the fight against cancer and dementia? *EMBO Reports*, 10(3), 194–200.

Jia, M., Chen, L., Xin, H. L., Zheng, C. J., Rahman, K., Han, T., & Qin, L. P. (2016). A friendly relationship between endophytic fungi and medicinal plants: A systematic review. *Frontiers in Microbiology*, 7, 906.

Jin, H. Q., Liu, H. B., Xie, Y. Y., Zhang, Y. G., Xu, Q. Q., Mao, L. J., . . . Zhang, C. L. (2018). Effect of the dark septate endophytic fungus Acrocalymma vagum on heavy metal content in tobacco leaves. *Symbiosis*, 74, 89–95.

Joe, M. M., Islam, M. R., Karthikeyan, B., Bradeepa, K., Sivakumar, P. K., & Sa, T. (2012). Resistance responses of rice-to-rice blast fungus after seed treatment with the endophytic Achromobacter xylosoxidans AUM54 strains. *Crop Protection*, 42, 141–148.

Jogawat, A., Saha, S., Bakshi, M., Dayaman, V., Kumar, M., Dua, M., Varma, A., Oelmüller, R., Tuteja, N., & Johri, A. K. (2013). Piriformospora indica reduces the growth diminution of rice seedlings during high salt stress. *Plant Signaling & Behavior*, 8(10), e26891.

Ju, Z., Wang, J., & Pan, S. (2009). Isolation and preliminary identification of the endophytic fungi that produce Hupzine A from four species in Hupziaceae and determination of Huperzine A by HPLC. *Fudan University Journal of Medical Sciences*, 445–449.

Kapoor, N., & Saxena, S. (2018). Endophytic fungi of Tinospora cordifolia with anti-gout properties. *3 Biotech*, 8(6), 264.

Karthikeyan, B., Joe, M. M., Islam, M. R., & Sa, T. (2012). ACC deaminase containing diazotrophic endophytic bacteria ameliorates salt stress in Catharanthus roseus through reduced ethylene levels and induction of antioxidative defense systems. *Symbiosis*, 56, 77–86.

Katoch, M., Singh, D., Kapoor, K. K., & Vishwakarma, R. A. (2019). Trichoderma lixii (IIIM-B4), an endophyte of Bacopa monnieri L. producing peptaibols. *BMC Microbiology*, 19(1), 1–10.

Khaksar, G., Siswanto, D., Treesubsuntorn, C., & Thiravetyan, P. (2016). Euphorbia milii-endophytic bacteria interactions affect the hormonal levels of the native host differently under various airborne pollutants. *Molecular Plant-Microbe Interactions*, 29(9), 663–673.

Khan, A. L., Shinwari, Z. K., Kim, Y.-H., Waqas, M., Hamayun, M., Kamran, M., & Lee, I.-J. (2012). Role of endophyte Chaetomium globosum LK4 in the growth of Capsicum annuum by the production of gibberellins and indole acetic acid. *Pakistan Journal of Botany*, 44(5), 1601–1607.

Khan, N., Afroz, F., Begum, M. N., Rony, S. R., Sharmin, S., Moni, F., Hasan, C. M., Shaha, K., & Sohrab, M. H. (2018). Endophytic Fusarium solani: A rich source of cytotoxic and antimicrobial napthaquinone and aza-anthraquinone derivatives. *Toxicology Reports*, 5, 970–976.

Khan, Z., Rho, H., Firrincieli, A., Hung, S. H., Luna, V., Masciarelli, O., . . . Doty, S. L. (2016). Growth enhancement and drought tolerance of hybrid poplar upon inoculation with endophyte consortia. *Current Plant Biology*, 6, 38–47.

Köberl, M., Ramadan, E. M., Adam, M., Cardinale, M., Hallmann, J., Heuer, H., Smalla, K., & Berg, G. (2013). Bacillus and Streptomyces were selected as broad-spectrum antagonists against soilborne pathogens from arid areas in Egypt. *FEMS Microbiology Letters*, 342(2), 168–178.

Kumar, C. G., Godumagadda, N. R., Mongolla, P., Pombala, S., Badrinarayan, P., Reddy, C. R., Das, B., Ramakrishna, K. V. S., Sridhar, B., & Kamal, A. (2017). Novel bioactive compounds from an endophytic Chaetomium arcuatum strain Saf-2 Isolated from Semecarpus anacardium. *Microbial Biotechnology: Technological Challenges and Developmental Trends*, 63.

Kumar, V., & Nautiyal, C. S. (2022). Plant abiotic and biotic stress alleviation: From an endophytic microbial perspective. *Current Microbiology*, 79(10), 311.

Kumara, P. M., Shweta, S., Vasanthakumari, M. M., Sachin, N., Manjunatha, B. L., Jadhav, S. S., Ravikanth, G., Ganeshaiah, K. N., & Shaanker, R. U. (2014). Endophytes and plant secondary metabolite synthesis: A molecular and evolutionary perspective. *Advances in Endophytic Research*, 177–190.

Kumaran, R. S., Muthumary, J., & Hur, B.-K. (2008). Taxol from *Phyllosticta citricarpa*, a leaf spot fungus of the angiosperm Citrus medica. *Journal of Bioscience and Bioengineering*, 106(1), 103–106.

Kusari, S., Hertweck, C., & Spiteller, M. (2012). Chemical ecology of endophytic fungi: Origins of secondary metabolites. *Chemistry & Biology*, 19(7), 792–798.

Kusari, S., Lamshöft, M., Zühlke, S., & Spiteller, M. (2008). An endophytic fungus from Hypericum perforatum that produces hypericin. *Journal of Natural Products*, 71(2), 159–162.

Kusari, S., Zühlke, S., & Spiteller, M. (2009). An endophytic fungus from Camptotheca acuminata that produces camptothecin and analogs. *Journal of Natural Products*, 72(1), 2–7.

Kusari, S., Zühlke, S., & Spiteller, M. (2011). Effect of artificial reconstitution of the interaction between the plant Camptotheca acuminata and the fungal endophyte Fusarium solani on camptothecin biosynthesis. *Journal of Natural Products*, 74(4), 764–775.

Lacava, P. T., & Azevedo, J. L. (2013). Biological control of insect-pest and diseases by endophytes. In *Advances in Endophytic Research* (pp. 231–256). Springer India.

Lata, R., Chowdhury, S., Gond, S. K., & White Jr, J. F. (2018). Induction of abiotic stress tolerance in plants by endophytic microbes. *Letters in Applied Microbiology*, 66(4), 268–276.

Leong, C. R., Rashid, S. A., Ang, S. N., Tan, W.-N., Tong, W. Y., & Ibrahim, D. (2018). Antimicrobial activity of Aspergillus sp. IBRL MP15 CCL, an endophytic fungus isolated from *Swietenia macrophylla* leaf against human pathogens. *Malaysian Journal of Microbiology*, 49–54.

Li, J. Y., Zhao, G. Z., Varma, A., Qin, S., Xiong, Z., Huang, H. Y., . . . Li, W. J. (2012). An endophytic Pseudonocardia species induces the production of artemisinin in Artemisia annua. *PLoS One*, 7(12), e51410.

Li, W. K., Zhou, J. Y., Lin, Z. W., & Hu, Z. B. (2007). Study on fermentation condition for the production of huperzine A from endophytic fungus 2F09P03B of Huperzia serrata. *Chinese Medicine Biotechnology*, 2(4), 254–259.

Liu, K., Ding, X., Deng, B., & Chen, W. (2009). Isolation and characterization of endophytic taxol-producing fungi from Taxus chinensis. *Journal of Industrial Microbiology and Biotechnology*, 36(9), 1171.

Liu, T., Zhang, X., Tian, S., Chen, L., & Yuan, J. (2020). Bioinformatics analysis of endophytic bacteria related to berberine in the Chinese medicinal plant *Coptis teeta* Wall. *3 Biotech*, 10, 1–12.

Liu, X., Dong, M., Chen, X., Jiang, M., Lv, X., & Zhou, J. (2008). Antimicrobial activity of an endophytic Xylaria sp. YX-28 and identification of its antimicrobial compound 7-amino-4-methylcoumarin. *Applied Microbiology and Biotechnology*, 78, 241–247.

Liu, Y., Liu, W., & Liang, Z. (2015). Endophytic bacteria from *Pinellia ternata*, a new source of purine alkaloids and bacterial manure. *Pharmaceutical Biology*, 53(10), 1545–1548.

Ludwig-Müller, J. (2015). Plants and endophytes: Equal partners in secondary metabolite production? *Biotechnology Letters*, 37, 1325–1334.

Ma, L., Cao, Y. H., Cheng, M. H., Huang, Y., Mo, M. H., Wang, Y., . . . Yang, F. X. (2013). Phylogenetic diversity of bacterial endophytes of Panax notoginseng with antagonistic characteristics towards pathogens of root-rot disease complex. *Antonie Van Leeuwenhoek*, 103, 299–312.

Maehara, S., Simanjuntak, P., Maetani, Y., Kitamura, C., Ohashi, K., & Shibuya, H. (2012). Ability of endophytic filamentous fungi associated with Cinchona ledgeriana to produce Cinchona alkaloids. *Journal of Natural Medicines*, 67, 421–423. https://doi.org/10.1007/s11418-012-0701-8

Malinowski, D. P., & Belesky, D. P. (1999). *Neotyphodium coenophialum* - endophyte infection affects the ability of tall fescue to use sparingly available phosphorus. *Journal of Plant Nutrition*, 22(4–5), 835–853.

Marques-Pereira, C., Proença, D. N., & Morais, P. V. (2022). The role of serratomolide-like amino lipids produced by bacteria of genus Serratia in nematicidal activity. *Pathogens*, 11(2), 198.

Mastouri, F., Björkman, T., & Harman, G. E. (2010). Seed treatment with Trichoderma harzianum alleviates biotic, abiotic, and physiological stresses in germinating seeds and seedlings. *Phytopathology*, 100(11), 1213–1221.

Maulidia, V., Soesanto, L., Syamsuddin, S., Khairan, K., Hamaguchi, T., Hasegawa, K., & Sriwati, R. (2020). Secondary metabolites produced by endophytic bacteria against the Root-Knot Nematode (Meloidogyne sp.). Biodiversitas Journal of Biological Diversity, 21(11), 5270–5.

Meng, J. J., & He, X. L. (2011). Effects of AM fungi on growth and nutritional contents of Salvia miltiorrhiza Bge. under drought stress. *Journal of Agriculture University Hebei*, 34(1), 51–61.

Meshram, V., Saxena, S., & Paul, K. (2016). Xylarinase: A novel clot-busting enzyme from an endophytic fungus Xylaria curta. *Journal of Enzyme Inhibition and Medicinal Chemistry*, 31(6), 1502–1511.

Miller, G. A. D., Suzuki, N., Ciftci-Yilmaz, S. U. L. T. A. N., & Mittler, R. O. N. (2010). Reactive oxygen species homeostasis and signaling during drought and salinity stresses. *Plant, Cell & Environment*, 33(4), 453–467.

Ming, Q., Han, T., Li, W., Zhang, Q., Zhang, H., Zheng, C., . . . Qin, L. (2012). Tanshinone IIA and tanshinone I are produced by Trichoderma atroviride D16, an endophytic fungus in Salvia miltiorrhiza. *Phytomedicine*, 19(3–4), 330–333.

Mishra, P. K., Bisht, S. C., Ruwari, P., Subbanna, A. R. N. S., Bisht, J. K., Bhatt, J. C., & Gupta, H. S. (2017). Genetic diversity and functional characterization of endophytic Bacillus thuringiensis isolates from the North Western Indian Himalayas. *Annals of Microbiology*, 67(2), 143–155.

Mohamad, O. A. A., Li, L., Ma, J.-B., Hatab, S., Xu, L., Guo, J.-W., Rasulov, B. A., Liu, Y.-H., Hedlund, B. P., & Li, W.-J. (2018). Evaluation of the antimicrobial activity of endophytic bacterial populations from Chinese traditional medicinal plant licorice and characterization of the bioactive secondary metabolites produced by Bacillus atrophaeus against Verticillium dahliae. *Frontiers in Microbiology*, 9, 924.

Moslehi, S., Pourmehr, S., Shirzad, A., & Khakvar, R. (2021). Potential of some endophytic bacteria in biological control of root-knot nematode Meloidogyne incognita. *Egyptian Journal of Biological Pest Control*, 31, 1–11.

Muangthong, A., Youpensuk, S., & Rerkasem, B. (2015). Isolation and characterization of endophytic nitrogen-fixing bacteria in sugarcane. *Tropical Life Sciences Research*, 26(1), 41.

Muhae-ud-Din, G., Moosa, A., Ghummen, U. F., Jabran, M., Abbas, A., Naveed, M., Jabbar, A., & Ali, M. A. (2018). Host status of commonly planted ornamentals to Meloidogyne incognita and management through endophytic bacteria. *Pakistan Journal of Zoology*, 50(4).

Müller, H., Westendorf, C., Leitner, E., Chernin, L., Riedel, K., Schmidt, S., Eberl, L., & Berg, G. (2009). Quorum-sensing effects in the antagonistic rhizosphere bacterium Serratia plymuthica HRO-C48. *FEMS Microbiology Ecology*, 67(3), 468–478.

Munif, A., Hallmann, J., & Sikora, R. A. (2013). The influence of endophytic bacteria on Meloidogyne incognita infection and tomato plant growth. *Journal of the ISSAAS*, 19(2), 68–74.

Murphy, B. R., Doohan, F. M., & Hodkinson, T. R. (2014). Yield increase induced by the fungal root endophyte Piriformospora indica in barley grown at low temperatures is nutrient-limited. *Symbiosis*, 62, 29–39.

Nagata, S., Yamaji, K., Nomura, N., & Ishimoto, H. (2015). Root endophytes enhance the stress-tolerance of Cicuta virosa L. growing in a mining pond in eastern Japan. *Plant Species Biology*, 30(2), 116–125.

Namdar, D., Charuvi, D., Ajjampura, V., Mazuz, M., Ion, A., Kamara, I., & Koltai, H. (2019). LED lighting affects the composition and biological activity of Cannabis sativa secondary metabolites. *Industrial Crops and Products*, 132, 177–185.

Namwongsa, J., Jogloy, S., Vorasoot, N., Boonlue, S., Riddech, N., & Mongkolthanaruk, W. (2019). Endophytic bacteria improve root traits, biomass, and yield of Helianthus tuberosus L. under normal and deficit water conditions *Journal of Microbiology and Biotechnology*, 29(11), 1777–1789.

Naveed, M., Mitter, B., Reichenauer, T. G., Wieczorek, K., & Sessitsch, A. (2014). Increased drought stress resilience of maize through endophytic colonization by Burkholderia phytofirmans PsJN and Enterobacter sp. FD17. *Environmental and Experimental Botany*, 97, 30–39.

Newman, L. A., & Reynolds, C. M. (2005). Bacteria and phytoremediation: New uses for endophytic bacteria in plants. *TRENDS in Biotechnology*, 23(1), 6–8.

Nie, W., Wang, M., Xia, X., Zhou, Y., Shi, K., Chen, Z., & Yu, J. Q. (2013). Silencing of tomato RBOH1 and MPK2 abolishes brassinosteroid-induced H2O2 generation and stress tolerance. *Plant, Cell & Environment*, 36(4), 789–803.

Ogbe, A. A., Finnie, J. F., & Van Staden, J. (2020). The role of endophytes in secondary metabolite accumulation in medicinal plants under abiotic stress. *South African Journal of Botany*, 134, 126–134.

Pandi, M., Kumaran, R. S., Choi, Y.-K., Kim, H. J., & Muthumary, J. (2011). Isolation and detection of taxol, an anticancer drug produced from *Lasiodiplodia theobromae*, an endophytic fungus of the medicinal plant Morinda citrifolia. *African Journal of Biotechnology*, 10(8), 1428–1435.

Pang, C.-H., & Wang, B.-S. (2010). Role of ascorbate peroxidase and glutathione reductase in ascorbate – glutathione cycle and stress tolerance in plants. *Ascorbate-Glutathione Pathway and Stress Tolerance in Plants*, 91–113.

Papp-Rupar, M., Karlstrom, A., Passey, T., Deakin, G., & Xu, X. (2022). The influence of host genotypes on the endophytes in the leaf scar tissues of apple trees and correlation of the endophytes with apple canker (*Neonectria ditissima*) development. *Phytobiomes Journal*, 6(2), 127–138.

Parniske, M. (2008). Arbuscular mycorrhiza: The mother of plant root endosymbioses. *Nature Reviews Microbiology*, 6(10), 763–775.
Passari, A. K., Mishra, V. K., Leo, V. V., Gupta, V. K., & Singh, B. P. (2016). Phytohormone production endowed with antagonistic potential and plant growth promoting abilities of culturable endophytic bacteria isolated from *Clerodendrum colebrookianum* Walp. *Microbiological Research*, 193, 57–73.
Pasternak, T., Potters, G., Caubergs, R., & Jansen, M. A. K. (2005). Complementary interactions between oxidative stress and auxins control plant growth responses at plant, organ, and cellular levels. *Journal of Experimental Botany*, 56(418), 1991–2001.
Patil, M. P., Patil, R. H., & Maheshwari, V. L. (2015). Biological activities and identification of bioactive metabolite from endophytic Aspergillus flavus L7 isolated from Aegle marmelos. *Current Microbiology*, 71, 39–48.
Piccoli, P., & Bottini, R. (2013). Abiotic stress tolerance induced by endophytic PGPR. *Symbiotic Endophytes*, 151–163.
Pompeng, P., Sommit, D., Sriubolmas, N., Ngamrojanavanich, N., Matsubara, K., & Pudhom, K. (2013). Antiangiogenetic effects of anthranoids from Alternaria sp., an endophytic fungus in a Thai medicinal plant Erythrina variegata. *Phytomedicine*, 20(10), 918–922.
Priti, V., Ramesha, B. T., Singh, S., Ravikanth, G., Ganeshaiah, K. N., Suryanarayanan, T. S., & Uma Shaanker, R. (2009). How promising are endophytic fungi as alternative sources of plant secondary metabolites? *Current Science*, 97(4), 477–478.
Qader, M. M., Kumar, N. S., Jayasinghe, L., Araya, H., & Fujimoto, Y. (2017). Bioactive sesquiterpenes from an endophytic fungus Bipolaris sorokiniana isolated from a popular medicinal plant *Costus speciosus*. *Mycology*, 8(1), 17–20.
Qiu, M., Xie, R. S., Shi, Y., Zhang, H., & Chen, H. M. (2010). Isolation and identification of two flavonoid-producing endophytic fungi from Ginkgo biloba L. *Annals of Microbiology*, 60, 143–150.
Raina, D., Singh, B., Bhat, A. K., Satti, N. K., & Singh, V. K. (2018). Antimicrobial activity of endophytes isolated from *Picrorhiza kurroa*. *Indian Phytopathology*, 71, 103–113.
Rajan, D., George, T., Divakaran, D., Rebello, S., & Jisha, M. S. (2016). Biomodulatory role of Enterobacter sp: A novel bacterial endophyte of Sida cordifolia and its comparative analysis with plant extract. *Journal of Biologically Active Products from Nature*, 6(5–6), 373–382.
Rakotoniriana, E. F., Rafamantanana, M., Randriamampionona, D., Rabemanantsoa, C., Urveg-Ratsimamanga, S., El Jaziri, M., Munaut, F., Corbisier, A.-M., Quetin-Leclercq, J., & Declerck, S. (2013). Study in vitro of the impact of endophytic bacteria isolated from Centella asiatica on the disease incidence caused by the hemibiotrophic fungus *Colletotrichum higginsianum*. *Antonie Van Leeuwenhoek*, 103, 121–133.
Rathnayake, G. R. N., Kumar, N. S., Jayasinghe, L., Araya, H., & Fujimoto, Y. (2018). Chemical investigation of metabolites produced by endophytic fungi Phialemonium curvatum from the leaves of Passiflora edulis. *Natural Product Research*, 32(20), 2483–2486.
Rehman, S., Shawl, A. S., Kour, A., Andrabi, R., Sudan, P., Sultan, P., Verma, V., & Qazi, G. N. (2008). An endophytic Neurospora sp. from Nothapodytes foetida producing camptothecin. *Applied Biochemistry and Microbiology*, 44, 203–209.
Ren, C.-G., & Dai, C.-C. (2012). Jasmonic acid is involved in the signaling pathway for fungal endophyte-induced volatile oil accumulation of Atractylodes lancea plantlets. *BMC Plant Biology*, 12(1), 1–11.
Riga, R., Happyana, N., & Hakim, E. H. (2019). Chemical constituents of Pestalotiopsis microspora HF 12440. *Journal of Applied Pharmaceutical Science*, 9(1), 108–124.
Rodriguez, R. J., & Redman, R. (2008). More than 400 million years of evolution and some plants still can't make it on their own: Plant stress tolerance via fungal symbiosis. *Journal of Experimental Botany*, 59(5), 1109–1114.
Rodriguez, R. J., White Jr, J. F., Arnold, A. E., & Redman, A. R. S. (2009). Fungal endophytes: Diversity and functional roles. *New Phytologist*, 182(2), 314–330.
Rodriguez, R. J., Woodward, C. J., & Redman, R. S. (2012). Fungal influence on plant tolerance to stress. *Biocomplexity of Plant – Fungal Interactions*, 155–163.
Rosenblueth, M., & Martínez-Romero, E. (2006). Bacterial endophytes and their interactions with hosts. *Molecular Plant-Microbe Interactions*, 19(8), 827–837.
Roy, S. K., & Roy, D. K. (2016). Use of medicinal plants and their vulnerability due to climate change in the northern part of Bangladesh. *American Journal of Plant Sciences*, 7(13), 1782.
Russo, M. L., Scorsetti, A. C., Vianna, M. F., Cabello, M., Ferreri, N., & Pelizza, S. (2019). Endophytic effects of Beauveria bassiana on corn (Zea mays) and its herbivore, Rachiplusia nu (Lepidoptera: Noctuidae). *Insects*, 10(4), 110.

Rustamova, N., Wubulikasimu, A., Mukhamedov, N., Gao, Y., Egamberdieva, D., & Yili, A. (2020). Endophytic bacteria associated with medicinal plant Vernonia anthelmintica: Diversity and characterization. *Current Microbiology*, 77, 1457–1465.

Ryan, R. P., Germaine, K., Franks, A., Ryan, D. J., & Dowling, D. N. (2008). Bacterial endophytes: Recent developments and applications. *FEMS Microbiology Letters*, 278(1), 1–9.

Sabu, R., Soumya, K. R., & Radhakrishnan, E. K. (2017). Endophytic Nocardiopsis sp. from Zingiber officinale with both anti-phytopathogenic mechanisms and antibiofilm activity against clinical isolates. *3 Biotech*, 7, 1–13.

Sahoo, N., Manchikanti, P., & Dey, S. (2010). Herbal drugs: Standards and regulation. *Fitoterapia*, 81(6), 462–471.

Saikkonen, K., Faeth, S. H., Helander, M., & Sullivan, T. J. (1998). Fungal endophytes: A continuum of interactions with host plants. *Annual Review of Ecology and Systematics*, 29(1), 319–343.

Saikkonen, K., Saari, S., & Helander, M. (2010). Defensive mutualism between plants and endophytic fungi? *Fungal Diversity*, 41, 101–113.

Saraf, M., Jha, C. K., & Patel, D. (2011). The role of ACC deaminase-producing PGPR in sustainable agriculture. *Plant Growth and Health Promoting Bacteria*, 365–385.

Sasan, R. K., & Bidochka, M. J. (2013). Antagonism of the endophytic insect pathogenic fungus Metarhizium robertsii against the bean plant pathogen Fusarium solani f. sp. phaseoli. *Canadian Journal of Plant Pathology*, 35(3), 288–293.

Savary, S., Willocquet, L., Pethybridge, S. J., Esker, P., McRoberts, N., & Nelson, A. (2019). The global burden of pathogens and pests on major food crops. *Nature Ecology & Evolution*, 3(3), 430–439.

Scherwinski, K., Grosch, R., & Berg, G. (2008). Effect of bacterial antagonists on lettuce: Active biocontrol of Rhizoctonia solani and negligible, short-term effects on nontarget microorganisms. *FEMS Microbiology Ecology*, 64(1), 106–116.

Shaharoona, B., Arshad, M., Zahir, Z. A., & Khalid, A. (2006). Performance of Pseudomonas spp. containing ACC-deaminase for improving the growth and yield of maize (Zea mays L.) in the presence of nitrogenous fertilizer. *Soil Biology and Biochemistry*, 38(9), 2971–2975.

Sharma, A., Malhotra, B., Kharkwal, H., Kulkarni, G. T., & Kaushik, N. (2020). Therapeutic agents from endophytes harbored in Asian medicinal plants. *Phytochemistry Reviews*, 19, 691–720.

Sharma, P., & Pandey, R. (2009). Biological control of root-knot nematode; Meloidogyne incognita in the medicinal plant; Withania somnifera and the effect of biocontrol agents on plant growth. Afr J Agric Res, 4(6), 564–567.

Sharma, N., Kushwaha, M., Arora, D., Jain, S., Singamaneni, V., Sharma, S., Shankar, R., Bhushan, S., Gupta, P., & Jaglan, S. (2018). New cytochalasin from Rosellinia sanctae-cruciana, an endophytic fungus of Albizia lebbeck. *Journal of Applied Microbiology*, 125(1), 111–120.

Sherameti, I., Shahollari, B., Venus, Y., Altschmied, L., Varma, A., & Oelmuller, R. (2005). The endophytic fungus Piriformospora indica stimulates the expression of nitrate reductase and the starch-degrading enzyme glucan-water dikinase in tobacco and Arabidopsis roots through a homeodomain transcription factor that binds to a conserved motif in their promoters. *Journal of Biological Chemistry*, 280(28), 26241–26247.

Shi, P., Zhu, K., Zhang, Y., & Chai, T. (2016). Growth and cadmium accumulation of Solanum nigrum L. seedling was enhanced by heavy metal-tolerant strains of Pseudomonas aeruginosa. *Water, Air, & Soil Pollution*, 227, 1–11.

Shoba, S., & Mythili, S. (2018). Biological activities of endophytic fungus Cochliobolus sp. AL24 isolated from *Aerva lanata* L. *Indian Journal of Pharmaceutical Education*, 52, 277–283.

Shu, S., Zhao, X., Wang, W., Zhang, G., Cosoveanu, A., Ahn, Y., & Wang, M. (2014). Identification of a novel endophytic fungus from Huperzia serrata which produces huperzine A. *World Journal of Microbiology and Biotechnology*, 30, 3101–3109.

Shwab, E. K., & Keller, N. P. (2008). Regulation of secondary metabolite production in filamentous ascomycetes. *Mycological Research*, 112(2), 225–230.

Siddiqui, Z. S., Wei, X., Umar, M., Abideen, Z., Zulfiqar, F., Chen, J., Hanif, A., Dawar, S., Dias, D. A., & Yasmeen, R. (2022). Scrutinizing the application of saline endophyte to enhance salt tolerance in rice and maize plants. *Frontiers in Plant Science*, 12, 770084.

Sieber, T. N. (2007). Endophytic fungi in forest trees: Are they mutualists? *Fungal Biology Reviews*, 21(2–3), 75–89.

Singh, R., & Dubey, A. K. (2018). Diversity and applications of endophytic actinobacteria of plants in special and other ecological niches. *Frontiers in Microbiology*, 9, 1767.

Sinha, A., Priya, R., Nimisha, M., & Osborne, W. J. (2015). Impact of endophytic Ralstonia sp. from Aloe vera gel and its antimicrobial activity. *Asian Journal of Pharmaceutical and Clinical Research*, 8(1), 259–262.

Song, X., Wu, H., Yin, Z., Lian, M., & Yin, C. (2017). Endophytic bacteria isolated from Panax ginseng improves ginsenoside accumulation in adventitious ginseng root culture. *Molecules*, 22(6), 837.

Spiers, D. E., Wax, L. E., Eichen, P. A., Rottinghaus, G. E., Evans, T. J., Keisler, D. H., & Ellersieck, M. R. (2012). Use of different levels of ground endophyte-infected tall fescue seed during heat stress to separate characteristics of fescue toxicosis. *Journal of Animal Science*, 90(10), 3457–3467.

Stierle, A., Strobel, G., Stierle, D., Grothaus, P., & Bignami, G. (1995). The search for a taxol-producing microorganism among the endophytic fungi of the Pacific yew, Taxus brevifolia. *Journal of Natural Products*, 58(9), 1315–1324.

Strobel, G. A., Hess, W. M., Ford, E., Sidhu, R. S., & Yang, X. (1996). Endophytic fungi in grasses and woody plants. *Journal of Industrial Microbiology*, 17, 417–423.

Strobel, G. A., Miller, R. V., Martinez-Miller, C., Condron, M. M., Teplow, D. B., & Hess, W. M. (1999). Cryptocandin, a potent antimycotic from the endophytic fungus *Cryptosporiopsis* cf. quercina. *Microbiology*, 145(8), 1919–1926.

Sugijanto, N. E., Diesel, A., Rateb, M., Pretsch, A., Gogalic, S., Zaini, N. C., Ebel, R., & Indrayanto, G. (2011). Lecythomycin, a new macrolactone glycoside from the endophytic fungus Lecythophora sp. *Natural Product Communications*, 6(5), https://doi.org/10.1934578X1100600521

Sumarah, M. W., Puniani, E., Sørensen, D., Blackwell, B. A., & Miller, J. D. (2010). Secondary metabolites from anti-insect extracts of endophytic fungi isolated from Picea rubens. *Phytochemistry*, 71(7), 760–765.

Sun, X., & Guo, L.-D. (2012). Endophytic fungal diversity: A review of traditional and molecular techniques. *Mycology*, 3(1), 65–76.

Taechowisan, T., Chanaphat, S., Ruensamran, W., & Phutdhawong, W. S. (2014). Antibacterial activity of new flavonoids from Streptomyces sp. BT01; an endophyte in Boesenbergia rotunda (L.) Mansf. *Journal of Applied Pharmaceutical Science*, 4(4), 8–13.

Taechowisan, T., Lu, C., Shen, Y., & Lumyong, S. (2007). Antitumor activity of 4-aryl coumarins from endophytic *Streptomyces aureofaciens* CMUAc130. *Journal of Cancer Research and Therapeutics*, 3(2), 86–91.

Takahashi, F., Kuromori, T., Urano, K., Yamaguchi-Shinozaki, K., & Shinozaki, K. (2020). Drought stress responses and resistance in plants: From cellular responses to long-distance intercellular communication. *Frontiers in Plant Science*, 11, 1407.

Tan, R. X., & Zou, W. X. (2001). Endophytes: A rich source of functional metabolites. *Natural Product Reports*, 18(4), 448–459.

Tan, X. M., Zhou, Y. Q., Zhou, X. L., Xia, X. H., Wei, Y., He, L. L., . . . Yu, L. Y. (2018). Diversity and bioactive potential of culturable fungal endophytes of *Dysosma versipellis*; a rare medicinal plant endemic to China. *Scientific Reports*, 8(1), 5929.

Tanaka, A., Christensen, M. J., Takemoto, D., Park, P., & Scott, B. (2006). Reactive oxygen species play a role in regulating a fungus – perennial ryegrass mutualistic interaction. *The Plant Cell*, 18(4), 1052–1066.

Tejesvi, M. V., Kini, K. R., Prakash, H. S., Subbiah, V., & Shetty, H. S. (2008). Antioxidant, antihypertensive, and antibacterial properties of endophytic Pestalotiopsis species from medicinal plants. *Canadian Journal of Microbiology*, 54(9), 769–780.

Ting, A. S. Y., Mah, S. W., & Tee, C. S. (2012). Evaluating the feasibility of induced host resistance by endophytic isolate Penicillium citrinum BTF08 as a control mechanism for Fusarium wilt in banana plantlets. *Biological Control*, 61(2), 155–159.

Tiwari, R., Kalra, A., Darokar, M. P., Chandra, M., Aggarwal, N., Singh, A. K., & Khanuja, S. P. S. (2010). Endophytic bacteria from Ocimum sanctum and their yield-enhancing capabilities. *Current Microbiology*, 60, 167–171.

Tofern, B., Kaloga, M., Witte, L., Hartmann, T., & Eich, E. (1999). Occurrence of loline alkaloids in Argyreia mollis (Convolvulaceae). *Phytochemistry*, 51(8), 1177–1180.

Torres, M. S., Singh, A. P., Shah, S., Herrera-Carrillo, Z., Gianfagna, T., White Jr, J. F., & Vorsa, N. (2009, January). LC-MS-MS identification and quantification of phenolics in symbiotic tall fescue. In *Proceedings of the 18th annual Rutgers Turfgrass Symposium* (Vol. 12). Rutgers University, Center for Interdisciplinary Studies in Turfgrass Science.

Torres, M. S., White Jr, J. F., Zhang, X., Hinton, D. M., & Bacon, C. W. (2012). Endophyte-mediated adjustments in host morphology and physiology and effects on host fitness traits in grasses. *Fungal Ecology*, 5(3), 322–330.

Tyburski, J., Dunajska, K., Mazurek, P., Piotrowska, B., & Tretyn, A. (2009). Exogenous auxin regulates H2O2 metabolism in roots of tomato (Lycopersicon esculentum Mill.) seedlings affecting the expression and activity of CuZn-superoxide dismutase, catalase, and peroxidase. *Acta physiologiae plantarum*, 31, 249–260.

Upreti, R., & Thomas, P. (2015). Root-associated bacterial endophytes from Ralstonia solanacearum resistant and susceptible tomato cultivars and their pathogen antagonistic effects. *Frontiers in Microbiology*, 6, 255.

Vaishnav, A., Shukla, A. K., Sharma, A., Kumar, R., & Choudhary, D. K. (2019). Endophytic bacteria in plant salt stress tolerance: Current and prospects. *Journal of Plant Growth Regulation*, 38, 650–668.

Valluru, R., Link, J., & Claupein, W. (2012). Consequences of early chilling stress in two Triticum species: Plastic responses and adaptive significance. *Plant Biology*, 14(4), 641–651.

Vargas, L., Santa Brigida, A. B., Mota Filho, J. P., de Carvalho, T. G., Rojas, C. A., Vaneechoutte, D., Van Bel, M., Farrinelli, L., Ferreira, P. C. G., & Vandepoele, K. (2014). Drought tolerance conferred to sugarcane by association with Gluconacetobacter diazotrophicus: A transcriptomic view of hormone pathways. *PLoS One*, 9(12), e114744.

Vendan, R. T., Yu, Y. J., Lee, S. H., & Rhee, Y. H. (2010). Diversity of endophytic bacteria in ginseng and their potential for plant growth promotion. *The Journal of Microbiology*, 48, 559–565.

Venkatachalam, R., Subban, K., & Paul, M. J. (2008). Taxol from Botryodiplodia theobromae (BT 115) – AN endophytic fungus of Taxus baccata. *Journal of Biotechnology*, 136, S189–S190.

Verma, H., Kumar, D., Kumar, V., Kumari, M., Singh, S. K., Sharma, V. K., Droby, S., Santoyo, G., White, J. F., & Kumar, A. (2021a). The potential application of endophytes in the management of stress from drought and salinity in crop plants. *Microorganisms*, 9(8), 1729.

Verma, S. K., Sahu, P. K., Kumar, K., Pal, G., Gond, S. K., Kharwar, R. N., & White, J. F. (2021b). Endophyte roles in nutrient acquisition, root system architecture development, and oxidative stress tolerance. *Journal of Applied Microbiology*, 131(5), 2161–2177.

Verma, S. K., Kharwar, R. N., & White, J. F. (2019). The role of seed-vectored endophytes in seedling development and establishment. *Symbiosis*, 78, 107–113.

Waller, F., Achatz, B., Baltruschat, H., Fodor, J., Becker, K., Fischer, M., . . . Kogel, K. H. (2005). The endophytic fungus Piriformospora indica reprograms barley to salt-stress tolerance, disease resistance, and higher yield. *Proceedings of the National Academy of Sciences*, 102(38), 13386–13391.

Wang, C., Wang, C., Gao, Y. L., Wang, Y. P., & Guo, J. H. (2016). A consortium of three plant growth-promoting rhizobacterium strains acclimates Lycopersicon esculentum and confers a better tolerance to chilling stress. *Journal of Plant Growth Regulation*, 35, 54–64.

Wang, J. W., Zheng, L. P., & Tan, R. X. (2006). The preparation of an elicitor from a fungal endophyte to enhance artemisinin production in hairy root cultures of Artemisia annua L. *Sheng Wu Gong Cheng Xue bao= Chinese Journal of Biotechnology*, 22(5), 829–834.

Wang, L., Liu, L., & Han, S. Z. (2009). Screening and identification of antimicrobe activity of endophytic fungus in Glycyrrhiza uralensis. *Biotechnology Bulletin*, 6, 34.

Wang, W. X., Barak, T., Vinocur, B., Shoseyov, O., & Altman, A. (2003). Abiotic resistance and chaperones: Possible physiological role of SP1, a stable and stabilizing protein from Populus. *Plant Biotechnology 2002 and Beyond: Proceedings of the 10th IAPTC&B Congress*, June 23–28, 2002 Orlando, Florida, USA, 439–443.

Wang, Y., & Frei, M. (2011). Stressed food – the impact of abiotic environmental stresses on crop quality. *Agriculture, Ecosystems & Environment*, 141(3–4), 271–286.

Wang, Y., Xu, L., Ren, W., Zhao, D., Zhu, Y., & Wu, X. (2012). Bioactive metabolites from Chaetomium globosum L18, an endophytic fungus in the medicinal plant Curcuma wenyujin. *Phytomedicine*, 19(3–4), 364–368.

Wani, S. H., Sandhu, J. S., & Gosal, S. S. (2008). Genetic engineering of crop plants for abiotic stress tolerance. *Advanced Topics in Plant Biotechnology and Plant Biology*, 149–183.

Wani, Z. A., Ashraf, N., Mohiuddin, T., & Riyaz-Ul-Hassan, S. (2015). Plant-endophyte symbiosis, an ecological perspective. *Applied Microbiology and Biotechnology*, 99, 2955–2965.

Waqas, M., Khan, A. L., Kamran, M., Hamayun, M., Kang, S.-M., Kim, Y.-H., & Lee, I.-J. (2012). Endophytic fungi produce gibberellins and indoleacetic acid and promote host-plant growth during stress. *Molecules*, 17(9), 10754–10773.

Waqas, M., Khan, A. L., Kang, S.-M., Kim, Y.-H., & Lee, I.-J. (2014). Phytohormone-producing fungal endophytes and hardwood-derived biochar interact to ameliorate heavy metal stress in soybeans. *Biology and Fertility of Soils*, 50, 1155–1167.

Weyens, N., van der Lelie, D., Taghavi, S., & Vangronsveld, J. (2009). Phytoremediation: Plant-endophyte partnerships take the challenge. *Current Opinion in Biotechnology*, 20(2), 248–254.

Wink, M. (2003). Evolution of secondary metabolites from an ecological and molecular phylogenetic perspective. *Phytochemistry*, 64(1), 3–19.

Wu, L., Han, T., Li, W., Jia, M., Xue, L., Rahman, K., & Qin, L. (2013). Geographic and tissue influence on endophytic fungal communities of Taxus chinensis var. mairei in China. *Current Microbiology*, 66, 40–48.

Wu, W., Chen, W., Liu, S., Wu, J., Zhu, Y., Qin, L., & Zhu, B. (2021). Beneficial relationships between endophytic bacteria and medicinal plants. *Frontiers in Plant Science*, 12, 646146.

Xu, J. Y., Han, Y. H., Chen, Y., Zhu, L. J., & Ma, L. Q. (2016). Arsenic transformation and plant growth promotion characteristics of As-resistant endophytic bacteria from As-hyperaccumulator Pteris vittata. *Chemosphere*, 144, 1233–1240.

Xu, L., Wang, A., Wang, J., Wei, Q., & Zhang, W. (2017). Piriformospora indica confers drought tolerance on Zea mays L. through enhancement of antioxidant activity and expression of drought-related genes. *The Crop Journal*, 5(3), 251–258.

Yan, H., Jin, H., Fu, Y., Yin, Z., & Yin, C. (2019). Production of rare ginsenosides Rg3 and Rh2 by endophytic bacteria from Panax ginseng. *Journal of Agricultural and Food Chemistry*, 67(31), 8493–8499.

Yang, H.-R., Yuan, J., Liu, L.-H., Zhang, W., Chen, F., & Dai, C.-C. (2019). Endophytic Pseudomonas fluorescens induced sesquiterpenoid accumulation mediated by gibberellic acid and jasmonic acid in *Atractylodes macrocephala* Koidz plantlets. *Plant Cell, Tissue and Organ Culture (PCTOC)*, 138, 445–457.

Yang, N. Y., Jiang, S., Shang, E. X., Tang, Y. P., & Duan, J. A. (2012). A new phenylpentanamine alkaloid produced by an endophyte Bacillus subtilis isolated from Angelica sinensis. *Journal of Chemical Research*, 36(11), 647–647.

Yang, R., Fan, X., Cai, X., & Hu, F. (2015). The inhibitory mechanisms by mixtures of two endophytic bacterial strains isolated from Ginkgo biloba against pepper Phytophthora blight. *Biological Control*, 85, 59–67.

Yang, S., Vanderbeld, B., Wan, J., & Huang, Y. (2010). Narrowing down the targets: Towards successful genetic engineering of drought-tolerant crops. *Molecular Plant*, 3(3), 469–490.

Yang, Y.-H., Yang, D.-S., Li, G.-H., Liu, R., Huang, X.-W., Zhang, K.-Q., & Zhao, P.-J. (2018). New secondary metabolites from an engineering mutant of endophytic Streptomyces sp. CS. *Fitoterapia*, 130, 17–25.

Yin, D. D., Wang, Y. L., Yang, M., Yin, D. K., Wang, G. K., & Xu, F. (2019). Analysis of Chuanxiong Rhizoma substrate on the production of ligustrazine in endophytic Bacillus subtilis by ultra-high performance liquid chromatography with quadrupole time-of-flight mass spectrometry. *Journal of Separation Science*, 42(19), 3067–3076.

Youssef, M., El-Nagdi, W., & Lotfy, D. E. M. (2020). Evaluation of the fungal activity of Beauveria bassiana, Metarhizium anisopliae, and Paecilomyces lilacinus as biocontrol agents against root-knot nematode, Meloidogyne incognita on cowpea. *Bulletin of the National Research Centre*, 44(1), 1–11.

Yuan, H., Ma, Q., Ye, L., & Piao, G. (2016). Traditional medicine and modern medicine from natural products. *Molecules*, 21(5), 559.

Zhang, H. W., Song, Y. C., & Tan, R. X. (2006). Biology and chemistry of endophytes. *Natural Product Reports*, 23(5), 753–771.

Zhang, H. W., Zhu, J., Gong, Z., & Zhu, J.-K. (2022). Abiotic stress responses in plants. *Nature Reviews Genetics*, 23(2), 104–119.

Zhang, L. Q., Guo, B., & Li, H. Y. (1998). Isolation of a fungus producing Vinbrastine. *Journal of Yunnan University: Natural Sciences*, 20, 214–215.

Zhang, L. Q., Zhang, W., Li, Q., Cui, R., Wang, Z., Wang, Y., Zhang, Y.-Z., Ding, W., & Shen, X. (2020). Deciphering the root endosphere microbiome of the desert plant Alhagi sparsifolia for drought resistance-promoting bacteria. *Applied and Environmental Microbiology*, 86(11), e02863–19.

Zhang, X., Li, C., & Nan, Z. (2012). Effects of cadmium stress on seed germination and seedling growth of Elymus dahuricus infected with the Neotyphodium endophyte. *Science China Life Sciences*, 55, 793–799.

Zhang, X., Li, G., Ma, J., Zeng, Y., Ma, W., & Zhao, P. (2010). Endophytic fungus Trichothecium roseum LZ93 antagonizing pathogenic fungi in vitro and its secondary metabolites. *The Journal of Microbiology*, 48, 784–790.

Zhang, Y., Yu, X., Zhang, W., Lang, D., Zhang, X., Cui, G., & Zhang, X. (2019). Interactions between endophytes and plants: Beneficial effect of endophytes to ameliorate biotic and abiotic stresses in plants. *Journal of Plant Biology*, 62, 1–13.

Zhao, L., Xu, Y., Lai, X.-H., Shan, C., Deng, Z., & Ji, Y. (2015). Screening and characterization of endophytic Bacillus and Paenibacillus strains from medicinal plant Lonicera japonica for use as potential plant growth promoters. *Brazilian Journal of Microbiology*, 46, 977–989.

Zhao, S., Zhou, N., Zhao, Z.-Y., Zhang, K., Wu, G.-H., & Tian, C.-Y. (2016). Isolation of endophytic plant growth-promoting bacteria associated with the halophyte Salicornia europaea and evaluation of their promoting activity under salt stress. *Current Microbiology*, 73, 574–581.

Zhou, J. Y., Li, X., Zhao, D., Deng-Wang, M. Y., & Dai, C. C. (2016). Reactive oxygen species and hormone signaling cascade in endophytic bacterium-induced essential oil accumulation in Atractylodes lancea. *Planta*, 244, 699–712.

Zhou, J. Y., Sun, K., Chen, F., Yuan, J., Li, X., & Dai, C. (2018). Endophytic Pseudomonas induces metabolic flux changes that enhance medicinal sesquiterpenoid accumulation in Atractylodes lancea. *Plant Physiology and Biochemistry*, 130, 473–481.

Zhou, S. L., Yang, F., Lan, S. L., Xu, N., & Hong, Y. H. (2009). Huperzine A producing conditions from endophytic fungus in SHB Huperzia serrata. *Journal of Microbiology*, 3, 32–36.

Zhou, X., Wang, Z., Jiang, K., Wei, Y., Lin, J., Sun, X., & Tang, K. (2007). Screening of taxol-producing endophytic fungi from Taxus chinensis var. mairei. *Applied Biochemistry and Microbiology*, 43, 439–443.

Zhu, Y. (2019). Isolation and identification of *Ammodendron bifolium* endophytic bacteria and the action mechanism of selected isolates-induced seed germination and their effects on host osmotic-stress tolerance. *Archives of Microbiology*, 201, 431–442.

Zia, R., Nawaz, M. S., Siddique, M. J., Hakim, S., & Imran, A. (2021). Plant survival under drought stress: Implications, adaptive responses, and integrated rhizosphere management strategy for stress mitigation. *Microbiological Research*, 242, 126626.

10 Role of Endophytes as Biostimulants in Medicinal Plant Protection and Resilience

Wiwiek Harsonowati, Dwinita Wikan Utami, Wiwin Setiawati, Sri Widawati, Suliasih, Deciyanto Soetopo, Susi Wuryantini, Sri Wahyuni Budiarti, and Mizu Istianto

10.1 THE IMPORTANCE OF MEDICINAL PLANTS AND THEIR RESPONSE TO ENVIRONMENTAL STRESS

The growing world population, insufficient drug supply, prohibitive treatment costs, side effects of several synthetic drugs, and resistance to infectious disease drugs have all increased the emphasis on plant material as a source of medicines for various human ailments. Globally, secondary metabolites of higher plants are the primary source of many natural drugs. Medicinal plants possess pharmacological properties (antidiabetes, anticancer, antihypertension, anticardiovascular, antileprosy, antimicrobial, etc.) due to their ability to produce diverse bioactive secondary metabolites (Mahboubi *et al.* 2017; Galovičová *et al.* 2021). Natural products from medicinal plants have no side effects and are relatively safe, eco-friendly, and readily accessible. Traditionally, many herbs have been used to treat various human diseases. In contrast to synthetic drugs, regarded as dangerous to humans and the ecosystem, herbal products now symbolize safe alternative medicines. Therefore, there is an urgent need to develop a cultivation system of medicinal plants with high growth and production under any environmental conditions. There have been numerous definitions suggested for the term 'medicinal plant'. According to the World Health Organization (WHO), a medicinal plant is any plant that contains a compound(s) that can be used for therapeutic purposes or is a precursor for chemo-pharmaceutical semi-synthesis in one or more of its organs/parts (WHO 1977). Medicinal plants are considered a rich source of ingredients that can be used in the development of pharmaceutical, non-pharmaceutical, or synthetic drugs. Furthermore, some plants are considered essential sources of nutrition, so they are suggested for their therapeutic properties. These plants include ginger, green tea, hazelnuts, aloe, pepper, and turmeric. Some plants and their derivatives are considered significant sources of active ingredients used in aspirin and toothpaste, among other things. Therefore, these plants play an essential role in the development of human cultures all over the world. There has been a tremendous increase in the use of herbal medicine over the last two decades; however, more research in this area still needs to be done.

Recently, the WHO estimated that 80% of people worldwide use plant-derived medicines for some ingredients of their primary health care requirements, and approximately 21,000 plants, or more than 30% of all plant species, can be used as medicinal plants in alternative medicine (WHO 2013). According to available data, more than three-quarters of the world's population depends primarily on plants and plant extracts for their health care requirements. Medicinal plants are expected

DOI: 10.1201/b23155-10

to account for up to 25% of total drugs in developed countries such as the United States and up to 80% in fast-developing countries such as India and China (Hosseinzadeh *et al.* 2015). As a result, the economic importance of medicinal plants is much greater in developing countries than in the rest of the globe. These nations supply two-thirds of the plants used in modern medicine, and the rural population's health care system relies on them.

Like other living components of the biosphere, medicinal plants are not immune to environmental stresses; hence, ecological limits negatively impact their growth and yields. During their life cycle, medicinal plant species are typically subject to various environmental challenges that negatively affect their development, productivity, reproductive ability, and survival (Figure 10.1). Medicinal plants are susceptible to multiple tissue-damaging abiotic stressors threatening their survival and productivity (Farsaraei *et al.* 2020; Ashrafi *et al.* 2022). Various climatic abiotic stresses such as drought, salinity, low and heat temperatures, and heavy metal toxicity restrict their growth and productivity worldwide by affecting physiological and biochemical homoeostatic conditions (Moori and Ahmadi-Lahijani 2020; Ashrafi *et al.* 2022).

In the natural ecosystem, plants are frequently infected by more than one stressor, either biotic or abiotic, which occur simultaneously or in combination, resulting in significant yield and economic losses. Plants are generally exposed to biotic stress, such as pests and diseases. These plants create effective strategies for avoiding or tolerating all of these stresses, which aids in their adaptation to stressful conditions. Plants use epigenetic memory, molecular crosstalk, reactive oxygen species (ROS) signalling, plant hormone accumulation (such as abscisic acid, jasmonates, ethylene, and salicylic acid), redox status and inorganic ion flux changes, R-gene resistance, and systemic acquired resistance to adapt and protect against the environmental stressors (Abdelshafy Mohamad *et al.* 2020; Moori and Ahmadi-Lahijani 2020; Raza *et al.* 2022; Myers *et al.* 2023). In addition, plants produce a cluster of low-molecular-mass compounds called secondary metabolites, including quinones, anthocyanins, phenolics, flavonoids, alkaloids, lignans, steroids, and terpenoids. Terpenoids are a class of organic chemicals primarily synthesized by plants as secondary metabolites. They have been used extensively in the pharmaceutical, food, and cosmetics sectors due to their medicinal and nutritional benefits (Hill and Connolly 2020). Due to their potent antiviral properties, terpenoids have recently been used in the clinical treatment of COVID-19 (Murck 2020). Plants have developed complex physiological and biochemical adaptations to adjust and adapt to

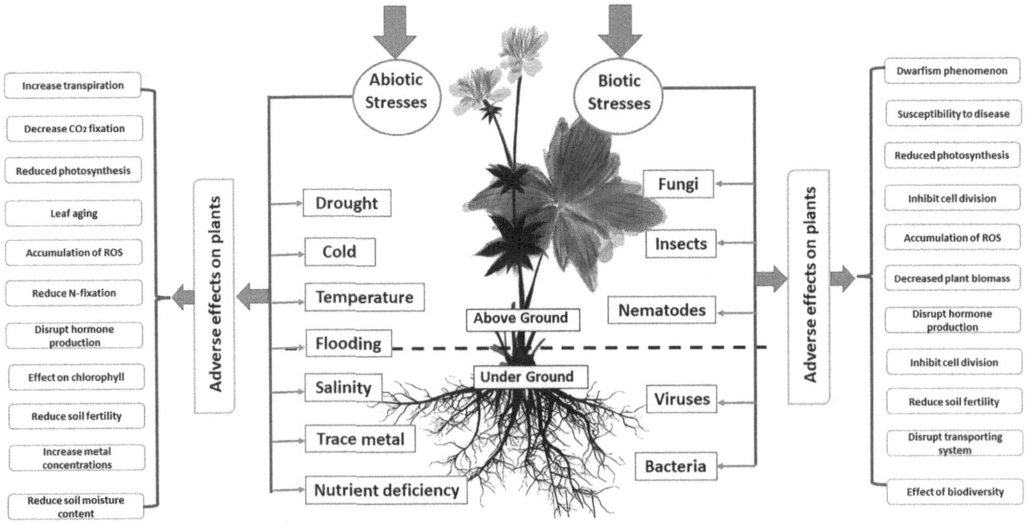

FIGURE 10.1 Adverse effects of diverse biotic and abiotic stresses on medicinal plant growth and development.

various environmental stresses. Plant responses can be observed at structural, anatomical, hormonal, molecular, and biochemical levels.

Amongst all biotic constraints, insect pests are one of the major global concerns (Saha *et al.* 2020). Around 20 to 40% of global crop production is lost to pests annually. Plant diseases cost the global economy around $220 billion each year and invasive insects around $70 billion, according to the Food and Agriculture Organization of the United Nations (USDA 2023). Plant diseases and pests such as insects, mites, and nematodes attack medicinal plants regularly and cause damage to these plants in all agroecosystems where small-, medium-, and large-scale farming is practiced. Despite severe damage, extensive studies on determining economic value (quality deterioration and yield reduction) for all invaders have not been conducted. However, some plant workers described the severity of the damage as major or minor. The status of these pests can alter when climate, cultural practices, and other control measures change, depending on the nature of the damage they inflict throughout plant development and storage of harvested material. Recently, many scholars have focused on the study of endophytes as biocontrol agents of pathogens and insects and successfully applied endophytes to plant protection (Fesel and Zuccaro 2016; Regaiolo *et al.* 2020; Popay *et al.* 2023; Rumyantsev *et al.* 2023). The potential mechanisms of endophyte inhibition of plant pathogens are by several means, including direct, indirect, and ecological effects. It can be seen that the main reasons for fungal endophytes in plant protection against pathogens might be phytoalexin production induced by fungal endophytes and endophytic ecological occupation

10.2 ALLEVIATION OF STRESS CONDITIONS IN MEDICINAL PLANTS BY ENDOPHYTES

Medicinal plants have become popular, with increasing market demand for plant-based products in health-related industries. Their supply is frequently insufficient due to lower production and inferior quality caused by pest and disease damage. Some chemical pesticides, including chlorpyriphos, dicofol, carbaryl, ethion, dimethoate, dichlorvos, triazophos, malathion, and quinalphos, are routinely used in medicinal plant protection. Residues of these toxic pesticides that remain on harvested plant parts may be transferred to final products consumed by consumers, which are hazardous to human health (Glare *et al.* 2012). Additionally, agrochemicals affect secondary metabolites' morphological, physiological, and biosynthesis (Karimmojeni *et al.* 2022). Therefore, medicinal plant protection is crucial to preventing or minimizing damage and producing healthy materials with high-quality yield.

Endophytic microbes are being studied extensively to improve agricultural production and environmental safety for sustainability. Endophytes, mainly bacteria, and fungi, are microbial symbionts of plants that inhabit the inter- and intracellular spaces within the host plant tissues without manifestation of causing any adverse or pathological effects. Medicinal plants harbour diverse endophyte domain populations with immense potential as biostimulants and stress alleviators. After being colonized by specific functional endophytes, medicinal plants develop symbiotic relationships, accompanied in recent years by an in-depth study and exploration of medicinal plant symbionts. These symbiotic relationships can regulate numerous host functions, including tolerance to various abiotic stressors (Chen *et al.* 2022; Rong *et al.* 2022; He *et al.* 2023; Kruasuwan *et al.* 2023; Long *et al.* 2023), promoting plant growth and development (Rho *et al.* 2020; Martínez-Arias *et al.* 2021; Harsonowati *et al.* 2022), stimulating defence systems against plant pathogens (Macedo-Raygoza *et al.* 2019; Harsonowati *et al.* 2020; Marian *et al.* 2022), and reducing pest penetration by producing secondary metabolites with insecticidal properties (Gao *et al.* 2010; Amutha 2021; Wang *et al.* 2022; Yerukala *et al.* 2022; Rumyantsev *et al.* 2023). In addition to playing a vital role in the growth and development of their host, endophytes also act as 'inducers' which assist their host plant species in accumulating some highly coveted plant secondary metabolites (PSMs) in response to stress conditions (Cherif-Silini *et al.* 2019; Singh *et al.* 2022).

Endophytes reside asymptomatically in plant cells and synthesize secondary metabolites or signalling molecules that serve as internal and external stimuli during mutualistic interaction (Tidke *et al.* 2019). The symbiotic association improves the host plants' physiology, fitness, and metabolite profile, while the plants provide nutrients and shelter for the endophytes. The colonization of endophytes can alter the mineral nutrient composition, phytohormonal balance, and the chemical constituents of root exudates and protect the plant against biotic and abiotic stressors. The endophyte-induced alterations of the plants offer numerous potentials for biotechnological, medicinal, and agricultural applications. Endophytes significantly improve medicinal plant growth under stress conditions such as nutrient deficiency, temperature (heat and cold), salinity, pests, and diseases.

Endophytes promote host plant growth and fitness by producing phytohormones or reducing abiotic and biotic stress. Enhancing the plant immune system and disease suppression are associated with having novel antibiotics, secondary metabolites, siderophores, and nitrogenous or other industrially critical chemical compounds. In addition, functional endophytes enhance the synthesis of important secondary metabolites in plants and improve the potential applicability of plants in agriculture, medicine, food, and horticulture. The search for microbial candidate sources with high application potential has become increasingly focused in recent years. Microbes associated with

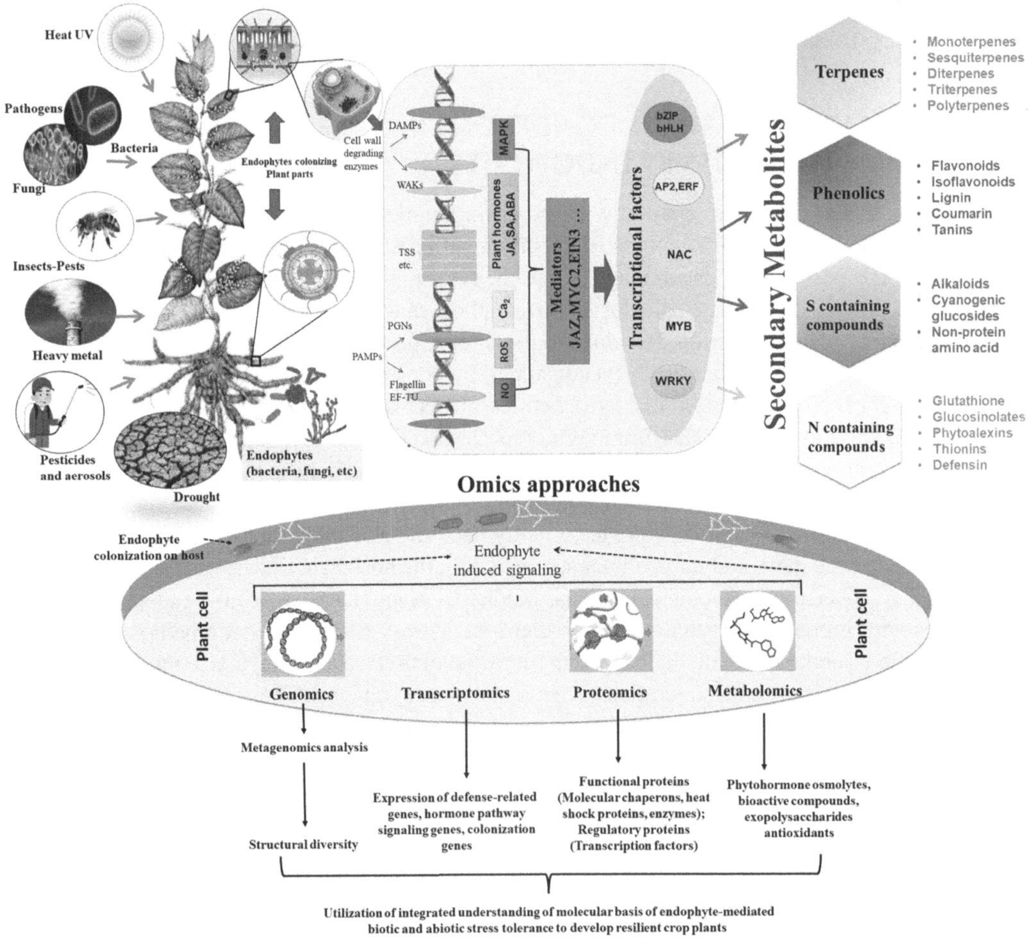

FIGURE 10.2 Adverse biotic and abiotic stresses influence plant growth and development; in response, plants employ endophytes to mitigate these stresses and the modulation of secondary metabolism by diverse transcription factors regulated by endophytes-mediated cell-signalling pathways in response to stress. Endophyte-mediated environmental stress tolerance can be developed using an integrated molecular basis, such as omics analysis.

plants, such as endophytes, are of particular interest because of their potential in poorly studied environments compared to soil microbes.

Genome-wide analysis of endophytes directly reflects the colonization preference and genetic characteristics of endophytes on various host plants. Genome-wide analysis significantly facilitates the identification of genes associated with host growth and development, antibiotic production, insertion elements, the endophytic secretory system, and other metabolic processes (Subudhi *et al.* 2019). Additionally, genome-wide analysis has been used to explain the survival and evolution of endophytes in their hosts (Singh *et al.* 2021). The complete genomes of several endophytes have been sequenced, with the number increasing. It can also serve as a model for systematically investigating interactions between endophytes and host plants. In addition, omics have been developed into 'macro-omics', which provides more robust technical support for the study of interactions between hosts and endophytes (Figure 10.2). In recent years, many researchers have conducted studies on the interaction between endophytes and hosts by using joint omics, but there is still a substantial knowledge deficit regarding this interaction (Li *et al.* 2022). There is still much work to be done in the subsequent phase to realize the ideal combination of information technology and data and to study the interactions between endophytes and hosts in a meaningful manner.

10.3 ENDOPHYTE-MEDIATED BIOSYNTHESIS AND ACCUMULATION OF MEDICINAL PLANT SECONDARY METABOLITES

Medicinal plants host diverse endophytes, including prokaryotic (archaea and bacteria) and eukaryotic (fungi) endophytes. Recent advances in medicinal plant–endophyte symbiotic interaction suggest that some of the reported plant tolerance or resistance to environmental stress comes from the host plant and the obscure plant symbionts known as endophytes that live in the plant's tissues. Endophytes are well distributed in all parts of host plants. Their biological functions promote growth, development, and propagation, as well as the biosynthesis of secondary metabolites of their host medicinal plants (Gupta *et al.* 2020). Plant secondary metabolites have gained widespread attention in recent decades due to their potential as either natural medicinal compounds or precursors for producing new medications and products.

Endophytes are essential in regulating plant growth and development and are essentially used as a promising strategy to enhance the biosynthesis of the host plant's secondary metabolites. Plant secondary metabolites are substances or bioactive compounds synthesized during plant growth and development due to environmental interactions. Secondary metabolites are mainly produced from primary metabolites and have no direct role in plant growth and development. They are well known for their role in defensive functions and medicinal properties. The biosynthesis of these metabolites is regulated by the interplay of signalling molecules comprising phytohormones. Most secondary metabolites are alkaloids, flavonoids, terpenoids, peptides, phenols, sterols, and other minor molecular organic compounds (Tian *et al.* 2017). As essential substances plants use for self-protection to cope with environmental conditions, secondary metabolites have various physiological functions, such as biological defence, and are involved in the plant response to abiotic stresses (Jin *et al.* 2017; Xie *et al.* 2018a).

Research on medicinal plants' SMs has advanced rapidly in recent years, particularly in human health and agricultural production. However, plant production of SMs is very limited and far from satisfying the growing market demand. As a result, finding technology to improve secondary metabolite biosynthesis in medicinal plants is critical. Recently, several studies have reported on how endophytes affect the synthesis of secondary metabolites. The synthesis and accumulation of secondary metabolites are the most important strategies of medicinal plants to protect against environmental stresses.

Endophytes regulate genes related to the biosynthesis of secondary metabolites by plant cells via signal transduction pathways. Endophytes also alter the rate and intensity of these genes' expression and induce the production and accumulation of corresponding SMs in medicinal plants (Table 10.1). Sun *et al.* (2022) reported the co-cultivation and the symbiotic association

TABLE 10.1
Mechanisms of Action: Medicinal Plant-Associated Endophytes Regulate Gene Expression Related to the Biosynthesis and Accumulation of Secondary Metabolites

Endophyte Classification	Endophyte species	Medicinal Plant Host	Mechanisms of Action	Functions	References
Bacteria	*Enterobacter hormaechei* H2A3 and H5A2	*Stevia rebaudiana* (candy leaf, sweetleaf, or sugar leaf)	Upregulating the transcript levels of the *KO*, *KAH*, *UGT74G1*, and *UGT76G1* genes	Increased steviol glycoside (SG) synthesis, flavonoid content, and flavonoid accumulation	(Oviedo-Pereira *et al.* 2022)
	Pseudomonas fluorescens	*Atractylodes macrocephala* (Baizhu in Chinese)	Mediating gibberellic and jasmonic acid signals to increase the activities of key enzymes: 3-hydroxy-3-methylglutaryl-CoA reductase (HMGR) and 1-deoxy-D-xylulose 5-phosphate reductoisomerase (DXR)	Induced sesquiterpenoid biosynthesis	(Yang *et al.* 2019)
	Bacillus subtilis	*Ligusticum chuanxiong* (Chuanxiong rhizoma)	Reducing sugar and polysaccharides in host plant with the stimulating effect of ammonium sulphate	Promoted the biosynthesis of ligustrazine and acetoin (biosynthetic precursor of ligustrazine)	(Yin *et al.* 2019)
	Paenibacillus polymyxa; Flavobacterium sp.	*Panax ginseng* (ginseng)	Mediating jasmonic acid signals	Induced biosynthesis of ginsenoside Rg3	(Gao *et al.* 2015; Fu 2019)
Fungi	*Funneliformis mosseae* and *Piriformospora indica*	*Polygonum cuspidatum* (Japanese knotweed)	Upregulating the expression of *PcCHS1, PcCRS1, PcRS11, PcRS*, and *PcSTS* genes	Accelerated biomass production, leaf physiological activity, and resveratrol accumulation	(Sun *et al.* 2022)
	Biscogniauxia sp. and *Didymella* sp.	*Vitis amurensis* Rupr (amur grape)	Activation of phenylalanine ammonia-lyase (PAL) and stilbene synthase (STS) gene expression	Stimulated the production of stilbenes	(Aleynova *et al.* 2021)
	Glomus hoi	*Ocimum basilicum* L. (basil)	Stimulate contents of total phenols (TPC) and phenylalanine ammonia-lyase (PAL) activity	Increasing the content of phenolic/active compounds like caffeic acid (CA) and rosmarinic (RA) acid	(Zare *et al.* 2021)

(*Continued*)

TABLE 10.1 (*Continued*)
Mechanisms of Action: Medicinal Plant-Associated Endophytes Regulate Gene Expression Related to the Biosynthesis and Accumulation of Secondary Metabolites

Endophyte Classification	Endophyte species	Medicinal Plant Host	Mechanisms of Action	Functions	References
	Rhizophagus irregularis	*Glycyrrhiza uralensis* (Chinese liquorice)	Upregulated the expression of aquaporin gene *PIP*; glycyrrhizin biosynthesis genes *SQS1*, *β-AS*, *CYP88D6*, and *CYP72A154*; and liquiritin biosynthesis gene *CHS*	Increasing the accumulation of glycyrrhizin and liquiritin	(Xie *et al.* 2018b)
	Gilmaniella sp. AL 12	*Atractylodes lancea* (Cangzhu in Chinese)	Mediating ethylene signalling pathways for jasmonic acid and salicylic acid (upstream signal), NO, and H_2O_2 (downstream signal)	Induced important signal mediating sesquiterpenoids, a primary active compound in biosynthesis	(Yuan *et al.* 2016)
	Rhizophagus intraradices	*Artemisia annua* (sweet wormwood or sweet sagewort)	Upregulating the expression of key biosynthesis genes via enhanced jasmonic acid levels	Increasing the accumulation of secondary metabolite artemisinin	(Mandal *et al.* 2015)
	Claroideoglomus claroideum and *Racocetra coralloidea*	*Paris polyphylla* var. yunnanensis	Increasing the activity of succinic dehydrogenase (SDH) and alkaline phosphatase (ALP)	Increasing the accumulation of steroid saponin and the quality of medicine	(Nong *et al.* 2015)

of endophytic fungi *Funneliformis mosseae* and *Piriformospora indica* to their host medicinal plant *Polygonum cuspidatum* significantly increased biosynthesis of resveratrol, a major medicinal component for the extraction of the drug. The application of the endophytic fungus *F. mosseae* accelerates physiological activity. It upregulates the expression of associated genes in resveratrol biosynthesis in the medicinal plant *Polygonum cuspidatum*, a perennial herb with medicinal components (Sun *et al.* 2022).

10.4 MECHANISMS OF ENDOPHYTES AS BIOSTIMULANTS IN IMPROVING THE PRODUCTIVITY AND QUALITY OF MEDICINAL PLANTS

Recently the effects of endophytes on the accumulation of active ingredients by medicinal plants have received increasing attention. Endophytes are typically found in host plants as multicellular aggregation communities, which have attracted significant interest as a unique 'microecological' system (Bziuk *et al.* 2021). Numerous plant biological functions depend on endophytes; each plant is a complex micro-ecosystem (Compant *et al.* 2021). Plants actively seek the help of their microbial symbionts to survive harsh environments. The endophytic microbiome provides several life-support functions for its host. It can assist host plants in sustaining basic life activities such as nutrient acquisition, immunological control, and biological stress tolerance by expanding the plant genome and metabolic capacity. Numerous research findings indicate that when plants recruit beneficial endophytes, the symbiotic effects can increase the release of essential molecular compounds during endophyte colonization. At various stages of successful endophytic colonization, certain plants can modulate the absorption of nutrients and the expression of related genes. Endophytes that successfully colonize plants rapidly use plant bodies to obtain the nutrients they require and establish a stable community of microbes, also known as the microbiome (Pande *et al.* 2021). Although endophytic microbial communities have been proposed as the next Green Revolution platform, basic research on microbial community construction and activity mechanics is still in its early stages (Cordovez *et al.* 2019).

Biosynthetic mechanisms of endophytes on secondary metabolite synthesis in plants and the factors that influence it are major research topics. Previous research has suggested that the synthesis of numerous secondary metabolites in plants is affected by environmental factors such as temperature, light, and plant–microbe interaction. Many studies have been conducted on how endophytes increase secondary metabolite synthesis in host plants. However, there is no consensus among scientists. There are four primary postulated mechanisms: (1) endophytes promote the photosynthesis rate and accumulation of photosynthesis-related compounds in plants affected by various photoperiod or aboveground light signals (Harsonowati *et al.* 2020; Wang *et al.* 2021), (2) endophytes regulate the expression of genes related to plant secondary metabolites (Sun *et al.* 2022), (3) the transfer of endophytes leads to the formation of compounds that alter the genetic material within plants, and (4) the synthesis of unique secondary metabolites by endophytes affects plant material synthesis pathways.

Signal transduction regulates the biosynthesis of secondary metabolites in the interaction between endophytes and medicinal plants. Signaling molecules such as elicitors and signalling hormones are crucial to endophytic recruitment, infestation, and colonization in the symbiotic interaction system. Plant endophyte elicitors are signalling molecules derived from a fungus or bacterium. These elicitors can induce plant disease resistance and promote the plant to produce various secondary metabolites, such as terpenoids, alkaloids, saponins, phenols, and flavonoids (Tian *et al.* 2012).

Elicitation is recognized as the most effective biotechnological technique for extensive production of plant secondary metabolites both in vitro and in vivo conditions as a result of the responses of plant tissues to elicitors (Figure 10.3). Recently, endophytes have been employed as biological elicitor sources due to their capacity to replicate disease responses in plant cells. Plant endophyte elicitors activated the signal transduction network for related gene activity, thus increasing bioactive compound accumulation in the host (Tidke *et al.* 2019). Alternatively, it is suggested that colonization of endophytes triggers plant hydrolases that fragment the endophytes, producing elicitors such as lipopolysaccharides, polysaccharides, and glycoproteins that stimulate plant defence responses and the production of secondary metabolites (Sharma *et al.* 2023). Findings revealed that the endophytic fungus *Gilmaniella* sp. AL12 can stimulate *Atractylodes lancea* to produce volatile oils. The active component from AL12 was identified as an extracellular mannan which ultimately promoted volatile oil accumulation (Chen *et al.* 2016). More metabolites from endophytes are yet to be exploited as elicitors, and these bioactive components could provide a promising alternative resource for medicinal plant protection and resilience.

Endophytes comprehensively demonstrate numerous medicinal plants' secondary metabolite biosynthesis signalling pathways. These metabolites are equally important to humanity; hence, endophytism and medicinal plant stress physiology offer promising and fascinating opportunities that can be exploited or manipulated to optimize or improve the production of valuable biologically active plant secondary metabolites under various conditions. This chapter outlines the biotic and abiotic stressors that affect secondary metabolites (quality and quantity) production and summarizes the beneficial effects of endophytes on their host of medicinal plants. The plant cell signalling

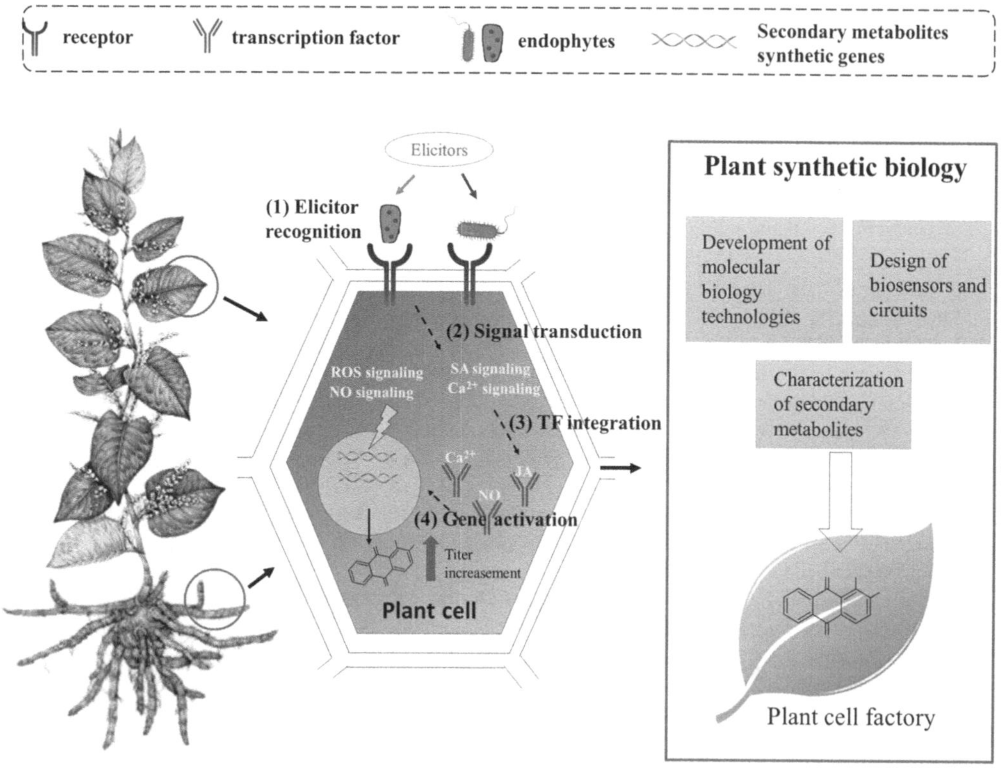

FIGURE 10.3 The endophyte-induced secondary metabolite synthesis mechanisms in host plant cells. (1) Recognition of plant endophytes elicitor; (2) signal transduction; (3) integration with transcription factors (TF); (4) gene activation; ROS, reactive oxygen species; JA, jasmonic acid; SA, salicylic acid; NO, nitric oxide.

pathways stimulated by endophytic action may lead to strategic improvements and biotechnology production of plant chemicals with medical and industrial utility. This information provides fundamental and new perspectives for future research and can be applied to obtain improved bioactive compound(s) from medicinal plants.

10.5 CONCLUSIONS AND FUTURE PERSPECTIVES

Medicinal plants in symbiotic association with endophytes based on mutual exploitation have more opportunities to survive in the environment. As critical components of the plant micro-ecosystem, endophytes significantly impact the production and accumulation of numerous secondary metabolites in the host plant. Their exploitation in agroecosystems as an eco-friendly and cost-effective alternative to agrochemical inputs may positively affect agricultural productivity and environmental sustainability. Numerous extensive studies have demonstrated endophytes' function in enhancing the accumulation of secondary metabolites in medicinal plants. Medicinal plants benefit extensively from their endophyte symbionts, and it is evident that their interaction plays significant roles in improving plant survival and adaptations. Technological applications of biostimulant-derived endophytes significantly enhance the quality of medicinal plants' growth, development, and secondary metabolite biosynthesis. Prospectively, the biotechnological application of functional plant endophytes serves as an adaptive strategy to enhance productivity, quality, and plant biosynthesis of secondary metabolites with health-promoting activities.

REFERENCES

Abdelshafy Mohamad OA, Ma JB, Liu YH, et al (2020) Beneficial endophytic bacterial populations associated with medicinal plant *Thymus vulgaris* alleviate salt stress and confer resistance to *Fusarium oxysporum*. *Front Plant Sci* 11:1–17. https://doi.org/10.3389/fpls.2020.00047

Aleynova OA, Suprun AR, Nityagovsky NN, et al (2021) The influence of the grapevine bacterial and fungal endophytes on biomass accumulation and stilbene production by the in vitro cultivated cells of *Vitis amurensis* Rupr. *Plants* 10:1–13. https://doi.org/10.3390/plants10071276

Amutha M (2021) In planta colonization of *Beauveria bassiana* in cotton plant and its effect against insect pests. *J Biol Control* 35:137–145. https://doi.org/10.18311/jbc/2021/28489

Ashrafi M, Azimi-Moqadam MR, Mohsenifard E, et al (2022) Physiological and molecular aspects of two *Thymus* species differently sensitive to drought stress. *BioTech* 11:1–11. https://doi.org/10.3390/biotech11020008

Bziuk N, Maccario L, Straube B, et al (2021) The treasure inside barley seeds: microbial diversity and plant beneficial bacteria. *Environ Microbiomes* 16:1–21. https://doi.org/10.1186/s40793-021-00389-8

Chen F, Ren CG, Zhou T, et al (2016) A novel exopolysaccharide elicitor from endophytic fungus *Gilmaniella* sp. AL12 on volatile oils accumulation in *Atractylodes lancea*. *Sci Rep* 6:1–17. https://doi.org/10.1038/srep34735

Chen Z, White JF, Malik K, et al (2022) Soil nutrient dynamics relate to *Epichloë* endophyte mutualism and nitrogen turnover in a low nitrogen environment. *Soil Biol Biochem* 174:108832. https://doi.org/10.1016/j.soilbio.2022.108832

Cherif-Silini H, Thissera B, Bouket AC, et al (2019) Durum wheat stress tolerance induced by endophyte *Pantoea agglomerans* with genes contributing to plant functions and secondary metabolite Arsenal. *Int J Mol Sci* 20. https://doi.org/10.3390/ijms20163989

Compant S, Cambon MC, Vacher C, et al (2021) The plant endosphere world – bacterial life within plants. *Environ Microbiol* 23:1812–1829. https://doi.org/10.1111/1462-2920.15240

Cordovez V, Dini-Andreote F, Carrión VJ, Raaijmakers JM (2019) Ecology and evolution of plant microbiomes. *Annu Rev Microbiol* 73:69–88. https://doi.org/10.1146/annurev-micro-090817-062524

Farsaraei S, Moghaddam M, Pirbalouti AG (2020) Changes in growth and essential oil composition of sweet basil in response to salinity stress and superabsorbents application. *Sci Hortic (Amsterdam)* 271:109465. https://doi.org/10.1016/j.scienta.2020.109465

Fesel PH, Zuccaro A (2016) Dissecting endophytic lifestyle along the parasitism/mutualism continuum in *Arabidopsis*. *Curr Opin Microbiol* 32:103–112. https://doi.org/10.1016/j.mib.2016.05.008

Fu Y (2019) Biotransformation of ginsenoside Rb1 to Gyp-XVII and minor ginsenoside Rg3 by endophytic bacterium *Flavobacterium* sp. GE 32 isolated from *Panax ginseng*. *Lett Appl Microbiol* 68:134–141. https://doi.org/10.1111/lam.13090

Galovičová L, Borotová P, Valková V, et al (2021) *Thymus vulgaris* essential oil and its biological activity. *Plants* 10:1–17. https://doi.org/10.3390/plants10091959

Gao FK, Dai CC, Liu XZ (2010) Mechanisms of fungal endophytes in plant protection against pathogens. *African J Microbiol Res* 4:1346–1351. https://doi.org/10.1111/j.0307-6946.2004.00642.x

Gao Y, Liu Q, Zang P, et al (2015) An endophytic bacterium isolated from *Panax ginseng* C.A. Meyer enhances growth, reduces morbidity, and stimulates ginsenoside biosynthesis. *Phytochem Lett* 11:132–138. https://doi.org/10.1016/j.phytol.2014.12.007

Glare T, Caradus J, Gelernter W, et al (2012) Have biopesticides come of age? *Trends Biotechnol* 30:250–258. https://doi.org/10.1016/j.tibtech.2012.01.003

Gupta S, Chaturvedi P, Kulkarni MG, Van Staden J (2020) A critical review on exploiting the pharmaceutical potential of plant endophytic fungi. *Biotechnol Adv* 39:107462. https://doi.org/10.1016/j.biotechadv.2019.107462

Harsonowati W, Marian M, Surono M, Narisawa K (2020) The effectiveness of a dark septate endophytic fungus, *Cladophialophora chaetospira* SK51, to mitigate strawberry fusarium wilt disease and with growth promotion activities. *Front Microbiol* 11:1–11. https://doi.org/10.3389/fmicb.2020.00585

Harsonowati W, Masrukhin, Narisawa K (2022) Prospecting the unpredicted potential traits of *Cladophialophora chaetospira* SK51 to alter photoperiodic flowering in strawberry, a perennial SD plant. *Sci Hortic (Amsterdam)* 295:110835. https://doi.org/10.1016/j.scienta.2021.110835

He Z, Cui K, Wang R, et al (2023) Multi-omics joint analysis reveals how *Streptomyces albidoflavus* OsiLf-2 assists *Camellia oleifera* to resist drought stress and improve fruit quality. *Front Microbiol* 14:1–12. https://doi.org/10.3389/fmicb.2023.1152632

Hill RA, Connolly JD (2020) Triterpenoids. *Nat Prod Rep* 37:962–998. https://doi.org/10.1039/c9np00067d

Hosseinzadeh S, Jafarikukhdan A, Hosseini A, Armand R (2015) The application of medicinal plants in traditional and modern medicine: A review of *ymus vulgaris*. *Int J Clin Med* 6:635–642. https://doi.org/10.4236/ijcm.2015.69084

Jin Z, Gao L, Zhang L, et al (2017) Antimicrobial activity of saponins produced by two novel endophytic fungi from *Panax notoginseng*. *Nat Prod Res* 31:2700–2703. https://doi.org/10.1080/14786419.2017.1292265

Karimmojeni H, Rezaei M, Tseng TM, Mastinu A (2022) Effects of metribuzin herbicide on some morpho-physiological characteristics of two *Echinacea* species. *Horticulturae* 8. https://doi.org/10.3390/horticulturae8020169

Kruasuwan W, Lohmaneeratana K, Munnoch JT, et al (2023) Transcriptome landscapes of salt-susceptible rice cultivar IR29 associated with a plant growth promoting endophytic *Streptomyces*. *Rice* 16. https://doi.org/10.1186/s12284-023-00622-7

Li Z, Wen W, Qin M, et al (2022) Biosynthetic mechanisms of secondary metabolites promoted by the interaction between endophytes and plant hosts. *Front Microbiol* 13: https://doi.org/10.3389/fmicb.2022.928967

Long F, Hu M-F, Chen S, et al (2023) Endophytic fungi regulate *HbNHX1* expression and ion balance in *Hordeum bogdanii* under alkaline stress. *J Fungi* 9:331. https://doi.org/10.3390/jof9030331

Macedo-Raygoza GM, Valdez-Salas B, Prado FM, et al (2019) *Enterobacter cloacae*, an endophyte that establishes a nutrient-transfer symbiosis with banana plants and protects against the black sigatoka pathogen. *Front Microbiol* 10:1–15. https://doi.org/10.3389/fmicb.2019.00804

Mahboubi M, Heidarytabar R, Mahdizadeh E, Hosseini H (2017) Antimicrobial activity and chemical composition of Thymus species ad Zataria multiflora essential oils. *Agric Nat Resour* 51:395–401. https://doi.org/10.1016/j.anres.2018.02.001

Mandal S, Upadhyay S, Wajid S, et al (2015) Arbuscular mycorrhiza increase artemisinin accumulation in *Artemisia annua* by higher expression of key biosynthesis genes via enhanced jasmonic acid levels. *Mycorrhiza* 25:345–357. https://doi.org/10.1007/s00572-014-0614-3

Marian M, Takashima Y, Harsonowati W, et al (2022) Biocontrol of Pythium root rot on lisianthus using a new dark septate endophytic fungus *Hyaloscypha variabilis* J1PC1. *Eur J Plant Pathol* 163:97–112. https://doi.org/10.1007/s10658-022-02459-0

Martínez-Arias C, Sobrino-Plata J, Medel D, et al (2021) Stem endophytes increase root development, photosynthesis, and survival of elm plantlets (*Ulmus minor* Mill.). *J Plant Physiol* 261. https://doi.org/10.1016/j.jplph.2021.153420

Moori S, Ahmadi-Lahijani MJ (2020) Hormopriming instigates defense mechanisms in Thyme (*Thymus vulgaris* L.) seeds under cadmium stress. *J Appl Res Med Aromat Plants* 19:100268. https://doi.org/10.1016/j.jarmap.2020.100268

Murck H (2020) Symptomatic protective action of glycyrrhizin (Licorice) in COVID-19 infection? *Front Immunol* 11:1–5. https://doi.org/10.3389/fimmu.2020.01239

Myers RJ, Fichman Y, Zandalinas SI, Mittler R (2023) Jasmonic acid and salicylic acid modulate systemic reactive oxygen species signaling during stress responses. *Plant Physiol* 191:862–873. https://doi.org/10.1093/plphys/kiac449

Nong Z, Ding B, Feng Y, et al (2015) Effects of mycorrhizal colonization and medicine quality of *Paris polyphylla* var yunnanensis inoculated by different foreign AM fungi species. *China J Chinese Mater Medica* 40:3158–3167.

Oviedo-Pereira DG, López-Meyer M, Evangelista-Lozano S, et al (2022) Enhanced specialized metabolite, trichome density, and biosynthetic gene expression in *Stevia rebaudiana* (Bertoni) Bertoni plants inoculated with endophytic bacteria *Enterobacter hormaechei*. *PeerJ* 10. https://doi.org/10.7717/peerj.13675

Pande A, Mun BG, Lee DS, et al (2021) NO network for plant-microbe communication underground: A review. *Front Plant Sci* 12:1–10. https://doi.org/10.3389/fpls.2021.658679

Popay AJ, Jensen JG, Simpson WR, et al (2023) Translocation of loline alkaloids in *Epichloë*-infected cereal and pasture grasses: What the insects tell us. *J Fungi* 9. https://doi.org/10.3390/jof9010096

Raza A, Salehi H, Rahman MA, et al (2022) Plant hormones and neurotransmitter interactions mediate antioxidant defenses under induced oxidative stress in plants. *Front Plant Sci* 13. https://doi.org/10.3389/fpls.2022.961872

Regaiolo A, Dominelli N, Andresen K, Heermann R (2020) The biocontrol agent and insect pathogen *Photorhabdus luminescens* interacts with plant roots. *Appl Environ Microbiol* 86. https://doi.org/10.1128/AEM.00891-20

Rho H, Van Epps V, Kim SH, Doty SL (2020) Endophytes increased fruit quality with higher soluble sugar production in Honeycrisp apple (*Malus pumila*). *Microorganisms* 8:1–17. https://doi.org/10.3390/microorganisms8050699

Rong ZY, Jiang DJ, Cao JL, et al (2022) Endophytic fungus *Serendipita indica* accelerates the ascorbate-glutathione cycle of white clover in response to water stress. *Front Microbiol* 13:1–9. https://doi.org/10.3389/fmicb.2022.967851

Rumyantsev SD, Alekseev VY, Sorokan AV, et al (2023) Additive effect of the composition of endophytic bacteria *Bacillus subtilis* on systemic resistance of wheat against greenbug aphid *Schizaphis graminum* due to lipopeptides. *Life* 13. https://doi.org/10.3390/life13010214

Saha T, Chandran N, Anu BC (2020) Major insect pests of vegetable crops in Bihar and their management. *Sustain Agric* 395–429.

Sharma G, Agarwal S, Verma K, et al (2023) Therapeutic compounds from medicinal plant endophytes: Molecular and metabolic adaptations. *J Appl Microbiol* 134:1–14. https://doi.org/10.1093/jambio/lxad074

Singh D, Thapa S, Mahawar H, et al (2022) *Prospecting potential of endophytes for modulation of biosynthesis of therapeutic bioactive secondary metabolites and plant growth promotion of medicinal and aromatic plants*. Springer International Publishing.

Singh P, Singh RK, Guo DJ, et al (2021) Whole genome analysis of sugarcane root-associated endophyte *Pseudomonas aeruginosa* B18 – a plant growth-promoting bacterium with antagonistic potential against *Sporisorium scitamineum*. *Front Microbiol* 12:1–21. https://doi.org/10.3389/fmicb.2021.628376

Subudhi E, Sahoo RK, Dey S, et al (2019) Unraveling plant-endophyte interactions: An omics insight. *Ref Ser Phytochem* 249–267. https://doi.org/10.1007/978-3-319-90484-9_2

Sun RT, Zhang ZZ, Feng XC, et al (2022) Endophytic fungi accelerate leaf physiological activity and resveratrol accumulation in *Polygonum cuspidatum* by up-regulating expression of associated genes. *Agronomy* 12:1–12. https://doi.org/10.3390/agronomy12051220

Tian CF, Garnerone AM, Mathieu-Demazière C, et al (2012) Plant-activated bacterial receptor adenylate cyclases modulate epidermal infection in the *Sinorhizobium meliloti-Medicago* symbiosis. *Proc Natl Acad Sci U S A* 109:6751–6756. https://doi.org/10.1073/pnas.1120260109

Tian N, Liu F, Wang P, et al (2017) The molecular basis of glandular trichome development and secondary metabolism in plants. *Plant Gene* 12:1–12. https://doi.org/10.1016/j.plgene.2017.05.010

Tidke SA, Kiran S, Giridhar P, Gokare RA (2019) Current understanding and future perspectives of endophytic microbes vis-a-vis production of secondary metabolites. In: Jha S (ed) *Endophytes and secondary metabolites*. Springer International Publishing, pp. 459–474.

USDA (2023) *Researchers helping protect crops from pests*. National Institute of Food and Agriculture. www.nifa.usda.gov/about-nifa/blogs/researchers-helping-protect-crops-pests. Accessed 11 Apr 2023.

Wang CF, Yang XQ, Sun J, et al (2022) New metabolites, antifeedant, insecticidal activities, and reciprocal relationship between insect and fungus from endophyte *Schizophyllum commune*. *Chem Biodivers* 19. https://doi.org/10.1002/cbdv.202200130

Wang T, Guo J, Peng Y, et al (2021) Light-induced mobile factors from shoots regulate rhizobium-triggered soybean root nodulation. *Science* 374(80):65–71. https://doi.org/10.1126/science.abh2890

WHO (1977) *World health promotion and development of training and research in traditional medicine*. https://apps.who.int/iris/bitstream/handle/10665/93212/WHA30.49_eng.pdf?sequence=1&isAllowed=y. Accessed 1 Apr 2023.

WHO (2013) WHO traditional medicine strategy 2014–2023. *World Heal Organ* 1–76. https://doi.org/2013

Xie J, Wu YY, Zhang TY, et al (2018a) New antimicrobial compounds produced by endophytic *Penicillium janthinellum* isolated from *Panax notoginseng* as potential inhibitors of FtsZ. *Fitoterapia* 131:35–43. https://doi.org/10.1016/j.fitote.2018.10.006

Xie W, Hao Z, Zhou X, et al (2018b) Arbuscular mycorrhiza facilitates the accumulation of glycyrrhizin and liquiritin in *Glycyrrhiza uralensis* under drought stress. *Mycorrhiza* 28:285–300. https://doi.org/10.1007/s00572-018-0827-y

Yang HR, Yuan J, Liu LH, et al (2019) Endophytic *Pseudomonas fluorescens* induced sesquiterpenoid accumulation mediated by gibberellic acid and jasmonic acid in *Atractylodes macrocephala* Koidz plantlets. *Plant Cell Tissue Organ Cult* 138:445–457. https://doi.org/10.1007/s11240-019-01640-4

Yerukala S, Butler DM, Bernard EC, et al (2022) Colonization efficacy of the endophytic insect-pathogenic fungus, *Beauveria bassiana*, across the plant kingdom: A meta-analysis. *CRC Crit Rev Plant Sci* 41:241–270. https://doi.org/10.1080/07352689.2022.2109287

Yin DD, Wang YL, Yang M, et al (2019) Analysis of Chuanxiong Rhizoma substrate on the production of ligustrazine in endophytic *Bacillus subtilis* by ultra-high-performance liquid chromatography with quadrupole time-of-flight mass spectrometry. *J Sep Sci* 42:3067–3076. https://doi.org/10.1002/jssc.201900030

Yuan J, Sun K, Deng-Wang MY, Dai CC (2016) The mechanism of ethylene signaling induced by endophytic fungus *Gilmaniella* sp. AL12 mediating sesquiterpenoids biosynthesis in *Atractylodes lancea. Front Plant Sci* 7:1–11. https://doi.org/10.3389/fpls.2016.00361

Zare M, Ganjeali A, Lahouti M (2021) Rosmarinic and caffeic acids contents in Basil (*Ocimum basilicum* L.) are altered by different levels of phosphorus and mycorrhiza inoculation under drought stress. *Acta Physiol Plant* 43:1–10. https://doi.org/10.1007/s11738-020-03191-0

11 Plant Growth-Promoting Activities for Bacterial and Fungal Endophytes Isolated from a Medicinal Plant, *Teucrium polium* L.

Shivani Gupta, Sumit Kumar, Gulshan Negi, and Sakshi Suchita

11.1 INTRODUCTION

A report by the World Health Organization (WHO) highlighted that approximately 80% of individuals in developing countries rely on conventional medicines, primarily organic remedies, for their primary health care needs (Farnsworth 1990). Moreover, in the modern field of pharmacology, approximately 25% of medications are developed from plants, with many others being chemical-based analogs based on plant-originated compounds. The demand for medicinal plants is increasing worldwide, both in developing and developed countries, due to the recognition of their natural products, which are known for their non-toxic properties, lack of side effects, and affordability. However, the global population of crucial medicinal plants is declining as a result of habitat loss, deforestation, and excessive harvesting. As a result, it is becoming increasingly impractical to directly extract natural compounds from wild medicinal plants to satisfy the present demand. Domestically grown medicinal plants were thought to be a feasible option; however, there are drawbacks, including inconsistent yields and different phytochemical profiles (Kala *et al.* 2006). To increase crop output, long-term plans involving the application of both organic and inorganic fertilizers as well as pesticides are frantically required. However, according to Aktar *et al.* (2009), these traditional methods harm soil quality and increase environmental contamination. To investigate alternate strategies, this chapter will concentrate on the interactions between plants and endophytic microbes in the medicinal herb *Teucrium polium.* By utilizing this symbiotic relationship, these interactions can be researched to produce desired or improved amounts of phytochemicals, or the microbes themselves can act as suppliers of these compounds. Growing interest has been shown in novel approaches based on microbial inoculation to counteract the drawbacks of current treatments. Plant–microbe symbiosis plays a vital role in promoting plant growth, improving agricultural traits, and enhancing soil quality and nutrient cycling (Puri *et al.* 2016).

Wild *Teucrium polium*, which belongs to the Lamiaceae family, is found in abundance in the dry areas of hills and deserts in practically all Mediterranean nations, as well as in southwestern Asia, Europe, and North Africa. The antimicrobial and antiseptic characteristics of this plant have long been utilized in medicine, and more recent research has isolated bioactive substances, including terpenoids and flavonoids, with potential pharmaceutical uses, including antimicrobial, antioxidant, and anticancer effects. The microbial endophytes associated with *Teucrium polium* have not been adequately studied, despite the ecological and therapeutic relevance of this plant in this particular setting. There are no published studies examining the microbial endophytes connected to this medicinal plant in its native dry habitat that promote the development of plants.

 DOI: 10.1201/b23155-11

11.2 ISOLATION OF BACTERIAL AND FUNGAL ENDOPHYTES FROM *TEUCRIUM POLIUM*

In Hassan's 2017 study, seven bacterial isolates of *Bacillus cereus* and *Bacillus subtilis*, as well as five fungal endophyte isolates of *Penicillium chryogenum* and *Penicillium crustosum*, were obtained and purified from the shoot system of *Teucrium polium* L. These bacterial and fungal endophytes possess plant growth-promoting (PGP) activities that directly or indirectly enhance plant growth. Among the seven bacterial and five fungal endophyte isolates, all the fungal endophytes were capable of producing ammonia and exhibited a higher capacity for solubilizing inorganic phosphate, while three bacterial endophytes (*Bacillus cereus* and *Bacillus subtilis*) were unable to produce ammonia. Additionally, plants inoculated with fungal endophytes displayed greater biomass production compared to those inoculated with bacteria. The direct promotion of plant growth by microbial endophytes involves the production of plant hormones, particularly IAA, or the mobilization of phosphorus. Indirect promotion of plant growth by endophytes includes antimicrobial activities, ammonia production, and the synthesis of degrading enzymes, which significantly inhibit pathogenic microorganisms.

11.3 AN OVERVIEW OF PLANT GROWTH-PROMOTING ENDOPHYTES

PGPEs reside within plant tissues and establish a close relationship that facilitates nutrient exchange and enzymatic activity (Khan *et al.* 2015). Endophytic microorganisms produce growth-promoting hormones that positively influence plant growth. Additionally, they can mobilize insoluble phosphate and supply nitrogen to host plants. By colonizing plant tissues without causing visible symptoms, microbial endophytes compete with other pathogens and contribute to the plant's defence against microbial diseases. Additionally, PGPEs create a variety of bioactive substances with the ability to stimulate plant development.

11.4 PLANT GROWTH-PROMOTING BACTERIAL ENDOPHYTES

The term 'plant growth-promoting bacterial endophytes' (PGPB) refers to bacteria that support plant growth and are often found in the soil's rhizosphere, the region around a plant's roots. According to Zhang *et al.* (2006) and Glick (2020), these bacteria typically penetrate the plant through fissures and lesions in the roots. Significantly, bacterial endophytes can thrive within plant tissues in a partnership with the plant that benefits both parties without harming or limiting the plant's ability to expand. Moreover, fungal endophytes are often found inside plant cells (intracellularly), and endophytic bacteria are typically found between plant cells (intercellularly). Additionally, even though endophytic PGPB are drawn to particular root exudates of a particular plant and enter the plant through the roots, many of these bacteria are mobile and can migrate through the plant to other tissues like leaves and stems, although they are typically present in lower concentrations than the roots (Sasse *et al.* 2018). Additionally, different bacterial species are drawn to diverse sets of tiny organic compounds released by various plant species and developmental phases (Dwibedi *et al.* 2022).

TABLE 11.1
List of Bacterial and Fungal Endophytes Isolated from the Medicinal Plant *Teucrium polium*

Bacterial and Fungal Endophytes	Isolated from	Phylum	References
Bacillus cereus	Leaves	Firmicutes	Hassan 2017
B. subtilis	Leaves	Firmicutes	Hassan 2017
Penicillium chrysogenum	Leaves	Ascomycota	Hassan 2017
P. crustosum	Leaves	Ascomycota	Hassan 2017
Lasiodiplodia theobromae	Leaves	Ascomycota	Balbool *et al.* 2018

11.5 MECHANISMS OF PGPB FOR DIRECT PLANT GROWTH PROMOTION

Bacterial endophytes employ a diverse array of mechanisms to directly enhance plant growth and development following their interaction with plants. Similar to rhizospheric plant growth-promoting bacteria, endophytic PGPB utilize a comparable set of mechanisms, if not identical. These mechanisms encompass various activities, including but not limited to:

1. Production of molecules, such as low-molecular-weight organic acids, that aid in the solubilization of inorganic phosphate and potassium.
2. Synthesis of siderophores, which are chelating agents that sequester iron from the soil and make it available to plants.
3. Synthesis of phytohormones, specifically gibberellins, and cytokinins, which regulate various developmental processes in plants.
4. Synthesis of auxins, notably indole 3-acetic acid (the most common auxin), which promote elongation and proliferation of plant cells.
5. Synthesis of trehalose, an unusual and highly stable sugar molecule that helps plants mitigate salt and drought stress by structuring water.
6. The ability to convert atmospheric nitrogen into ammonia, which is crucial for protein and nucleic acid synthesis.
7. Synthesis of the enzyme 1-aminocyclopropane-1-caroxylate (ACC) deaminase, which reduces plant ethylene levels, thereby alleviating the inhibitory effects of various abiotic stresses (Olanrewaju *et al.* 2017).

Genes involved in the manufacture of plant hormones, including auxin, cytokinin, and gibberellin, are frequently identified in the microbiome of endophytic communities, although they may not always be found in the same bacteria. All of these methods help plants grow and develop, but the production of ACC deaminase is arguably the most important one used by PGPB (Narayanan and Glick 2022).

11.6 MECHANISMS OF PGPB FOR INDIRECT PLANT GROWTH PROMOTION

Indirect plant growth promotion occurs when plant pathogens, usually fungi but also some bacteria, insects, and nematodes, are prevented from or have their inhibitory effects reduced by PGPB. Some endophytic PGPBs use biocontrol techniques to prevent the actions of different phytopathogens. These are some of the indirect mechanisms they employ:

1. Antibiotic synthesis
2. Manufacturing hydrogen cyanide
3. The expulsion of enzymes that hydrolyse fungal cell walls
4. The formation of siderophores, which inhibits the amount of iron available for the growth of phytopathogens
5. The emission of VOCs that prevent the growth of phytopathogens
6. Production of chemical substances that make target plants develop induced systemic resistance (ISR)
7. Synthesis of ACC deaminase, which reduces the level of growth-inhibiting stress ethylene in plants (Glick 2020)

11.7 MECHANISM OF ENDOPHYTIC FUNGI

Endophytic fungi reside within plant tissues and possess the capability to function as regulatory agents, either as biological agents or elicitors, in facilitating induced resistance and mitigating abiotic stresses. These endophytic fungi enhance the adaptability of plants through various modes of operation.

Typical mechanisms employed by endophytic fungi involve direct pathogen inhibition through competition, antibiosis, and mycoparasitism. Additionally, they also induce resistance indirectly, thereby activating the plant's defence system to combat diseases.

11.8 COMPETITION

Endophytes employ competition as a possible mechanism to hinder the colonization of pathogens in host tissues. This mechanism primarily exerts direct effects, making systemic colonization of the endophyte crucial for more effective antagonism against pathogens. A notable example is the application of a mixture of endophytes derived from cacao leaves onto the foliage, which resulted in a significant reduction in *Phytophthora* infection through competitive interactions (Arnold and Herre 2003).

11.9 INDUCTION OF RESISTANCE

When a plant is being attacked by pests or pathogens, endophytic fungi can activate the plant's defence mechanisms. Induced systemic resistance (ISR) or defence priming, a phenomenon that increases the plant's defence responses and makes it more capable of functioning, is a result of this activation event (Poveda *et al.* 2020). ISR is frequently regulated by ethylene or jasmonate without leading to an increase in pathogenicity-related (PR) proteins. The increase of PR proteins is the outcome of systemic acquired resistance (SAR), which is typically linked to pathogen infection and is mediated by salicylic acid (Tripathi *et al.* 2019). Invading pathogens are specifically targeted and destroyed by these PR proteins, which also include the enzymes chitinase and β-1,3-glucanase, strengthening the cell walls to fend off infection and cell death. However, endophytes may also increase genes involved in disease to induce systemic resistance. For instance, it has been discovered that the tomato plant pathogen *Fusarium solani* increases ISR against the foliar pathogen *Septorialyco persici* by activating PR genes in the roots, such as PR7 and PR5 (Kavroulakis *et al.* 2007).

11.10 MYCOPARASITISM

Endophytes use the essential tactic of mycoparasitism to protect the ecology of their hosts by directly attacking the pathogen or its reproductive organs. The pathogenic fungi's cell walls are damaged by the endophytic fungi's invasion of their hyphae, which results in cell lysis. These four stages can be used to break down this process. Recognizing the pathogenic fungus and encouraging chemotropic growth of the endophytic fungal mycelium are the first measures. The third and fourth steps then entail direct physical contact between the pathogenic and endophytic fungi, which causes the target fungi to be pierced and the cell walls to be degraded (Ram *et al.* 2018). Various species of *Trichoderma*, which are utilized in biocontrol, have been observed to penetrate the hyphae of the pathogenic fungus *Rhizoctonia solani*, leading to its demise.

11.11 ANTIBIOSIS

A phenomenon wherein endophytes produce a variety of secondary metabolites, some of which have antimicrobial activities, aiding in the prevention of plant pathogenic microbial growth. Numerous metabolites that have antibiotic properties have been produced by endophytes (Adetunji *et al.* 2019). However, it is urgently necessary to investigate the possibilities for the commercialization of secondary metabolites generated by endophytic fungal organisms. Endophytes and hosts have a symbiotic interaction that encourages the creation and release of metabolites with a variety of uses, including pathogen suppression. For instance, the *Talaromyces flavus* isolate Tfl, also known as *Penicillium dangeardii*, demonstrates the capacity to reduce *Verticillium* wilt in eggplants when grown in the field. Furthermore, this isolate has demonstrated the capability to suppress *Verticillium* wilt in potatoes as well (Naraghi *et al.* 2010).

11.12 PRODUCTION OF SECONDARY METABOLITES BY ENDOPHYTES

Endophytes have been proven to be important sources of secondary metabolites with distinctive structural properties and biological activity, including terpenoids, steroids, alkaloids, and derivatives of isocoumarin. For instance, the endophytic fungus *Taxomyces andreanae* can create Taxol, a strong anticancer medicine found in the bark of the yew tree (*Taxus brevifolia*). Yew trees are scarce and slowly growing; thus it takes a lot of labour to process the bark to get a small amount of the medicine. Taxol is scarce in yews, making up only 0.01–0.03% of the dry weight, which has an impact on its high market price. In addition, it has been discovered that various endophytic fungi, present in plants other than Taxus species, such as *Pestalotiopsis microspora*, *Periconia* sp., and *Colletotrichum gloeosporioides*, produce taxol (Gangadevi and Muthumary 2008).

Podophyllotoxin, a lignan produced by the fungal endophyte *Trametes hirsuta*, which was isolated from *Podophyllum* sp., has anticancer properties. Etoposide and teniposide, two derivatives of podophyllotoxin, are being utilized in chemotherapy for cancer (Puri *et al.* 2006). Similar to camptothecin, a phytochemical with anticancer effects, fungus obtained from the inner bark of *Nothapodytes foetida* (Wight) Sleumer's.

Although there have been several reports highlighting the identification of novel endophytic bacteria, the complex chemical and biological mechanisms underlying endophytic processes are still poorly understood. It is becoming clearer that host specialization is important in the endophytic interactions between higher plants and microbes (Bacon and White 2000). Understanding these interactions can help in the selection of particular endophytes for the investigation of new natural medicines. Endophytes may aid in plant defence by creating antimicrobial substances, some of which may be useful medicinally due to their biological actions against fungi, bacteria, and other diseases.

11.13 HIDDEN TREASURES OF PLANT ENDOSPHERE IN *TEUCRIUM POLIUM* L.

Microbial endophytes produce plant hormones, mainly indole-3-acetic acid (IAA), and mobilize phosphorus, which both directly contribute to the promotion of plant growth. They also play a role in indirect plant growth promotion through processes like ammonia generation, antibacterial actions, and the creation of enzymes that efficiently suppress harmful microbes. Beyond controlling plant growth, microbial IAA plays a role in the symbiotic relationship between plants and microbes. The greatest IAA concentrations were found in the bacterial endophyte Tp.1B and the fungal endophyte Tp.5F, which had ranges of 19.9 to 63.5 mg mL^{-1} and 4.1 to 23.4 mg mL^{-1}, respectively. This shows that these endophytes can promote the growth of roots and the overall health of plants. The main phenotypic trait has been found as root development control, which is mediated by IAA. Notably, IAA plays a crucial role in promoting plant root formation and stimulating cell division, even under harsh environmental conditions.

Enzymatic and antibacterial functions are used by endophytes as a covert technique for promoting plant development. The research's conclusions showed that all isolated endophytes produced enzymes such as cellulose and protease. By aiding in hydrolysis, these enzymes enable endophytes to enter plant tissues and form symbiotic relationships with their host plants. Endophytes improve plant nutrition by acquiring nutrients from plants and secreting enzymes. By dissolving proteins and polysaccharides, these extracellular enzymatic activities also aid in the senescence of plants.

11.14 CONCLUSION AND FUTURE PROSPECTS

The cultivation of medicinal plants necessitates interdisciplinary research in rhizosphere biology, microbiology, ecology, and agricultural technology. These studies aim to develop efficient approaches for biomass production and the extraction of high-quality plant material enriched with phytochemicals. By prospecting and producing phytochemicals from microorganisms, it is possible to prevent the undesired destruction of medicinal plants, thus preserving the green cover. Moreover, this approach enables the large-scale production of products at economically favourable prices.

REFERENCES

Adetunji, C.O., D. Kumar, M. Raina, O. Arogundade, and N.B. Sarin. 2019. Endophytic microorganisms as biological control agents for plant pathogens: a panacea for sustainable agriculture. *Plant Biotic Interactions: State of the Art* 1–20.

Aktar, M.W., D. Sengupta, and A. Chowdhury. 2009. Impact of pesticides use in agriculture: their benefits and hazards. *Interdisciplinary Toxicology* 2(1): 1–12.

Arnold, A.E., and E.A. Herre. 2003. Canopy cover and leaf age affect colonization by tropical fungal endophytes: ecological pattern and process in *Theobroma cacao* (Malvaceae). *Mycologia* 95(3): 388–398.

Bacon, C.W., and J.F. White. 2000. *Microbial endophytes.* Marcel Deker Inc., New York.

Balbool, B.A., A.M. Abdel-Azeem, M.H. Moubasher, and E.A. Helmy. 2018. Production of L-asparaginase (L-ASN) from endophytic *Lasiodiplodia theobromae* hosted *Teucrium polium* in Egypt. *Microbial Biosystems* 3(2): 46–55.

Dwibedi, V., S.K. Rath, M. Joshi, R. Kaur, G. Kaur, D. Singh, G. Kaur, and S. Kaur. 2022. Microbial endophytes: application towards sustainable agriculture and food security. *Applied Microbiology and Biotechnology* 106(17): 5359–5384.

Farnsworth, N.R. 1990. The role of ethnopharmacology in drug development. *Ciba Foundation Symposium* 154: 2–11.

Gangadevi, V., and J. Muthumary. 2008. Taxol, an anticancer drug produced by an endophytic fungus *Bartalinia robillardoides* Tassi, isolated from a medicinal plant, *Aegle marmelos* Correa ex Roxb. *World Journal of Microbiology and Biotechnology* 24: 717–724.

Glick, B.R. 2020. *Beneficial plant-bacterial interactions*, 2nd ed. Springer, Berlin/Heidelberg, Germany, 383–398.

Hassan, S.E.D. 2017. Plant growth-promoting activities for bacterial and fungal endophytes isolated from a medicinal plant of *Teucrium polium* L. *Journal of Advanced Research* 8(6): 687–695.

Kala, C.P., P.P. Dhyani, and B.S. Sajwan. 2006. Developing the medicinal plants sector in northern India: challenges and opportunities. *Journal of Ethnobiology and Ethnomedicine* 2: 1–15.

Kavroulakis, N., S. Ntougias, G.I. Zervakis, C. Ehaliotis, K. Haralampidis, and K.K. Papadopoulou. 2007. Role of ethylene in the protection of tomato plants against soil-borne fungal pathogens conferred by an endophytic *Fusarium solani* strain. *Journal of Experimental Botany* 58(14): 3853–3864.

Khan, A.R., I. Ullah, M. Waqas, R. Shahzad, S.J. Hong, G.S. Park, B.K. Jung, I.J. Lee, and J.H. Shin. 2015. Plant growth-promoting potential of endophytic fungi isolated from *Solanum nigrum* leaves. *World Journal of Microbiology and Biotechnology* 31: 1461–1466.

Naraghi, L., A. Heydari, S. Rezaee, M. Razavi, and H. Jahanifar. 2010. Study on antagonistic effects of *Talaromyces flavus* on *Verticillium albo-atrum*, the causal agent of potato wilt disease. *Crop Protection* 29(7): 658–662.

Narayanan, Z., and B.R. Glick. 2022. Secondary metabolites produced by plant growth-promoting bacterial endophytes. *Microorganisms* 10(10): 2008.

Olanrewaju, O.S., B.R. Glick, and O.O. Babalola. 2017. Mechanisms of action of plant growth promoting bacteria. *World Journal of Microbiology and Biotechnology* 33: 1–16.

Poveda, J., P. Abril-Urias, and C. Escobar. 2020. Biological control of plant-parasitic nematodes by filamentous fungi inducers of resistance: *Trichoderma*, mycorrhizal and endophytic fungi. *Frontiers in Microbiology* 11: 992.

Puri, A., K.P. Padda, and C.P. Chanway. 2016. Seedling growth promotion and nitrogen fixation by a bacterial endophyte *Paenibacillus polymyxa* P2b-2R and its GFP derivative in corn in a long-term trial. *Symbiosis* 69(2): 123–129.

Puri, S.C., A. Nazir, R. Chawla, R. Arora, S. Riyaz-ul-Hasan, T. Amna, B. Ahmed, V. Verma, S. Singh, R. Sagar, and A. Sharma. 2006. The endophytic fungus *Trametes hirsuta* is a novel alternative source of podophyllotoxin and related aryl tetralin lignans. *Journal of Biotechnology* 122(4): 494–510.

Ram, R.M., C. Keswani, K. Bisen, R. Tripathi, S.P. Singh, and H.B. Singh. 2018. Biocontrol technology: eco-friendly approaches for sustainable agriculture. In *Omics technologies and bio-engineering*. Academic Press, 177–190.

Sasse, J., E. Martinoia, and T. Northen. 2018. Feed your friends: do plant exudates shape the root microbiome? *Trends in Plant Science* 23(1): 25–41.

Tripathi, D., G. Raikhy, and D. Kumar. 2019. Chemical elicitors of systemic acquired resistance – salicylic acid and its functional analogs. *Current Plant Biology* 17: 48–59.

Zhang, H.W., Y.C. Song, and R.X. Tan. 2006. Biology and chemistry of endophytes. *Natural Product Reports* 23(5): 753–771.

12 Endophytic Microbes and Their Inductive Role in Abiotic Stress Tolerance in Medicinal Plants

Bhavna Nigam, Dheeraj Rathore, and Indra Jeet Chaudhary

12.1 INTRODUCTION

Medicinal plants are useful flora because many of their parts consist of chemically active constituents that cause a physiological reaction when used to cure different illnesses. Medicinal plant use has been practiced in countries such as Sri Lanka, Thailand, China, Nepal, India, Japan, and Pakistan since ancient times (Astutik *et al.*, 2019; Basu *et al.*, 2023). The earliest evidence of medicinal plant use comes from Mesopotamian civilizations and dates back to 2600 BC (Ahirwar, 2023). Plants with medicinal qualities have been used in Babylonian, Assyrian, Greek, Chinese, and Hebrew civilizations throughout history (Schulz, 2023). A few medicinal plants must now be cultivated on a commercial level to satisfy the pharmaceutical industry's ever-growing demands (Jeelani *et al.*, 2018).

Environmental factors such as temperature, moisture, illumination, soil conditions, and the availability and identity of soil fauna all have a significant impact on the quality and yield of medicinal plants (Namdar *et al.*, 2019). It is becoming increasingly clear that medicinal plants can be heavily affected by their interactions with particular bacterial endophytes (Ek-Ramos *et al.*, 2019). Long-term, symbiotic relationships between endophytes and host plants can promote plant growth and are particularly beneficial in agricultural practice (Compant *et al.*, 2010). Extreme conditions (below or above the optimal levels) limit plant growth and development. The community structure of endophytic bacteria is affected by multiple abiotic factors, including drought, low or high temperature, salt stress and acidic conditions, heavy metal stress, nutrient stress, and starvation, which harm medicinal plants (Chiellini *et al.*, 2014; Yang *et al.*, 2017). Biotic stress results from damage done to plants by living organisms, such as bacteria, viruses, fungi, parasites, harmful insects, weeds, and cultivated or native plants. Therefore, this chapter focuses on the development of applications of endophytic microbes for plant growth under stress conditions.

12.2 EFFECTS OF BIOTIC AND ABIOTIC STRESSES ON MEDICINAL PLANTS

Biotic stresses mainly result in disease in plants. Microorganisms can cause plant wilt, leaf spots, root rot, seed damage, and many more problems. Insects can cause severe physical damage to plants, including to the leaves, stems, bark, and flowers. Insects can also act as vectors of viruses and bacteria from infected plants to healthy plants (Eigenbrode *et al.*, 2018). Invasive plants through allelopathy cause severe damage to native plants. Plants are vulnerable to several abiotic stresses as the conditions change, including drought, salinity, heavy metal toxicity, temperature (high and low), high light intensity, low light intensity, UVB radiation, nutrient deficiency, and ozone. Abiotic stresses are linked to decreases in yield and quality of crops (Chaudhary and Rathore, 2021a, 2021b, 2021c; Kale *et al.*, 2024). Some medicinal plants, their economic importance, and their abiotic

DOI: 10.1201/b23155-12

TABLE 12.1
Medicinal Plants under Abiotic Stress

Medicinal Plant	Uses	Abiotic Stress	References
Dioscorea bulbifera	Antispasmodic, analgesic, aphrodisiac, and diuretic	Metal	Narula *et al.* (2005)
Pluchea lanceolata	Bronchitis, dyspepsia, and rheumatoid arthritis	Heavy metals	Kumar *et al.* (2004)
Jatropha curcas	Skin diseases and rheumatism, piles	Salinity	Gao *et al.* (2008)
Melissa officinalis L	Insomnia and anxiety	Salinity and water	Ozturk *et al.* (2004)
Ocimum sp.	Cancers, antifertility, and adaptogenic	Drought	Khalid (2006)
Orthosiphon stamineus	Antiallergenic, antihypertensive, anti-inflammatory, and nephritis	Salinity	Ting *et al.* (2009)
Thymus maroccanus Ball.	Antitussive, antiseptic, antispasmodic, and anti-helminthic	Salinity	Belaqziz *et al.* (2009)
Populus euphratica	Anodyne, anti-inflammatory, febrifuge, and vermifuge	Salinity	Zhang *et al.* (2004)
Matricaria chamomilla	Sore stomach, irritable bowel syndrome, and oral mucositis	Salinity	Razmjoo *et al.* (2008)
Matricaria chamomilla	Mucositis and irritable bowel syndrome	$CuCl_2$	Eliasova *et al.* (2004)
Bacopa monnieri	Antioxidants, epilepsy, loss of memory, and asthma	Salinity and drought	Debnath (2008)
Datura innoxia Mill	Anodyne, antispasmodic, hallucinogenic, hypnotic, and narcotic	Light, dark, and HCHO deprivation	Hashem *et al.* (2022)
Ziziphora clinopodioides	Antibacterial, sedative, stomach ache, carminative, antiseptic, and wound healing	Salinity and defoliation	Koocheki *et al.* (2008)
Dioscorea dregeana	Sedative and anti-inflammation	Smoke, temperature	Debnath *et al.* (2011)

stresses are listed in Table 12.1. Abiotic stresses reduce crop productivity and are identified to be serious constraints to the world's food supply (Chaudhary and Rathore, 2021a, 2021b, 2021c; Kale *et al.*, 2024). Among the major abiotic stresses that reduce global growth are waterlogging, drought, soil salinity, and soil toxicity with heavy metals (Singh *et al.*, 2019; Kale *et al.*, 2024). Environmental and abiotic stresses cause a variety of physiological changes as well as changes in plant chemical properties (Chaudhary and Rathore, 2022). Abiotic stress is known to have a strong adverse impact on plant medicinal qualities (Terletskaya *et al.*, 2021). Plant responses to various types of stresses differ. Water stress caused by mannitol has been shown to decrease growth in plants, leading to a higher root-to-shoot ratio (Mechri *et al.*, 2020). Sodium chloride helps improve the response of water-stressed plants to mannitol-induced water stress (Gómez *et al.*, 2020). The application of sodium chloride increases the relative water content and growth of seedlings (which is lowered by water stress) (Waśkiewicz *et al.*, 2013; Kumar *et al.*, 2018). Salt stress has been shown to significantly reduce plant development, gas exchange, photochemical quenching, and potassium ion content in shoots and roots. Salinity is also linked to higher levels of leaf glycine betaine, free proline, and sodium ion content.

Research on medicinal plants discovered that as drought stress increases, biomass and seed yields drastically decrease. Reduced seed yield results in lower essential oil yield. Extreme drought stress results in a rise in total phenolic amounts and antioxidant enzymes (Saeidnejad *et al.*, 2013). Moderate and mild salt stresses cause a 36% and 76% upsurge in secondary metabolite content in *Bacopa monnieri*, respectively. Bacoside A (a saponin) is a significant secondary metabolite of *Bacopa monnieri* with numerous medicinal properties. It has been concluded that salt treatment increases biomass yield and saponin content in *B. monnieri* (Bharti *et al.*, 2013). When plants are subjected to environmental stress, endogenous melatonin is over-expressed (Zhang *et al.*, 2015). The toxicity of heavy metals in medicinal plants reduces plant growth and the amount of chlorophyll in the leaves. Elevated cadmium concentrations are linked to overall damage to leaf structure and cell membrane-cell wall fusion (Ali *et al.*, 2013; Cui *et al.*, 2013).

12.3 ENDOPHYTIC MICROBES

The term endophyte is derived from two Greek words: 'endoendon' meaning within, and 'phyte-phyton' meaning plant. This term applies to the organisms living within plants. Endophytes are microorganisms residing inside healthy plant tissues without causing any apparent harm to the host (Fadiji and Babalola, 2020). The term endophyte was first used by De Bary in 1866 to denote all the microbes residing inside living healthy plants. A few years before, Wani and Hakeem (2022) had recognized the fungi in wheat leaves and called them endophytic fungi. Kandasamy and Kathirvel (2023) proposed that endophytes are fungi or bacteria, which, for all or part of their life cycle, invade the tissues of living plants and cause unapparent and asymptomatic infections entirely within plant tissues but cause no symptoms of disease. Based on the nature of pathogenicity, endophytes may be of three types: (1) pathogens of another host that are non-pathogenic in their endophytic relationship, (2) non-pathogenic microbes, and (3) pathogens that have been rendered non-pathogenic but are still capable of colonization by selection methods or genetic alteration (Vaishnavi and Kumar, 2021; Collinge *et al.*, 2022).

12.3.1 Sources of Endophytic Colonization

The source of endophytic colonization ranges from transmission via seeds and vegetative planting material to entrance from the surrounding environment such as the rhizosphere and phyllosphere (Lata *et al.*, 2018). A study demonstrated bacterial transmission via vegetative planting material for *Acetobacter diazotrophicus* for two successive cuttings of sugarcane, while the occurrence of endophytic bacteria in healthy seeds has been reported for some plants (Srebot *et al.*, 2021). High similarity between microbes found internally in root tissue with those found in the rhizosphere has indicated that the rhizosphere may be one of the main sources of endophytic colonization.

12.3.2 Establishment of Endophytes in Medicinal Plant Tissues

Endophytes follow the same pattern as pathogenic microorganisms to enter the plants. The expression of defence-related molecules in the host is found to be less at the time of entry of endophytes than pathogens. Except for some endophytes transmitted by seed or vegetative propagation material, successful endophytic colonization goes through several important stages, including host finding, recognition, colonization of the plant surface, and entrance into the internal plant tissue. Endophytic microbes enter plants through natural openings such as hydathodes, stomata, and lenticels; wounds caused by abrasion with soil particles; pathogen damage; and formation of lateral roots, micropores, and abiotic mechanical damage (Table 12.2). As a consequence of

TABLE 12.2
Colonization Rates of Endophytic Fungi from Medicinal Plants

Plants and Their Parts	Colonization Rate	References
Rhazya stricta, stem	60.00	
Moringa peregrine, stem	62.50	Khan *et al.*, 2017
Caralluma acutangula, stem	21.50	
Rhazya stricta, leaves	100.0	
Moringa peregrine, leaves	0	
Caralluma acutangula, leaves	0	
Rhazya stricta, roots	37.50	
Moringa peregrine, roots	25.00	
Caralluma acutangula, roots	0	

Rhizoctonia damage of *Phaseolus vulgaris*, the two endophytes *Enterobacter asburiae* JM22 and *Pseudomonas fluorescens* 89 B-27 extensively colonized fungal necrotic tissue (Verma *et al.*, 2019). To penetrate the cuticle and cell walls, microbes produce cell wall-degrading enzymes such as cutinase, pectinase, cellulase, hemicellulase, protease, and lignin-peroxidases to enter the plant (Gond *et al.*, 2021). The expression of a proteinase was identified and purified during infection of endophytic *Acremonium typhinum* to the grass Poaampla (Lata *et al.*, 2018). This proteinase was abundant within fungal membrane vesicles and in the plant and/or fungal cell walls at the time of infection but absent in fungal pure culture. In the case of root endophytes, *Piriformospora indica* interferes with the host cell death program to form a mutualistic symbiosis with plants (Sarkar *et al.*, 2021). The already established plant endophytes may transmit from one generation to another generation via plant seeds. The rhizosphere bacteria may enter and establish as root endophytes through the emergence of lateral roots or root hair cells, primary and lateral root cracks, and diverse tissue wounds occurring as a result of plant growth (Verma *et al.*, 2021). The genotype of the host and fungus is an important factor in determining the establishment of symbiotic relations. A PCR-based genetic analysis of *Venturiaditricha*, a fungal endophyte of birch trees, showed that the infection frequency by a particular endophyte genotype is influenced by the host genotype (Griffiths *et al.*, 2020).

12.4 ROLE OF ENDOPHYTES IN MEDICINAL PLANTS UNDER STRESS CONDITIONS

The symbiotic interactions between plant and endophyte may result in several outcomes, as defined by fitness benefits to each of the partners (Hawkes *et al.*, 2020). The effects on host plants can be positive (mutualism), neutral (commensalism and neutralism), or negative (parasitism, competition, and amensalism). Variations in the outside environment put the plant metabolism out of homeostasis, which creates the necessity for the plant to harbour advanced genetic and metabolic mechanisms within its cellular system (Gill and Tuteja, 2010). Herein, the importance of microbes, especially the endophytes, increases immensely. Endophytic microbes aid in plant health by deterring herbivory and pathogenesis while also facilitating plant growth through nutrient uptake (modification of root morphology, alteration of nitrogen accumulation and metabolism), water use efficiency (osmotic adjustment, stomatal regulation), and curtailing of environmental stresses (White *et al.*, 2019; Mathur and Roy, 2021). The endophytes, in return, obtain access to the host plant's nutrients and disseminate them to the next generation. Endophytes can endorse secondary metabolite gathering in plants. The close association can progress host plant tolerance to the abiotic stresses of soil salinity, drought, and risky temperatures. Their symbiosis also sheds light on plant growth and active compound construction (Figure 12.1).

12.4.1 Stress Tolerance in Medicinal Plants by Endophytes

Symbiotically conferred abiotic stress tolerance involves at least two mechanisms: (1) activation of host stress response systems soon after exposure to stress, allowing the plants to avoid or mitigate the impacts of the stress, and (2) biosynthesis of antistrips biochemical by endophytes (El Sabagh *et al.*, 2022). Studies conducted on *Arthrobacter* sp. and *Bacillus* sp. isolated from pepper plants showed a significant reduction in upregulation and even downregulation of some stress-inducible genes when compared with gene expression in inoculated plants. *Phomaglomerata* and *Penicillium* sp. significantly increased plant biomass; related growth parameters; and assimilation of essential nutrients, such as potassium, calcium, and magnesium, and reduced the sodium toxicity in cucumber plants under sodium chloride and polyethylene glycol-induced salinity and drought stress when compared with control plants (Waqas *et al.*, 2012).

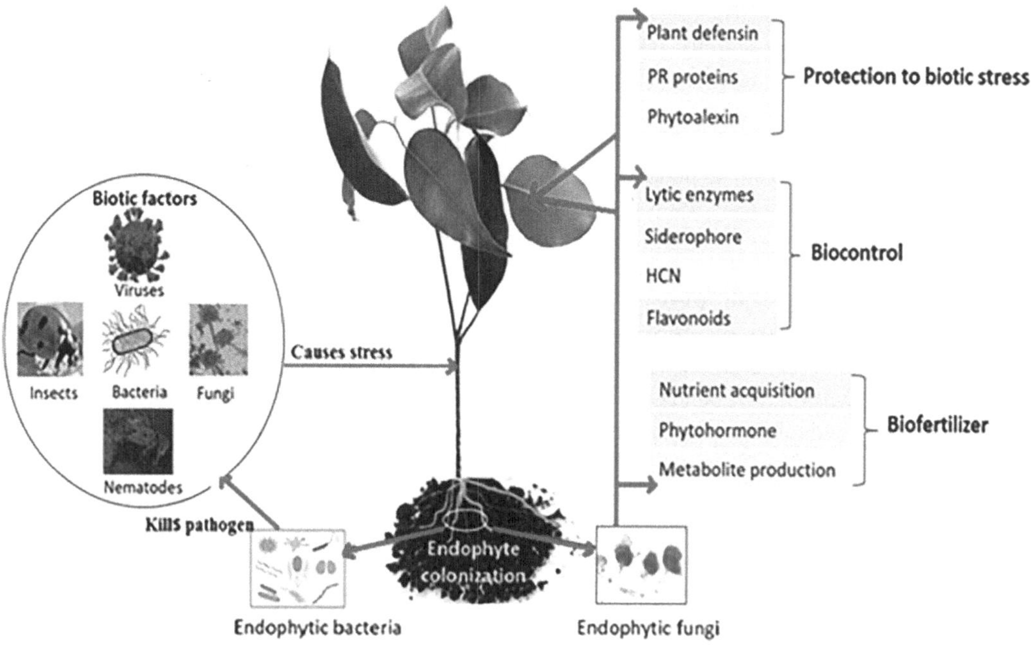

FIGURE 12.1 Role of endophytes in plants.

12.4.1.1 Water-Use Efficiency

Water-use efficiency (WUE) is an important trait in plants, including medicinal plants, as it directly relates to their ability to efficiently utilize water for growth and survival (Weerarathne, 2021). Endophytes are microorganisms that live inside plant tissues without causing apparent harm to the host. They can have beneficial effects on the plants, including enhancing their tolerance to various environmental stresses, including water stress. Endophytes can play a significant role in improving water-use efficiency in medicinal plants through several mechanisms (Fadiji and Babalola, 2020):

- Drought tolerance: Some endophytes have been found to promote drought tolerance in plants. They can enhance the water status of the host plant by facilitating water uptake, reducing transpiration rates, and improving the plant's ability to retain water.
- Stomatal regulation: Endophytes can modulate the stomatal behaviour of plants, which is crucial for regulating water loss through transpiration. They can help plants maintain proper stomatal closure under water-deficit conditions, thereby reducing water loss and improving water-use efficiency.
- Phytohormone production: Certain endophytes can produce phytohormones such as abscisic acid (ABA), which plays a key role in plant responses to water stress. ABA regulates stomatal closure and promotes the synthesis of osmoprotectants, enabling plants to cope with drought conditions.
- Nutrient acquisition: Endophytes can enhance nutrient uptake in plants, which indirectly affects water-use efficiency. Improved nutrient acquisition can lead to healthier plants with optimized physiological processes, including water uptake and utilization.
- Antioxidant activity: Water stress often leads to the generation of reactive oxygen species (ROS) in plants, causing oxidative damage. Some endophytes produce antioxidant enzymes and compounds that scavenge ROS, reducing cellular damage and improving plant survival under water-limited conditions.

By enhancing water-use efficiency and drought tolerance, endophytes can help medicinal plants withstand periods of water scarcity, which can be especially crucial in arid and semi-arid regions or during drought events (Hanaka *et al.*, 2021). This can result in improved growth, productivity, and overall health of medicinal plants, ensuring a reliable supply of high-quality plant-based medicinal products (Mofokeng *et al.*, 2022). However, it's important to note that the effects of endophytes on water-use efficiency can vary depending on the specific endophyte–plant interaction and environmental conditions. Further research is needed to explore and harness the potential of endophytes for improving water-use efficiency in medicinal plants.

12.4.1.2 Osmotic Stress Responses

Osmotic stress occurs when plants are exposed to high concentrations of solutes in the surrounding soil or when they experience water deficit conditions (Ozturk *et al.*, 2021). This can lead to water imbalances within the plant cells, affecting their normal physiological processes and overall health. However, endophytes play a crucial role in helping medicinal plants cope with osmotic stress (Ogbe *et al.*, 2020). Some endophytes promote water uptake and absorption by increasing the root surface area or secreting enzymes that improve soil water retention (Liu *et al.*, 2021). This helps the plants maintain their water balance even under osmotic stress conditions. Endophytes can synthesize compatible solutes or osmolytes, such as proline, glycine betaine, or trehalose, which accumulate in plant cells and help maintain osmotic balance (Kamran *et al.*, 2022). These osmolytes act as osmoprotectants, protecting cellular structures and macromolecules from damage caused by osmotic stress. Endophytes can produce phytohormones such as abscisic acid, which plays a crucial role in regulating plant responses to osmotic stress (Phurailatpam and Mishra, 2020). ABA helps plants close stomata, reduce water loss through transpiration, and promote the synthesis of stress-related proteins. Osmotic stress can lead to the generation of reactive oxygen species within plant cells, causing oxidative damage (Chaudhary and Rathore, 2018, 2019, 2020). Some endophytes produce antioxidant enzymes or compounds, such as superoxide dismutase, catalase, or flavonoids, which scavenge ROS and protect plants from oxidative stress (Sahu *et al.*, 2022). Endophytes can enhance nutrient uptake and assimilation in plants, enabling them to acquire essential minerals required for osmotic stress tolerance (Chandran *et al.*, 2020). For example, certain endophytes can solubilize phosphate or fix atmospheric nitrogen, providing additional nutrients to the host plant. Endophytes can stimulate the plant's defence mechanisms and induce systemic resistance against osmotic stress. They can trigger the production of defence-related compounds, activate stress-responsive genes, or prime the plant's immune system for better stress tolerance (Lu *et al.*, 2021).

Overall, endophytes play a multifaceted role in enhancing osmotic stress tolerance in medicinal plants. Their presence and interactions with the host plants can significantly improve water uptake, osmotic balance, antioxidant defences, and nutrient acquisition, ultimately leading to improved plant growth and survival under osmotic stress conditions.

12.4.1.3 Nutrient Uptake and Recycling

Endophytes form a mutualistic relationship with plants, providing various benefits, including nutrient uptake and recycling (Chandran *et al.*, 2020; Vishwakarma *et al.*, 2021). When it comes to medicinal plants, endophytes can play a significant role in enhancing nutrient availability and improving plant growth and medicinal compound production. Here's how nutrient uptake and recycling by endophytes occur in medicinal plants:

- **Nitrogen fixation:** Some endophytic bacteria can fix atmospheric nitrogen into a form that plants can use. Nitrogen fixation is an essential process for plants as nitrogen is a crucial nutrient required for their growth and development. Endophytic bacteria such as species of *Rhizobium*, *Azotobacter*, and *Azospirillum* can form symbiotic associations with medicinal plants and provide them with a source of fixed nitrogen (Vaghela and Gohel, 2023).

- **Phosphate solubilization:** Endophytic bacteria can solubilize insoluble forms of phosphorus in the soil, making it available for uptake by plants (Rawat *et al.*, 2021). These bacteria produce organic acids and enzymes that break down the complex phosphorus compounds into simpler forms that plants can utilize. By enhancing phosphate solubilization, endophytes improve the availability of phosphorus, an essential nutrient for plant growth and development.
- **Mineral uptake and transport:** Endophytes can enhance the uptake and transport of essential minerals such as iron, zinc, and manganese in medicinal plants (Singh *et al.*, 2018). These microorganisms can solubilize minerals from the soil and make them accessible to plants by producing organic acids and chelating compounds. They can also facilitate the movement of minerals within the plant tissues, ensuring their efficient distribution to different parts of the plant (Vishwakarma *et al.*, 2021).
- **Organic matter decomposition:** Endophytes contribute to the decomposition of organic matter within plant tissues, releasing essential nutrients (Johnston-Monje and Raizada, 2011). They secrete enzymes that break down complex organic compounds, making them available for plant uptake. This process helps in recycling nutrients and ensures their availability for ongoing plant growth and development.
- **Hormone production:** Some endophytic microorganisms can produce plant growth-promoting hormones such as auxins, cytokinins, and gibberellins (Ismail *et al.*, 2021). These hormones play a crucial role in regulating various plant processes, including nutrient uptake and utilization. By producing these hormones, endophytes can stimulate plant growth and enhance nutrient assimilation in medicinal plants.

Overall, endophytes contribute to nutrient uptake and recycling in medicinal plants by fixing atmospheric nitrogen, solubilizing phosphorus, facilitating mineral uptake and transport, decomposing organic matter, and producing plant growth-promoting hormones. These interactions between endophytes and medicinal plants promote plant growth, improve nutrient availability, and potentially enhance the production of medicinal compounds.

12.4.1.3 Salt Tolerance

Endophytes have been found to play a significant role in enhancing the salt tolerance of medicinal plants (Abdelshafy Mohamad *et al.*, 2020). High salt concentrations in the soil can disrupt the ion balance within plant cells and lead to water stress (Nigam *et al.*, 2022). Endophytes can produce enzymes or organic compounds that help maintain the ion balance and regulate osmotic adjustment in plants, allowing them to tolerate higher salt levels (Gupta *et al.*, 2021). Endophytes can produce phytohormones such as indole-3-acetic acid (IAA), cytokinins, and gibberellins, which regulate plant growth and development (Ismail *et al.*, 2021). These hormones can help plants cope with salt stress by promoting root growth, nutrient uptake, and overall plant vigour. Salt stress can lead to the accumulation of reactive oxygen species, which cause oxidative damage to plant cells (Nigam *et al.*, 2022). Endophytes can produce enzymes such as superoxide dismutase, catalase, and peroxidase that scavenge these ROS, reducing oxidative stress and enhancing salt tolerance (Siddiqui *et al.*, 2022). Salt stress can impair the uptake of essential nutrients by plants. Some endophytes can solubilize minerals and make them available to the host plant, thereby enhancing nutrient uptake and ensuring the proper growth and development of medicinal plants even under high salt conditions. Endophytes can interact with the plant's signalling pathways and influence hormone crosstalk (Phurailatpam and Mishra, 2020). This interaction can modulate the plant's response to salt stress and activate stress-related genes, leading to enhanced salt tolerance. Research studies have shown that specific endophytic strains isolated from medicinal plants, such as *Azospirillum*, *Bacillus*, and *Pseudomonas* species, can significantly improve the salt tolerance of their host plants (Mahgoub *et al.*, 2021). However, it's important to note that the effectiveness of endophytes in enhancing salt tolerance may vary depending on the specific plant species, endophyte strain, and environmental conditions. Overall, harnessing the beneficial effects of endophytes holds great promise for improving the salt tolerance of medicinal plants, enabling their cultivation in saline soils, and expanding their potential applications in agriculture and pharmaceutical industries.

12.4.1.4 Temperature Tolerance

The effect of endophytes on temperature tolerance may vary depending on the plant species and the characteristics of the endophytes involved, there are a few general ways in which endophytes can enhance temperature tolerance in medicinal plants (Wu *et al.*, 2021). Some endophytes produce compounds that help plants cope with high temperatures (Shaffique *et al.*, 2022). These compounds can act as antioxidants, osmoprotectants, or heat shock proteins, which protect plant cells from heat-induced damage. By colonizing medicinal plants, these endophytes can help enhance the heat tolerance of the host plants. Endophytes can improve the nutrient uptake efficiency of medicinal plants, which in turn can enhance their temperature tolerance (Vishwakarma *et al.*, 2021; Pandey *et al.*, 2023). Nutrient deficiencies can make plants more vulnerable to temperature stress. Endophytes can promote nutrient acquisition and assimilation within the plant, helping it better withstand temperature extremes. ABA is involved in regulating plant responses to various stresses, including temperature stress (Devireddy *et al.*, 2021). Endophytes can help optimize ABA levels in medicinal plants, enabling them to better adapt to temperature fluctuations. It's important to note that the specific mechanisms and effects of endophytes on temperature tolerance can vary depending on the plant species and the characteristics of the endophytes involved.

12.4.1.5 Heavy Metal Tolerance

When it comes to medicinal plants, the heat tolerance of their metal accumulation is a significant concern. Excessive accumulation of metals can be toxic to plants and hinder their growth and development (Edelstein and Ben-Hur, 2018). However, certain endophytic microorganisms have been shown to assist medicinal plants in coping with heat stress and metal tolerance simultaneously. Some endophytes possess the ability to bind and sequester metal ions within their cells (Sharma and Kumar, 2021). This activity helps in reducing the concentration of toxic metals within the plant tissues, minimizing their harmful effects. Endophytic microorganisms often produce enzymes that can detoxify heavy metals by converting them into less toxic forms. These enzymes, such as metallothioneins, phytochelatins, and glutathione, can help maintain metal homeostasis within plants, even under heat stress (Singh *et al.*, 2016). Endophytes can activate certain genes in medicinal plants that are involved in stress responses (Ogbe *et al.*, 2020). These genes may encode heat shock proteins, metal transporters, or other proteins involved in stress tolerance mechanisms. Endophytic microorganisms help medicinal plants cope with heat stress and metal toxicity simultaneously, improving their overall tolerance. Harnessing the potential of endophytes can be a promising strategy to enhance the cultivation of medicinal plants in metal-contaminated environments or regions with high temperatures.

12.5 CONCLUSION

Rhizobacteria and mycorrhizae are well known for plant growth promotion. Investigation is still ongoing to understand the endophytic nature of microorganisms. It is still under the investigation that how pathogenic microbes are changing their lifestyle as endophytic lifestyle over the period of time. It is necessary to understand physiological and molecular changes in endophytes as well as hosts during their interaction. Microbe-mediated stress tolerance in plants is an eco-friendly approach for better crop yield. It can increase cropland and species diversity. Endophytes may also be a good tool for the improvement of the quality and yield of plant products through genetic engineering and tissue culture. Induced drought tolerance was also observed in various studies. The commercial application of endophytic microbes for plant growth promotion and natural products is not as common as PGPR or mycorrhizae due to a range of scientific and technical challenges. Studies need to be carried out on how to overcome the challenges for commercial formulations of endophytic microorganisms.

REFERENCES

Abdelshafy Mohamad, O.A., Ma, J.B., Liu, Y.H., Zhang, D., Hua, S., Bhute, S., Hedlund, B.P., Li, W.J. and Li, L., 2020. Beneficial endophytic bacterial populations associated with the medicinal plant Thymus vulgaris alleviate salt stress and confer resistance to Fusarium oxysporum. *Frontiers in Plant Science*, 11, p. 47.

Ahirwar, R.K., 2023. An overview of medicinal plants: Drugs of tomorrow. *Stress-Responsive Factors and Molecular Farming in Medicinal Plants*, pp. 1–16.

Ali, B., Wang, B., Ali, S., Ghani, M.A., Hayat, M.T., Yang, C., Xu, L. and Zhou, W.J., 2013.5-Aminolevulinic acid ameliorates the growth, photosynthetic gas exchange capacity, and ultrastructural changes under cadmium stress in Brassica napus L. *Journal of Plant Growth Regulation*, 32, pp. 604–614.

Astutik, S., Pretzsch, J. and Ndzifon Kimengsi, J., 2019. Asian medicinal plants' production and utilization potentials: A review. *Sustainability*, 11(19), p. 5483.

Basu, R., Dasgupta, S., Babu, S.N. and Noor, A., 2023. Medicinal Plants in the Indian traditional medicine and current practices. *Bioprospecting of Tropical Medicinal Plants*, pp. 253–286.

Belaqziz, R., Romane, A. and Abbad, A., 2009. Salt stress effects on germination, growth, and essential oil content of an endemic thyme species in Morocco (Thymus maroccanus Ball.). *Journal of Applied Sciences Research*, July, pp. 858–863.

Bharti, N., Yadav, D., Barnawal, D., Maji, D. and Kalra, A., 2013. Exiguobacterium oxidotolerans, a halotolerant plant growth-promoting rhizobacteria, improves the yield and content of secondary metabolites in Bacopa monnieri (L.) Pennell under primary and secondary salt stress. *World Journal of Microbiology and Biotechnology*, 29, pp. 379–387.

Chandran, V., Shaji, H. and Mathew, L., 2020. Endophytic microbial influence on plant stress responses. In *Microbial Endophytes* (pp. 161–193). Woodhead Publishing.

Chaudhary, I.J. and Rathore, D., 2018. Suspended particulate matter deposition and its impact on urban trees. *Atmospheric Pollution Research*, 9(6), pp. 1072–1082.

Chaudhary, I.J. and Rathore, D., 2019. Dust pollution: Its removal and effect on foliage physiology of urban trees. *Sustainable Cities and Society*, 51, p. 101696.

Chaudhary, I.J. and Rathore, D., 2020. Relative effectiveness of ethylenediurea, phenyl urea, ascorbic acid, and urea in preventing groundnut (Arachis hypogaea L) crop from ground level ozone. *Environmental Technology & Innovation*, 19, p. 100963.

Chaudhary, I.J. and Rathore, D., 2021a. Assessment of ozone toxicity on cotton (Gossypium hirsutum L.) cultivars: Its defensive system and intraspecific sensitivity. *Plant Physiology and Biochemistry*, 166, pp. 912–927.

Chaudhary, I.J. and Rathore, D., 2021b. Assessment of the dose-response relationship between ozone dose and groundnut (Arachis hypogaea L) cultivars using open top chamber (OTC) and ethylenediurea (EDU). *Environmental Technology & Innovation*, 22, p. 101494.

Chaudhary, I.J. and Rathore, D., 2021c. Micro-morphological and anatomical response of groundnut (Arachis hypogaea L.) cultivars to ground-level ozone. *Journal of Applied Biology and Biotechnology*, 9(4), pp. 137–150.

Chaudhary, I.J. and Rathore, D., 2022. Effects of ambient and elevated ozone on morphophysiology of cotton (Gossypium hirsutum L.) and its correlation with yield traits. *Environmental Technology & Innovation*, 25, p. 102146.

Chiellini, C., Maida, I., Emiliani, G., Mengoni, A., Mocali, S., Fabiani, A., Biffi, S., Maggini, V., Gori, L., Vannacci, A. and Gallo, E.R., 2014. Endophytic and rhizospheric bacterial communities isolated from the medicinal plants Echinacea purpurea and Echinacea angustifolia. *International Microbiology*, 17, pp. 165–174.

Collinge, D.B., Jensen, B. and Jørgensen, H.J., 2022. Fungal endophytes in plants and their relationship to plant disease. *Current Opinion in Microbiology*, 69, p. 102177.

Compant, S., Clément, C. and Sessitsch, A., 2010. Plant growth-promoting bacteria in the rhizo- and endosphere of plants: Their role, colonization, mechanisms involved and prospects for utilization. *Soil Biology and Biochemistry*, 42(5), pp. 669–678.

Cui, W., Gao, C., Fang, P., Lin, G. and Shen, W., 2013. Alleviation of cadmium toxicity in Medicago sativa by hydrogen-rich water. *Journal of Hazardous Materials*, 260, pp. 715–724.

Debnath, M., 2008. Responses of Bacopa monnieri to salinity and drought stress in vitro. *Journal of Medicinal Plants Research*, 2(11), pp. 347–351.

Devireddy, A.R., Tschaplinski, T.J., Tuskan, G.A., Muchero, W. and Chen, J.G., 2021. Role of reactive oxygen species and hormones in plant responses to temperature changes. *International Journal of Molecular Sciences*, 22(16), p. 8843.

Edelstein, M. and Ben-Hur, M., 2018. Heavy metals and metalloids: Sources, risks, and strategies to reduce their accumulation in horticultural crops. *Scientia Horticulturae*, 234, pp. 431–444.

Eigenbrode, S.D., Bosque-Pérez, N.A. and Davis, T.S., 2018. Insect-borne plant pathogens and their vectors: Ecology, evolution, and complex interactions. *Annual Review of Entomology*, 63, pp. 169–191.

Ek-Ramos, M.J., Gomez-Flores, R., Orozco-Flores, A.A., Rodríguez-Padilla, C., González-Ochoa, G. and Tamez-Guerra, P., 2019. Bioactive products from plant-endophytic gram-positive bacteria. *Frontiers in Microbiology*, 10, p. 463.

EL Sabagh, A., Islam, M.S., Hossain, A., Iqbal, M.A., Mubeen, M., Waleed, M., Reginato, M., Battaglia, M., Ahmed, S., Rehman, A. and Arif, M., 2022. Phytohormones as growth regulators during abiotic stress tolerance in plants. *Frontiers in Agronomy*, 4, p. 765068.

Eliašová, A., Repčák, M. and Pastírová, A., 2004. Quantitative changes of secondary metabolites of Matricaria chamomilla by abiotic stress. *Zeitschrift für Naturforschung C*, 59(7–8), pp. 543–548.

Fadiji, A.E. and Babalola, O.O., 2020. Exploring the potentialities of beneficial endophytes for improved plant growth. *Saudi Journal of Biological Sciences*, 27(12), pp. 3622–3633.

Gao, X., Starmer, J. and Martin, E.R., 2008. A multiple testing correction method for genetic association studies using correlated single nucleotide polymorphisms. *Genetic Epidemiology: The Official Publication of the International Genetic Epidemiology Society*, 32(4), pp. 361–369.

Gill, S.S. and Tuteja, N., 2010. Reactive oxygen species and antioxidant machinery in abiotic stress tolerance in crop plants. *Plant Physiology and Biochemistry*, 48(12), pp. 909–930.

Gómez, D., Martínez, J., Hernández, L., Escalante, D., Yabor, L., Sershen and Lorenzo, J.C., 2020. Modifying sugarcane mineral levels through sodium chloride and mannitol exposure in temporary immersion bioreactors. *In Vitro Cellular & Developmental Biology-Plant*, 56, pp. 169–176.

Gond, S.K., Lata, R. and Singh, A., 2021. Endophytic microbes promote plant growth and alter host secondary metabolites. In *Microbiome Stimulants for Crops* (pp. 59–72). Woodhead Publishing.

Griffiths, S.M., Galambao, M., Rowntree, J., Goodhead, I., Hall, J., O'Brien, D., Atkinson, N. and Antwis, R.E., 2020. Complex associations between cross-kingdom microbial endophytes and host genotype in ash dieback disease dynamics. *Journal of Ecology*, 108(1), pp. 291–309.

Gupta, S., Schillaci, M., Walker, R., Smith, P.M., Watt, M. and Roessner, U., 2021. Alleviation of salinity stress in plants by endophytic plant-fungal symbiosis: Current knowledge, perspectives and future directions. *Plant and Soil*, 461, pp. 219–244.

Hanaka, A., Ozimek, E., Reszczyńska, E., Jaroszuk-Ściseł, J. and Stolarz, M., 2021. Plant tolerance to drought stress in the presence of supporting bacteria and fungi: An efficient strategy in horticulture. *Horticulturae*, 7(10), p. 390.

Hashem, S., Ali, T.A., Akhtar, S., Nisar, S., Sageena, G., Ali, S., Al-Mannai, S., Therachiyil, L., Mir, R., Elfaki, I. and Mir, M.M., 2022. Targeting cancer signaling pathways by natural products: Exploring promising anti-cancer agents. *Biomedicine & Pharmacotherapy*, 150, p. 113054.

Hawkes, C.V., Bull, J.J. and Lau, J.A., 2020. Symbiosis and stress: How plant microbiomes affect host evolution. *Philosophical Transactions of the Royal Society B*, 375(1808), p. 20190590.

Ismail, M.A., Amin, M.A., Eid, A.M., Hassan, S.E.D., Mahgoub, H.A., Lashin, I., Abdelwahab, A.T., Azab, E., Gobouri, A.A., Elkelish, A. and Fouda, A., 2021. Comparative study between exogenously applied plant growth hormones versus metabolites of microbial endophytes as plant growth-promoting for Phaseolus vulgaris L. *Cells*, 10(5), p. 1059.

Jeelani, S.M., Rather, G.A., Sharma, A. and Lattoo, S.K., 2018. In perspective: Potential medicinal plant resources of Kashmir Himalayas, their domestication and cultivation for commercial exploitation. *Journal of Applied Research on Medicinal and Aromatic Plants*, 8, pp. 10–25.

Johnston-Monje, D. and Raizada, M.N., 2011. 4.58 plant and endophyte relationships: Nutrient management. In *Comprehensive Biotechnology* (2nd ed., pp. 713–727) Academic.

Kale, S.S., Chauhan, R., Nigam, B., Gosavi, S., Chaudhary, I.J. 2024. Effectiveness of nanoparticles in improving soil fertility and eco-friendly crop resistance: A comprehensive review. *Biocatalysis and Agricultural Biotechnology*, p. 103066. https://doi.org/10.1016/j.bcab.2024.103066.

Kamran, M., Imran, Q.M., Ahmed, M.B., Falak, N., Khatoon, A. and Yun, B.W., 2022. Endophyte-mediated stress tolerance in plants: A sustainable strategy to enhance resilience and assist crop improvement. *Cells*, 11(20), p. 3292.

Kandasamy, G.D. and Kathirvel, P., 2023. Insights into bacterial endophytic diversity and isolation with a focus on their potential applications – a review. *Microbiological Research*, 266, p. 127256.

Khalid, H.M., 2006. Embracing diversity in user needs for effective design. *Applied Ergonomics*, 37(4), pp. 409–418.

Khan, A.L., Gilani, S.A., Waqas, M., Al-Hosni, K., Al-Khiziri, S., Kim, Y.H., Ali, L., Kang, S.M., Asaf, S., Shahzad, R. and Hussain, J., 2017. Endophytes from medicinal plants and their potential for producing indole acetic acid, improving seed germination, and mitigating oxidative stress. *Journal of Zhejiang University. Science B*, 18(2), p. 125.

Koocheki, A., Nassiri-Mahallati, M. and Azizi, G., 2008. Effect of drought, salinity, and defoliation on growth characteristics of some medicinal plants of Iran. *Journal of Herbs, Spices & Medicinal Plants*, 14(1–2), pp. 37–53.

Kumar, A.K., Sharma, S.K., Lata, C., Devi, R., Kulshrestha, N., Krishnamurthy, S.L., Singh, K. and Yadav, R.K., 2018. Impact of water deficit (salt and drought) stress on physiological, biochemical, and yield attributes on wheat (Triticum aestivum) varieties. *Indian Journal of Agricultural Sciences*, 88(10), pp. 1624–1632.

Kumar, S., Narula, A., Sharma, M.P. and Srivastava, P.S., 2004. In vitro propagation of Pluchea lanceolata, a medicinal plant, and effect of heavy metals and different aminopurines on quercetin content. *In Vitro Cellular & Developmental Biology-Plant*, 40, pp. 171–176.

Lata, R., Chowdhury, S., Gond, S.K. and White Jr, J.F., 2018. Induction of abiotic stress tolerance in plants by endophytic microbes. *Letters in Applied Microbiology*, 66(4), pp. 268–276.

Liu, H.Q., Lu, X.B., Li, Z.H., Tian, C.Y. and Song, J., 2021. The role of root-associated microbes in growth stimulation of plants under saline conditions. *Land Degradation & Development*, 32(13), pp. 3471–3486.

Lu, H., Wei, T., Lou, H., Shu, X. and Chen, Q., 2021. A critical review of communication mechanism within plant-endophytic fungi interactions to cope with biotic and abiotic stresses. *Journal of Fungi*, 7(9), p. 719.

Mahgoub, H.A., Fouda, A., Eid, A.M., Ewais, E.E.D. and Hassan, S.E.D., 2021. Biotechnological application of plant growth-promoting endophytic bacteria isolated from halophytic plants to ameliorate salinity tolerance of Vicia faba L. *Plant Biotechnology Reports*, 15, pp. 819–843.

Mathur, P. and Roy, S., 2021. Insights into the plant responses to drought and decoding the potential of the root-associated microbiome for inducing drought tolerance. *Physiologia Plantarum*, 172(2), pp. 1016–1029.

Mechri, B., Tekaya, M., Attia, F., Hammami, M. and Chehab, H., 2020. Drought stress improved the capacity of Rhizophagus irregularis for inducing the accumulation of oleuropein and mannitol in olive (Olea europaea) roots. *Plant Physiology and Biochemistry*, 156, pp. 178–191.

Mofokeng, M.M., du Plooy, C.P., Araya, H.T., Amoo, S.O., Mokgehle, S.N., Pofu, K.M. and Mashela, P.W., 2022. Medicinal plant cultivation for sustainable use and commercialization of high-value crops. *South African Journal of Science*, 118(7–8), pp. 1–7.

Namdar, D., Charuvi, D., Ajjampura, V., Mazuz, M., Ion, A., Kamara, I. and Koltai, H., 2019. LED lighting affects the composition and biological activity of Cannabis sativa secondary metabolites. *Industrial Crops and Products*, 132, pp. 177–185.

Narula, A., Kumar, S. and Srivastava, P.S., 2005. Abiotic metal stress enhances diosgenin yield in Dioscorea bulbifera L. cultures. *Plant Cell Reports*, 24, pp. 250–254.

Nigam, B., Dubey, R.S. and Rathore, D., 2022. Protective role of exogenously supplied salicylic acid and PGPB (Stenotrophomonas sp.) on spinach and soybean cultivars grown under salt stress. *Scientia Horticulturae*, 293, p. 110654.

Ogbe, A.A., Finnie, J.F. and Van Staden, J., 2020. The role of endophytes in secondary metabolite accumulation in medicinal plants under abiotic stress. *South African Journal of Botany*, 134, pp. 126–134.

Ozturk, C., Zhang, Y. and Trappe, W., 2004, October. Source-location privacy in energy-constrained sensor network routing. In *Proceedings of the 2nd ACM workshop on Security of Ad hoc and Sensor Networks* (pp. 88–93). ACM.

Ozturk, M., Turkyilmaz Unal, B., García-Caparrós, P., Khursheed, A., Gul, A. and Hasanuzzaman, M., 2021. Osmoregulation and its actions during the drought stress in plants. *Physiologia Plantarum*, 172(2), pp. 1321–1335.

Pandey, P., Tripathi, A., Dwivedi, S., Lal, K. and Jhang, T., 2023. Deciphering the mechanisms, hormonal signaling, and potential applications of endophytic microbes to mediate stress tolerance in medicinal plants. *Frontiers in Plant Science*, 14.

Phurailatpam, L. and Mishra, S., 2020. Role of plant endophytes in conferring abiotic stress tolerance. *Plant Ecophysiology and Adaptation under Climate Change: Mechanisms and Perspectives II: Mechanisms of Adaptation and Stress Amelioration*, pp. 603–628.

Rawat, P., Das, S., Shankhdhar, D. and Shankhdhar, S.C., 2021. Phosphate-solubilizing microorganisms: Mechanism and their role in phosphate solubilization and uptake. *Journal of Soil Science and Plant Nutrition*, 21, pp. 49–68.

Razmjoo, K.H., Heydarizadeh, P.A. and Sabzalian, M.R., 2008. Effect of salinity and drought stresses on growth parameters and essential oil content of Matricaria chamomile. *International Journal of Agriculture and Biology*, 10(4), pp. 451–454.

Saeidnejad, A.H., Kafi, M., Khazaei, H.R. and Pessarakli, M., 2013. Effects of drought stress on quantitative and qualitative yield and antioxidative activity of Bunium persicum. *Turkish Journal of Botany*, 37(5), pp. 930–939.

Sahu, P.K., Jayalakshmi, K., Tilgam, J., Gupta, A., Nagaraju, Y., Kumar, A., Hamid, S., Singh, H.V., Minkina, T., Rajput, V.D. and Rajawat, M.V.S., 2022. ROS generated from biotic stress: Effects on plants and alleviation by endophytic microbes. *Frontiers in Plant Science*, 13, p. 1042936.

Sarkar, S., Dey, A., Kumar, V., Batiha, G.E.S., El-Esawi, M.A., Tomczyk, M. and Ray, P., 2021. Fungal endophyte: An interactive endosymbiont with the capability of modulating host physiology in myriad ways. *Frontiers in Plant Science*, 12, p. 701800.

Schulz, H., 2023. 3 relationships between traditional. *Medicinal Plants Used in Traditional Persian Medicine*, p. 136.

Shaffique, S., Khan, M.A., Wani, S.H., Pande, A., Imran, M., Kang, S.M., Rahim, W., Khan, S.A., Bhatta, D., Kwon, E.H. and Lee, I.J., 2022. A review on the role of endophytes and plant growth promoting rhizobacteria in mitigating heat stress in plants. *Microorganisms*, 10(7), p. 1286.

Sharma, P. and Kumar, S., 2021. Bioremediation of heavy metals from industrial effluents by endophytes and their metabolic activity: Recent advances. *Bioresource Technology*, 339, p. 125589.

Siddiqui, Z.S., Wei, X., Umar, M., Abideen, Z., Zulfiqar, F., Chen, J., Hanif, A., Dawar, S., Dias, D.A. and Yasmeen, R., 2022. Scrutinizing the application of saline endophyte to enhance salt tolerance in rice and maize plants. *Frontiers in Plant Science*, 12, p. 770084.

Singh, D., Geat, N., Rajawat, M.V.S., Mahajan, M.M., Prasanna, R., Singh, S., Kaushik, R., Singh, R.N., Kumar, K. and Saxena, A.K., 2018. Deciphering the mechanisms of endophyte-mediated biofortification of Fe and Zn in wheat. *Journal of Plant Growth Regulation*, 37, pp. 174–182.

Singh, S., Chaudhary, I.J. and Kumar, P., 2019. Utilization of low-cost agricultural waste for removal of toxic metals from the environment: A review. *International Journal of Scientific Research in Biological Sciences*, 6, p. 4.

Singh, S., Parihar, P., Singh, R., Singh, V.P. and Prasad, S.M., 2016. Heavy metal tolerance in plants: Role of transcriptomics, proteomics, metabolomics, and ionomics. *Frontiers in Plant Science*, 6, p. 1143.

Srebot, M.S., Tano, J., Carrau, A., Ferretti, M.D., Martínez, M.L., Orellano, E.G. and Rodriguez, M.V., 2021. Bacterial wilt biocontrol by the endophytic bacteria Gluconacetobacter diazotrophicus in Río Grande tomato cultivar. *Biological Control*, 162, p. 104728.

Terletskaya, N.V., Korbozova, N.K., Kudrina, N.O., Kobylina, T.N., Kurmanbayeva, M.S., Meduntseva, N.D. and Tolstikova, T.G., 2021. The influence of abiotic stress factors on the morphophysiological and phytochemical aspects of the acclimation of the plant *Rhodiola semenowii* Boriss. *Plants*, 10(6), p. 1196.

Ting, D.Z.Y., Hill, C.J., Soibel, A., Keo, S.A., Mumolo, J.M., Nguyen, J. and Gunapala, S.D., 2009. A high-performance long wavelength superlattice complementary barrier infrared detector. *Applied Physics Letters*, 95(2).

Vaghela, N. and Gohel, S., 2023. Medicinal plant-associated rhizobacteria enhance the production of pharmaceutically important bioactive compounds under abiotic stress conditions. *Journal of Basic Microbiology*, 63(3–4), pp. 308–325.

Vaishnavi, K. and Kumar, M.H., 2021. Chapter-2 endophytes as bioherbicides, biopesticides, and bioinsecticides. *Chief Editor*, 151, p. 25.

Verma, D.K., Pandey, A.K., Mohapatra, B., Srivastava, S., Kumar, V., Talukdar, D., Yulianto, R., Zuan, A.T.K., Jobanputra, A.H. and Asthir, B., 2019. Plant growth-promoting rhizobacteria: An eco-friendly approach for sustainable agriculture and improved crop production. In *Microbiology for Sustainable Agriculture, Soil Health, and Environmental Protection* (pp. 3–80). Apple Academic Press.

Verma, S.K., Sahu, P.K., Kumar, K., Pal, G., Gond, S.K., Kharwar, R.N. and White, J.F., 2021. Endophyte roles in nutrient acquisition, root system architecture development, and oxidative stress tolerance. *Journal of Applied Microbiology*, 131(5), pp. 2161–2177.

Vishwakarma, K., Kumar, N., Shandilya, C. and Varma, A., 2021. Unraveling the role of endophytes in micronutrient uptake and enhanced crop productivity. *Symbiotic Soil Microorganisms: Biology and Applications*, pp. 63–85.

Wani, Z.A. and Hakeem, K.R., 2022. Endophytes: An overview. *Endophyte Biology*, pp. 1–19.

Waqas, M., Khan, A.L., Kamran, M., Hamayun, M., Kang, S.M., Kim, Y.H. and Lee, I.J., 2012. Endophytic fungi produce gibberellins and indoleacetic acid and promote host-plant growth during stress. *Molecules*, 17(9), pp. 10754–10773.

Waśkiewicz, A., Muzolf-Panek, M. and Goliński, P., 2013. Phenolic content changes in plants under salt stress. *Ecophysiology and Responses of Plants Under Salt Stress*, pp. 283–314.

Weerarathne, L.V.Y., 2021. *Water-Use Efficiency in Perennial Ryegrass (Lolium perenne L.)*. A Thesis Presented in Partial Fulfillment of the Requirements for the Degree of Doctor of Philosophy in Plant Science at Massey University, Manawatu, New Zealand (Doctoral dissertation, Massey University).

White, J.F., Kingsley, K.L., Zhang, Q., Verma, R., Obi, N., Dvinskikh, S., Elmore, M.T., Verma, S.K., Gond, S.K. and Kowalski, K.P., 2019. Endophytic microbes and their potential applications in crop management. *Pest Management Science*, 75(10), pp. 2558–2565.

Wu, W., Chen, W., Liu, S., Wu, J., Zhu, Y., Qin, L. and Zhu, B., 2021. Beneficial relationships between endophytic bacteria and medicinal plants. *Frontiers in Plant Science*, 12, p. 646146.

Yang, L., Danzberger, J., Schöler, A., Schröder, P., Schloter, M. and Radl, V., 2017. Dominant groups of potentially active bacteria shared by barley seeds become less abundant in root-associated microbiomes. *Frontiers in Plant Science*, 8, p. 1005.

Zhang, F., Yang, Y.L., He, W.L., Zhao, X. and Zhang, L.X., 2004. Effects of salinity on growth and compatible solutes of callus induced from Populus euphratica. *In Vitro Cellular & Developmental Biology-Plant*, 40, pp. 491–494.

Zhang, N., Sun, Q., Zhang, H., Cao, Y., Weeda, S., Ren, S. and Guo, Y.D., 2015. Roles of melatonin in abiotic stress resistance in plants. *Journal of Experimental Botany*, 66(3), pp. 647–656.

Index

Note: Page numbers in *italics* indicate a figure and page numbers in **bold** indicate a table on the corresponding page.

L

M